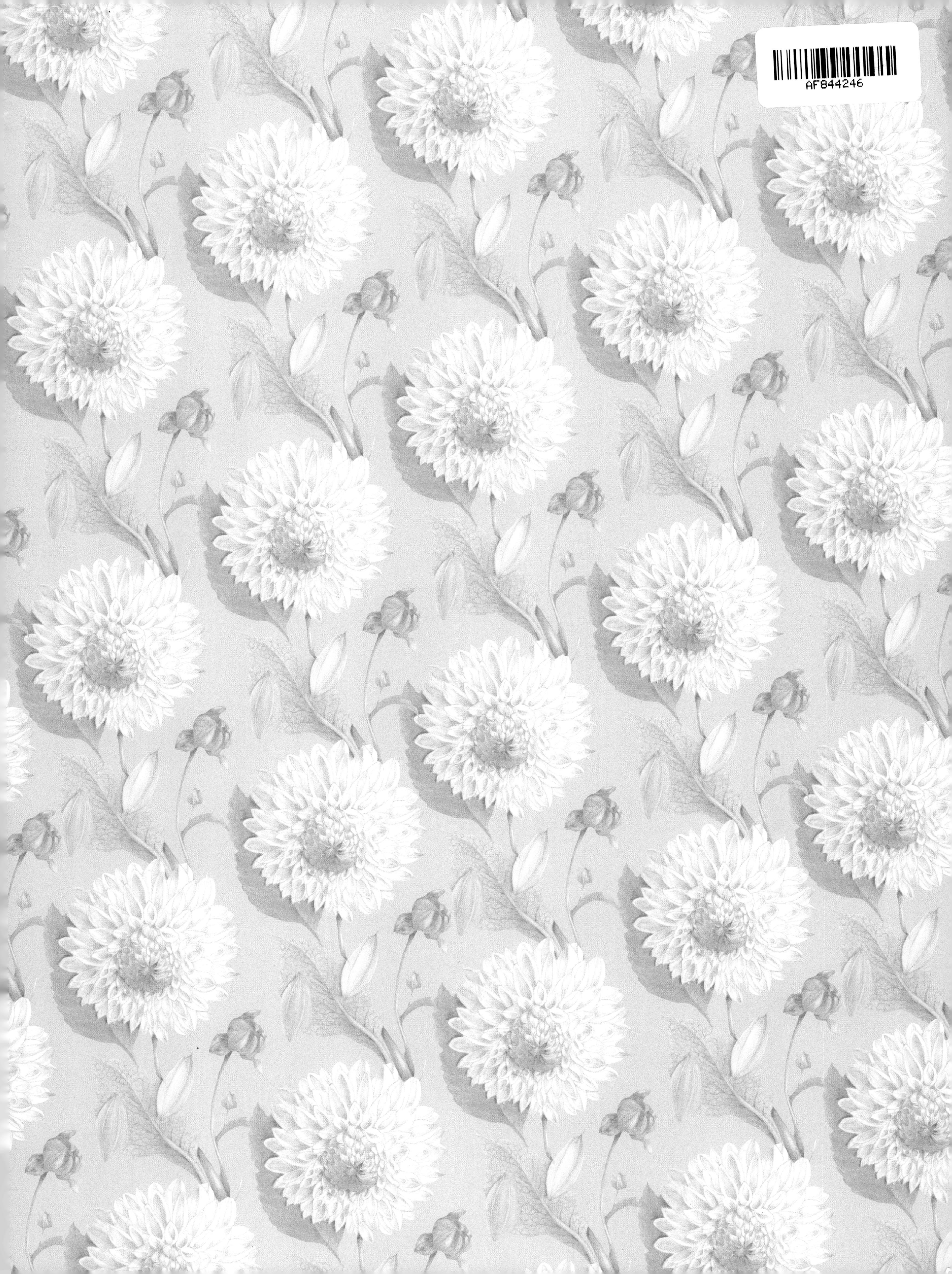

# THE ALBUM OF PLANT FAMILIES

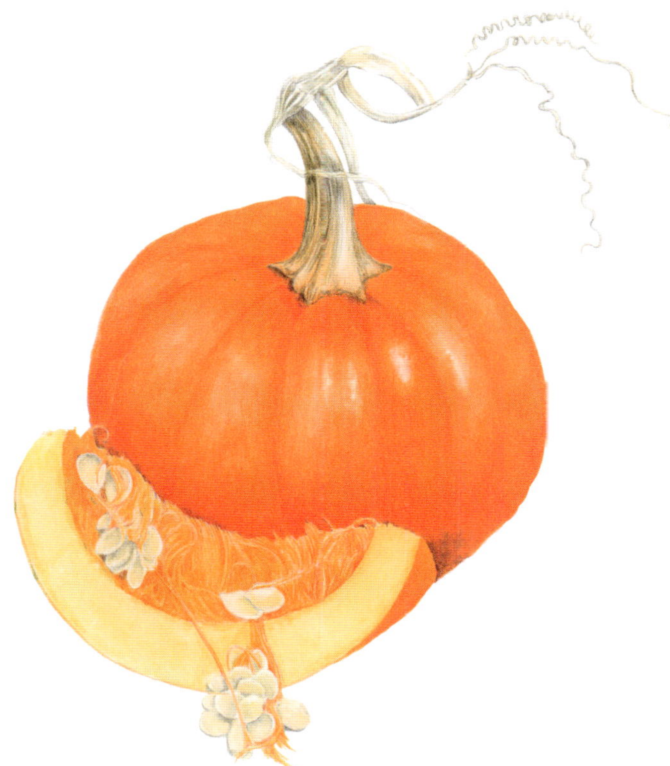

*An Illustrated Exploration*

*of Nature's*

*Beauty and Bounty*

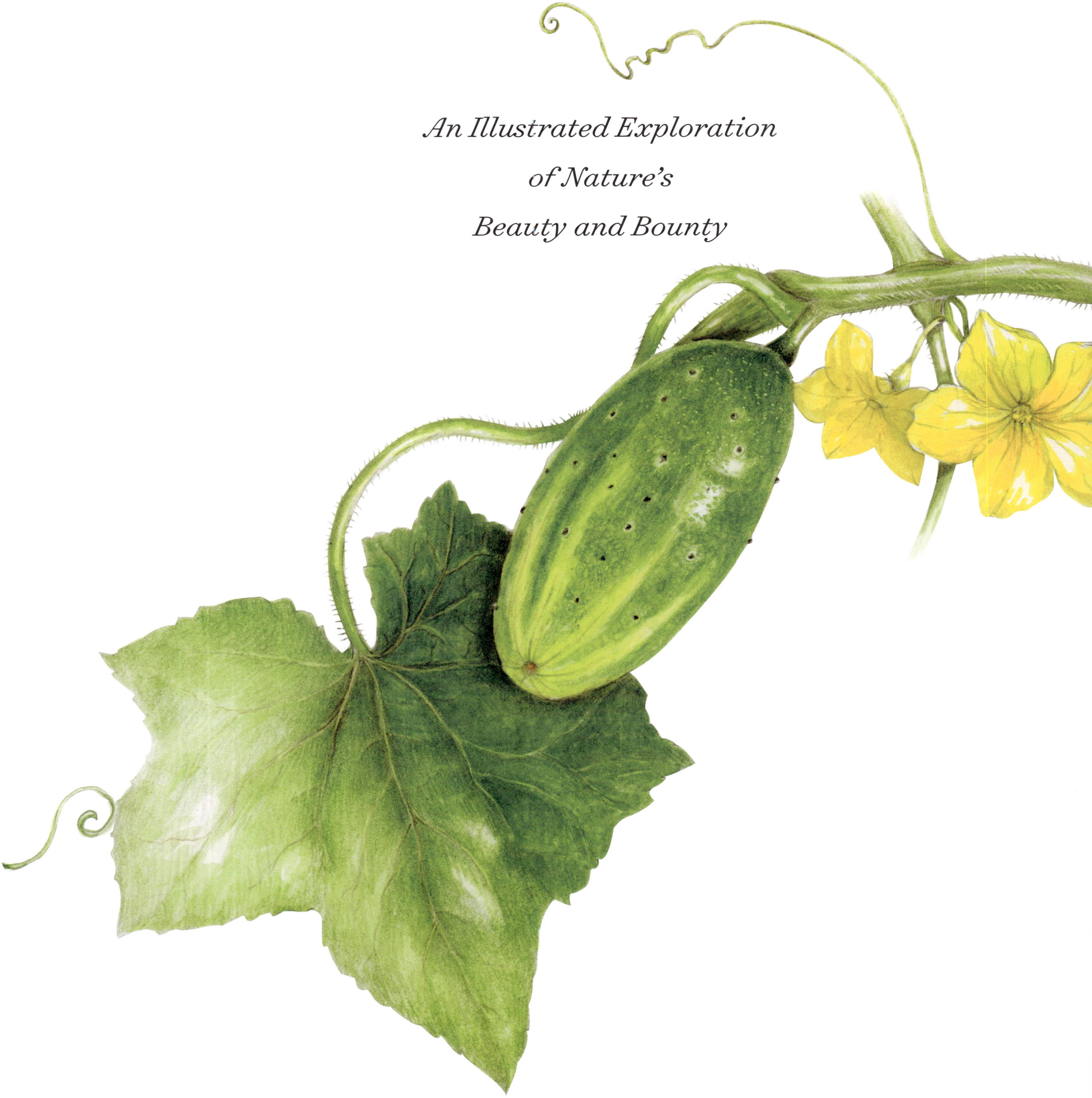

# The Album of Plant Families

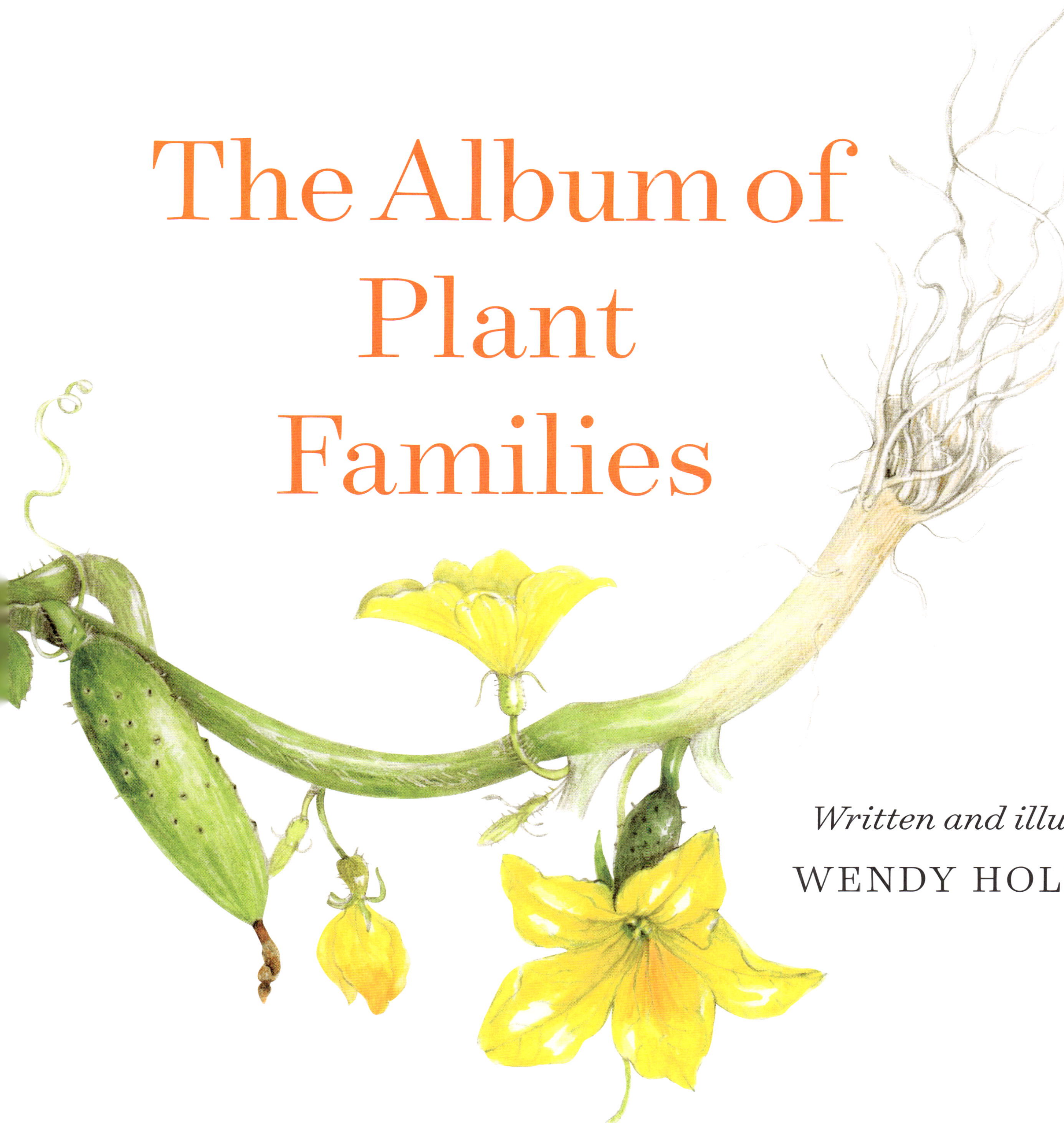

*Written and illustrated by*
WENDY HOLLENDER

ABBEVILLE PRESS
New York London

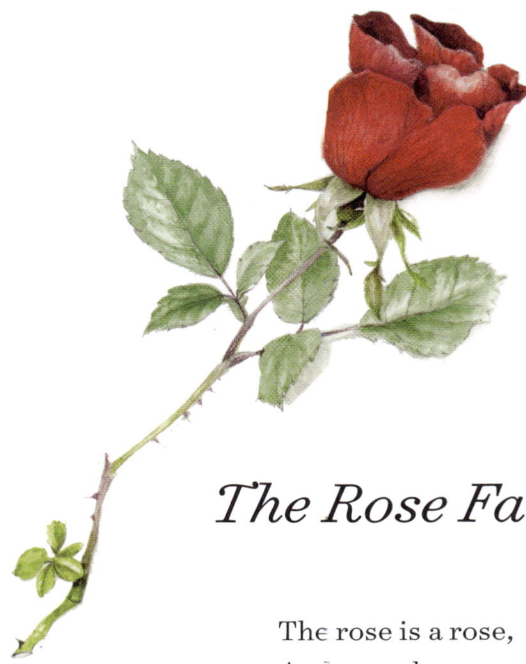

# *The Rose Family*

The rose is a rose,
And was always a rose.
But the theory now goes
That the apple's a rose,
And the pear is, and so's
The plum, I suppose.
The dear only knows
What will next prove a rose.
You, of course, are a rose—
But were always a rose.
—Robert Frost

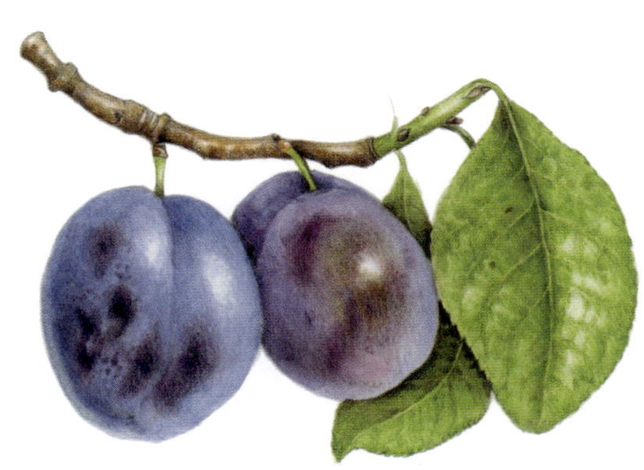

# Contents

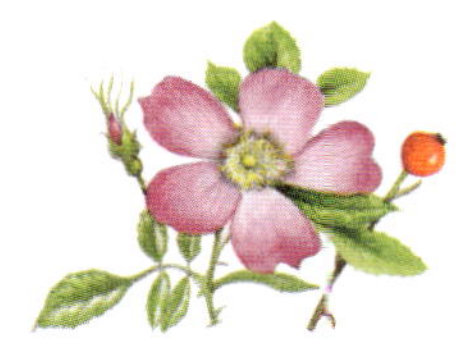

# Introduction

I grew up in a family of short people. We were low to the ground with strong muscles and curly or wavy hair. My family shares lots of traits, some of which are obvious when you see us together. Plants have shared traits too—think tendrils on vines instead of curly hair and square stems in the mint family helping the plant to stand up tall.

With my paintings of plant families, my goal is to help you understand plants in an empowering and joyful way. Since many adults and children are unfamiliar with the scientific side of plants, I have arranged my paintings to provide a visual, accessible gateway to the relationships between plants. Plant families share visual patterns, often in the structure of their flowers, leaves, and stems. The families also share features such as scent, taste, and medicinal qualities, and sometimes irritating and poisonous properties. Exploring these relationships is valuable, helping to deepen a connection to nature and assisting with plant identification.

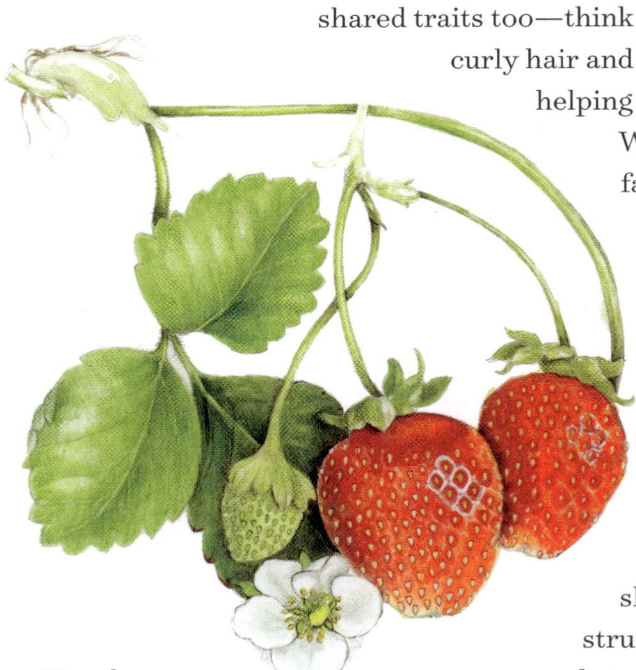

Strawberry (*Fragaria × ananassa*), from the rose family

As a botanical illustrator, I draw plants with as much detail as possible, making the paintings useful for identifying and understanding plants. I have been doing this work exclusively for twenty-six years, though I was an artist and textile designer before then. I have been fortunate to follow plants around the world to illustrate them. I always work from real plants so I can closely observe and fully experience the plant, even if characteristics such as smell and taste are not apparent in a drawing. I like to eat all the edible plants and observe but keep my distance from the poisonous ones. I consider myself a bit of a plant explorer and ethnobotanist. With each illustration, I feel as if I am discovering these plants for the first time, uncovering secrets that I never knew before. I could probably read about plants and find all this information, but it's more rewarding for me to learn from a personal relationship with each plant.

My earliest memory of plants is from when I was four years old. My family went on a summer vacation to visit relatives in Canada. Our great uncle Dave and his wife, Mary, lived in a small town near Toronto in an old wooden house with a dry goods store below and apartments for the family above. Mary was clad in an apron that tied at the waist and had ruffles on top. She took my hand and led me out back to see the garden where sweet peas were growing. Mary showed me how to pick a pod, split it open, and munch on the round, bright green peas inside. Next, she bent down over the bed of low green leaves in groups of three with bright red heart-shaped berries dangling below. I learned how to pick a strawberry and was encouraged to eat it. I was immediately enamored. All of a sudden, I was awakened to the magic of plants, which until that moment were nothing more than a sea of green color to me. Back home in our suburban backyard, I was now aware of my mother's pretty, colorful flowers and an occasional strawberry. Do you remember your earliest memory of plants?

Plants eventually drew me to the Hudson Valley of New York after living in Manhattan for thirty-five years. It all started with a trip to a local you-pick farm called Kelder's near the hamlet of Accord, New York, when fellow botanical artist Carol Woodin invited me to go strawberry picking. It wasn't until I visited Kelder's Farm again, seventeen years later, to gather sunflowers for this project, that my first visit there came back to me. The first time I went to this farm was in spring to pick strawberries and sweet peas, the same two plants I had learned to pick as a four-year-old! Coincidence? Or were those buried early memories still leading me along on my journey with plants? I kept going back all summer to Kelder's to pick the next ripe fruit or vegetable. It was a delightful experience to learn what my food looked like growing. I often had no idea. At the end of that summer I decided I needed to get myself a house right there, near Kelder's Farm, and that is what I did!

We generally read books looking for answers, but sometimes we have more questions as we read, each new insight allowing our curiosity to grow. I hope you will find in my paintings and the text in this book a way to allow your own curiosity and wonder to thrive, enticing you to go on your own path of discovery about nature and the plant world. This project of researching plants and developing these plant family paintings (over

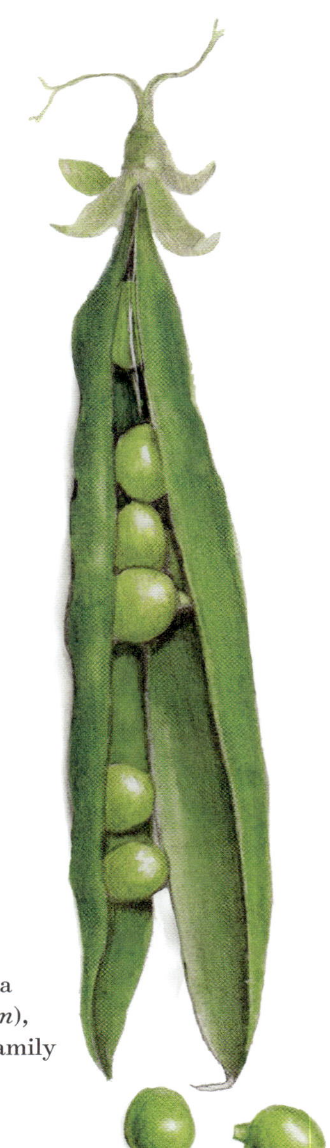

Sugar Snap Pea (*Pisum sativum*), from the pea family

a four-year period) is a culmination of my twenty-six years as a botanical illustrator, learning about plants as I go. In this book, you will find twenty-four original paintings, each of a different family to investigate. In each painting, I have illustrated from six to twenty-eight individual species or varieties of plants or mushrooms. I write about my process of developing each painting and what I discovered. There are also many guest essays written by plant experts in various fields, allowing you to hear different perspectives on these well-known and important plants. My hope is that when you finish looking at these paintings, you will come away with many questions that will spark your curiosity and put you on your own path to exploration.

## What It Means to Be Part of a Plant Family: Identifying Patterns

*What's in a name? That which we call a rose*
*by any other name would smell as sweet.*
—Juliet, William Shakespeare's *Romeo and Juliet*

The classification and naming of plants—their nomenclature—is helpful to understand. A common name is a simple name given to an individual plant species. Common names are not universal, and many plants have multiple common names, so a system of consistent scientific names was created that is the same anywhere in the world. A scientific name consists of two parts. The first part, the genus (plural: genera), may be likened to a person's family name, and the second part, the species, to a person's first name. The genus is always capitalized, but the species is lowercase, and the full scientific name is italicized. The scientific name for onion is *Allium cepa*. *Allium* is the genus, and *cepa* is the species. Garlic is also in the genus *Allium*, and its scientific name is *Allium sativum*.

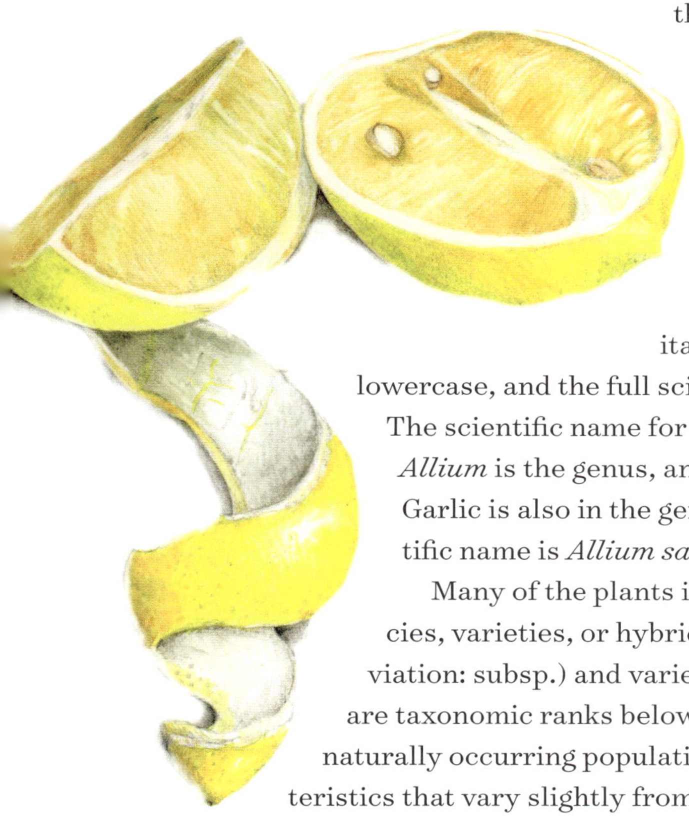

Meyer Lemon (*Citrus* × *meyeri*), from the citrus family

Many of the plants illustrated are subspecies, varieties, or hybrids. Subspecies (abbreviation: subsp.) and variety (abbreviation: var.) are taxonomic ranks below species, indicating a naturally occurring population with minor characteristics that vary slightly from what is typical for the species such as flower color or geographic location. A hybrid is a plant produced from two different species or varieties. Hybridization happens occasionally in nature, though some plants, such as those in the genus *Cucurbita* (gourd family, Cucurbitaceae), hybridize frequently. Humans create hybrids intentionally. A hybrid cross is indicated by the symbol ×, as in *Citrus* × *meyeri*, the Meyer lemon. A cultivar (cultivated variety) is a plant intentionally bred for particular characteristics that differ from those of a typical member of the species. Pumpkin, kabocha squash, and zucchini are all cultivated varieties of *Cucurbita* species. Cultivars often have fanciful or descriptive names, such as Granny Smith apple or Ruby Moon hyacinth bean.

Plant families also have common and scientific names. As with species, there are usually many common names but only one universal scientific name for each family. In plants, the scientific family name almost always ends in *-ceae*. Brassicaceae, for example, is the scientific name for the mustard family. Names are always changing in the natural world based on updated DNA findings and other considerations. Therefore, you may know different names for some of the species in this book; that is unavoidable.

There are more than 380,000 individual plant species in the world, and these plants fit into some 600 unique families, each with shared characteristics that can be learned to make it easier to recognize similarities in plants.

I have created twenty-four paintings for this book. The first twenty-two represent families in the plant kingdom. By learning these twenty-two plant families, you are on your way to being able to identify 140,000 plants! The final two paintings describe mushrooms I encountered over a growing season in my area of New York State. Mushrooms, part of the fungus kingdom, are not considered plants, but they are closely tied to the plant kingdom. The "roots" of mushrooms, known as mycelia, intertwine with the roots of trees and plants underground. These two paintings feature fifty-two species of wild mushrooms.

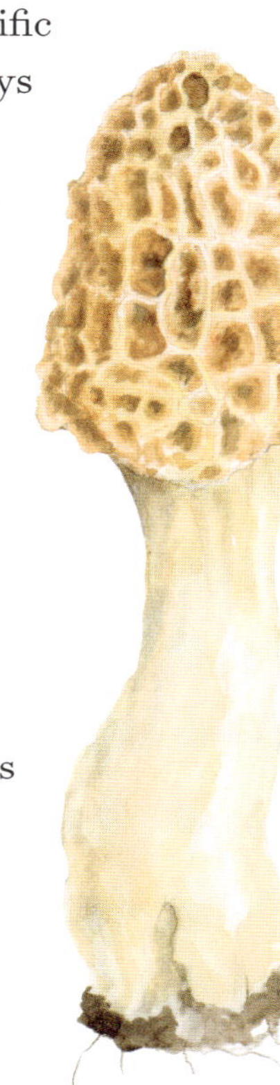

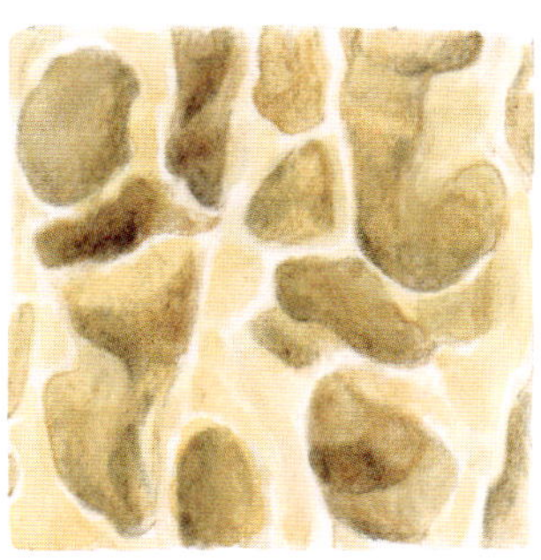

Yellow Morel (*Morchella americana*), from the fungus kingdom

Patterns have always fascinated me, whether I was doodling repetitive designs or learning to carefully craft letters as a child. In both drawing and writing, the repetition of motion brings me a sense of comfort and familiarity. As I began writing the text for my book on plant families, I found it intriguing that the word "familiar," sharing the same root as the word "family," naturally found its way into my thoughts. Family holds profound meaning for me, encompassing memories of laughter, smells, and shared experiences. It's a concept that extends to my botanical work, where understanding plant families has provided a framework for organizing and comprehending the vast diversity of flora. I've painted many individual plant portraits, starting with individual fruits and vegetables, then progressing to flowers and larger compositions of one plant in its various life-cycle stages. The collection of family paintings I have created for this book showcase the similarities and differences between plants in the same families. Rather than presenting one plant at a time, these paintings feature many plants in the same family relating to each other on the page, highlighting their similar patterns and structures as well as what makes each a unique species.

What does it mean to be part of a plant family? I was curious about that myself when I started to notice that some flowers were similar to other flowers. Once I learned the structure of one flower in a family, it was so much easier to draw or identify another plant in that same family. Plants in the same family share characteristics, not unlike humans who are related to each other. Flower parts, including number of petals and structure of reproductive parts, are among the primary shared characteristics, as are leaf arrangement, stem shape, various types of fruits, and roots. Scent and taste are often shared as well.

## Key Identifying Characteristics of Plants

Most of the plants in these plant family paintings are angiosperms, or flowering plants. Angiosperms are further divided into monocots and dicots. A monocot is a type of flowering plant characterized by having one cotyledon (the first leaf) in its seed, parallel-veined leaves, and flower parts arranged in multiples of three. The monocot families in this book are Amaryllidaceae (amaryllis), Poaceae (grass), Liliaceae (lily), Orchidaceae (orchid), and Arecaceae (palm). All the other plant families in this book are dicots, which have two first leaves (cotyledons) and branched or netlike veining in their mature leaves.

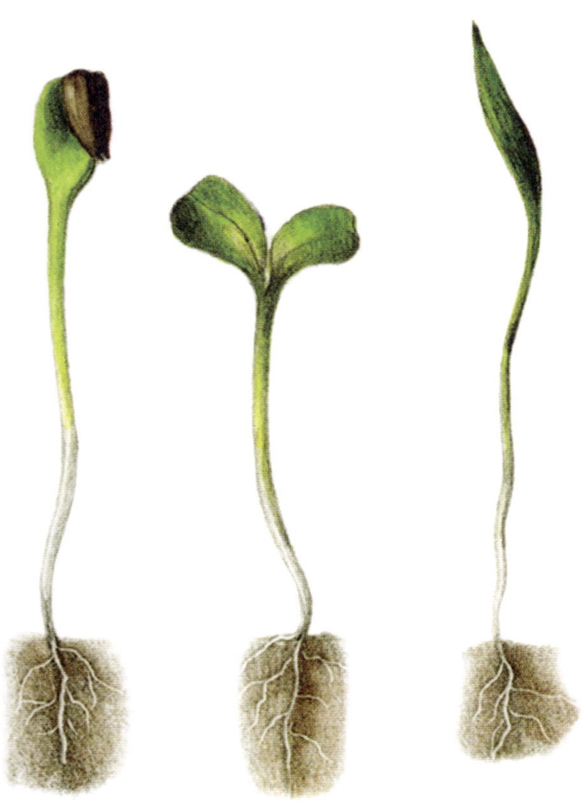

Germinating monocot seedlings have one leaflet, dicot seedlings have two

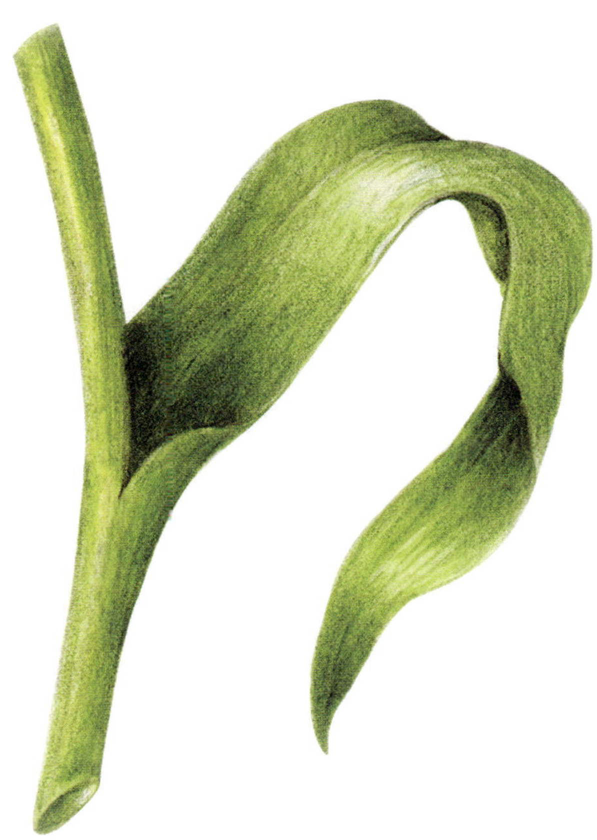

Mature leaves of monocots are straplike and have parallel veins

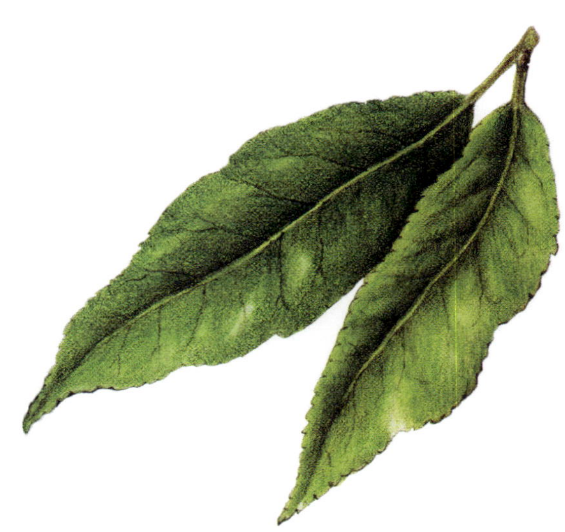

Mature leaves of dicots have branching or netlike veins.

Leaves may be simple (single and undivided, like a rose leaf) or compound (divided into separate leaflets, like a black locust leaf), and may be pinnate (with a central axis like a feather) or palmate (hand-shaped, like a maple leaf). Opposite leaves occur in pairs along the stem; alternate leaves occur singly along the stem.

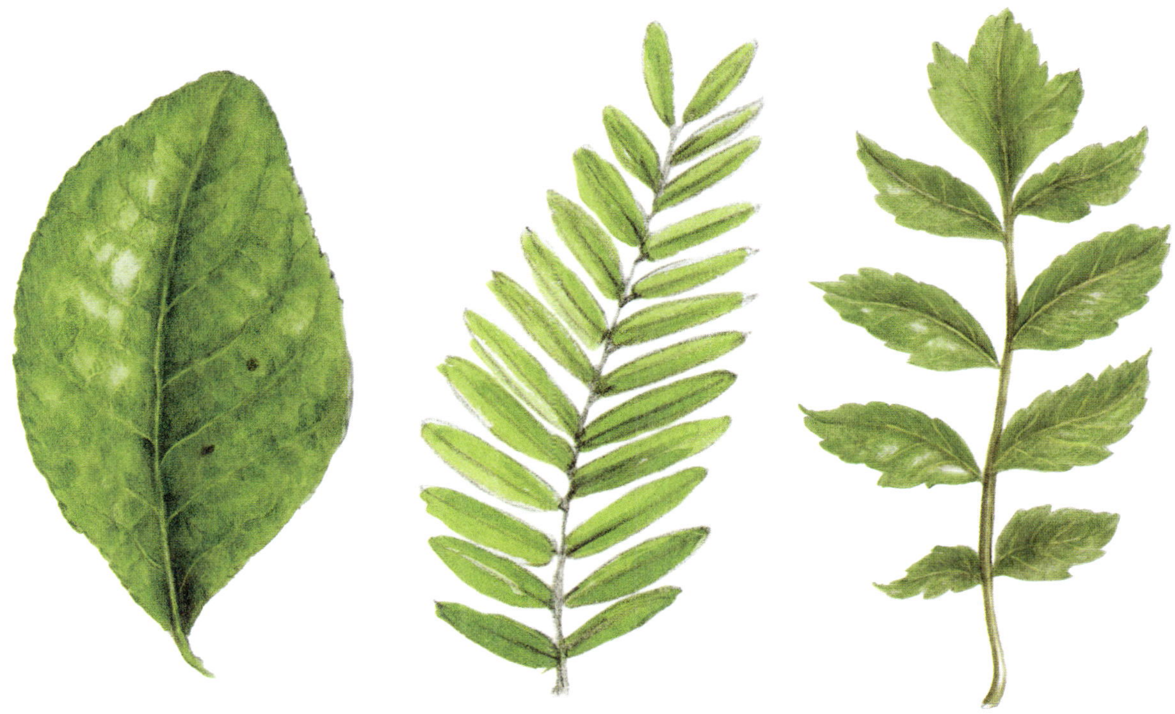

A simple leaf, such as a rose leaf (*left*) is single and undivided. A compound leaf is divided into leaflets (*center* and *right*). These three leaves are all pinnate, with a central axis, like the shaft of a feather.

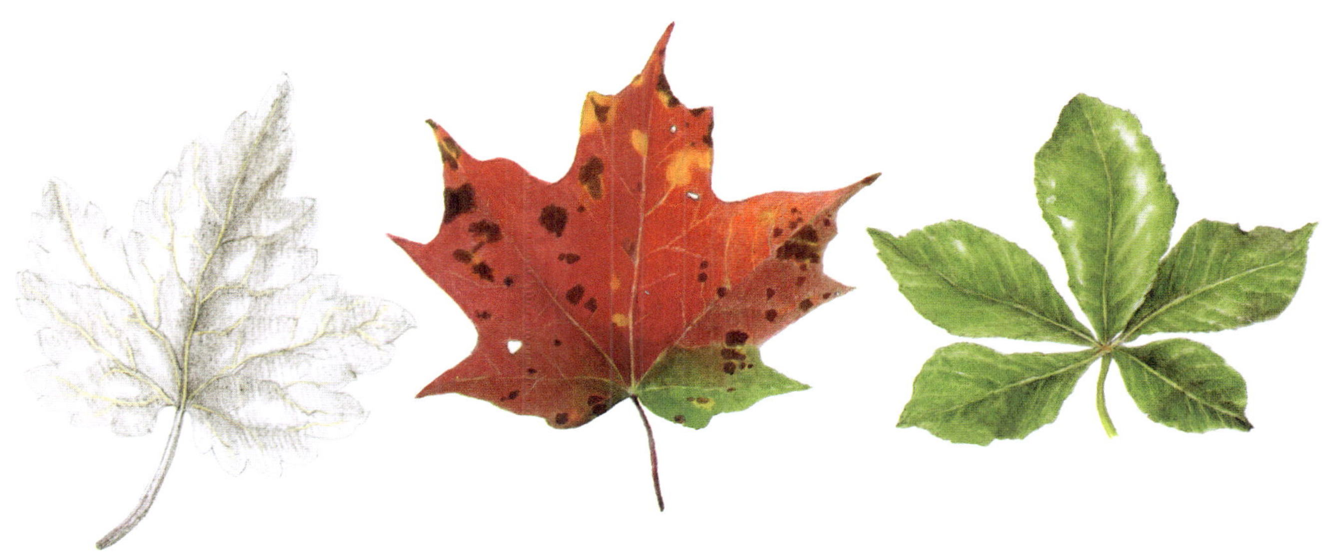

A palmate leaf is hand-shaped, like a maple leaf (*left* and *center*); a palmately compound leaf, such as a horse chestnut leaf, is divided into leaflets arising from a common point (*right*).

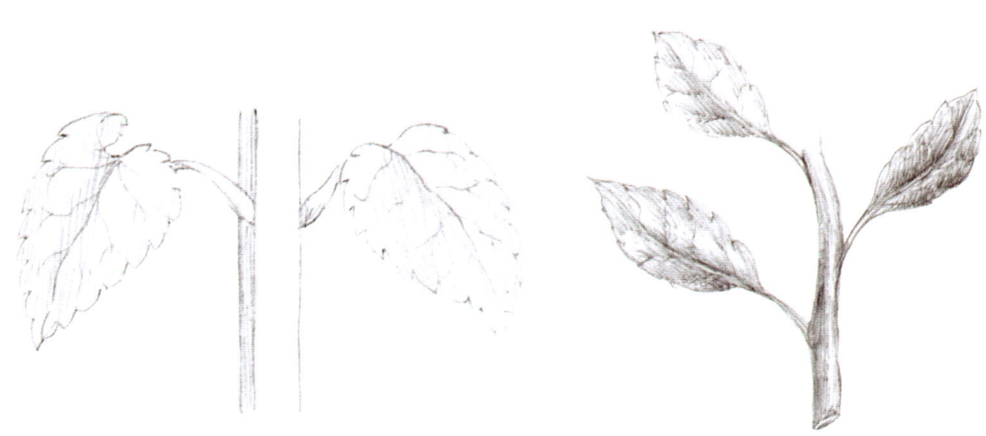

Leaves grow on the stem in opposite pairs (*left*), or grow singly in an alternate arrangement (*right*).

This illustration shows the parts of a complete (or bisexual) apple blossom, which has both male (stamens) and female (pistil) reproductive parts.

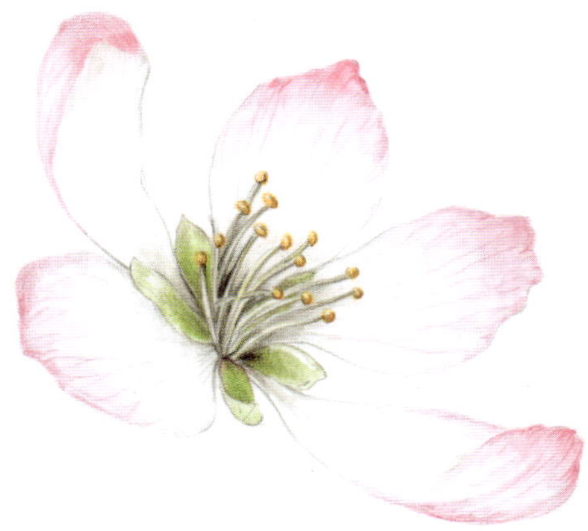

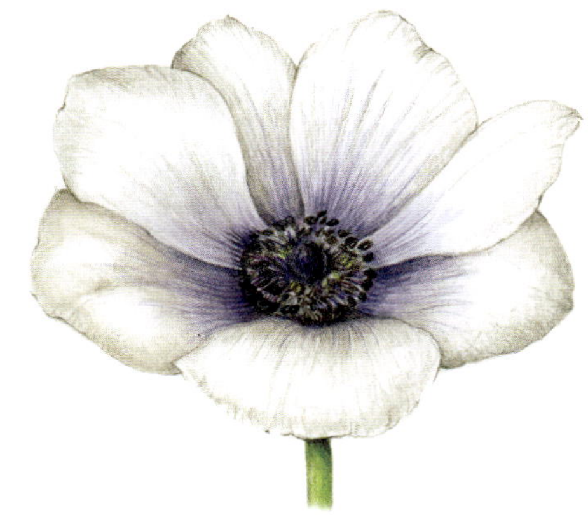

Radial symmetry in an anemone

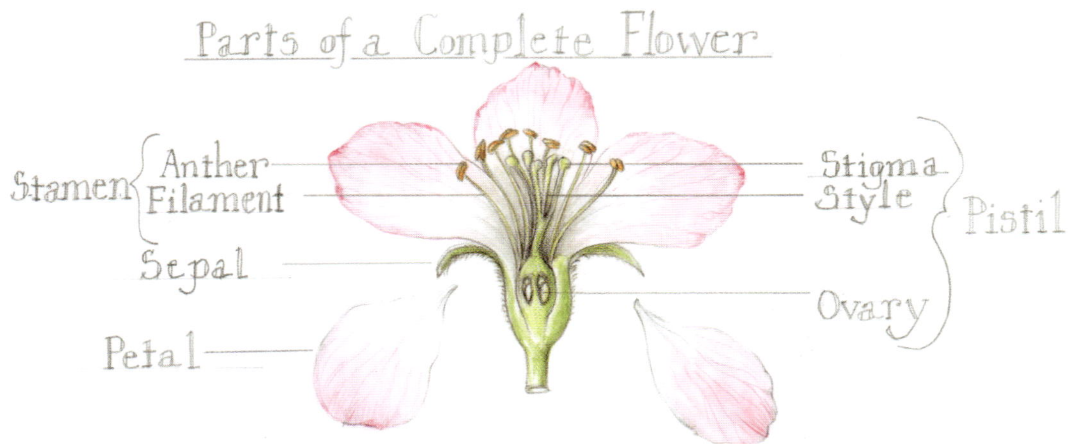


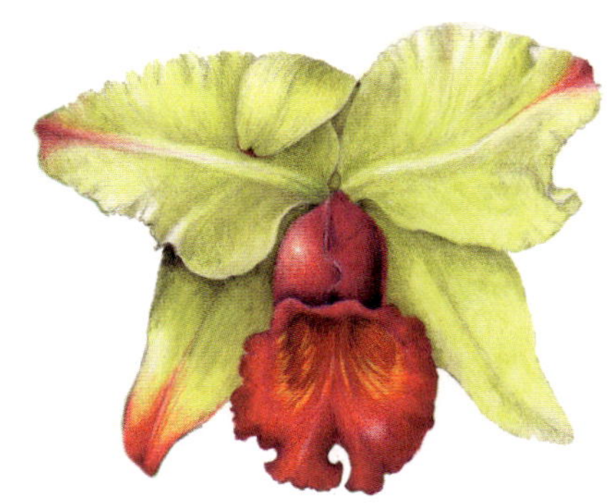

Bilateral symmetry in an orchid

Tulips have *tepals*, the word for when all of the petals and sepals on a flower look alike.

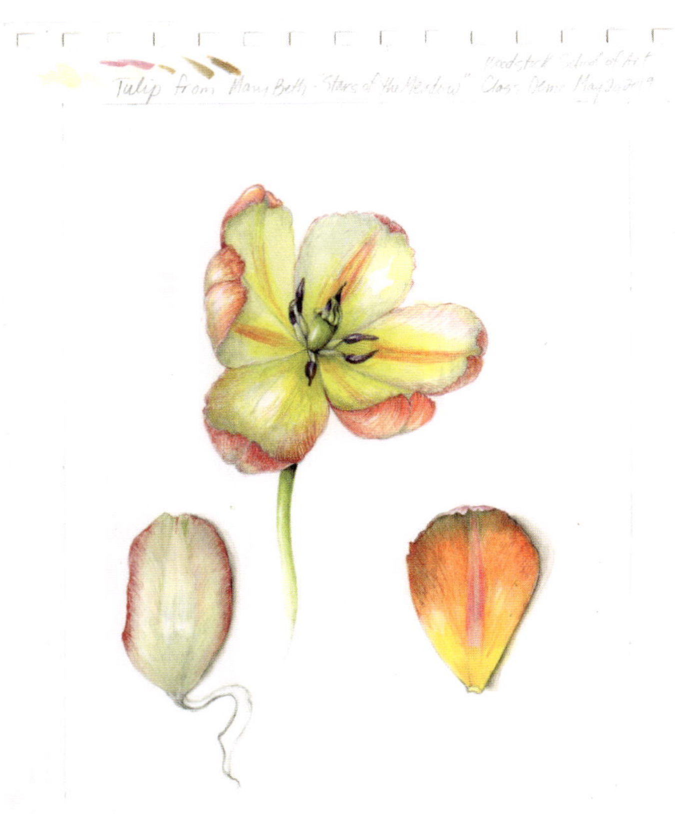

A major key to identifying and classifying plants into families is their flower structure, which can vary greatly. A typical flower structure with all its parts is shown above. Some flowers are bisexual, or complete, having both male and female parts, while others are unisexual, either male or female. The female reproductive part is the pistil; at the top of the pistil is the stigma, where pollen (from the male reproductive part) lands; a tube leads the pollen from the stigma to the ovary, which can develop into a fruit or seed capsule; the ovary may be divided into sections or chambers, called carpels. The male reproductive part is the stamen, comprising a stemlike filament that supports the anther at the top, which holds pollen.

Flowers usually have petals, the colorful modified leaves that we love so much. Often below the petals are sepals, leaflike structures that hold the flower together. Sometimes petals and sepals look alike, both often resembling colorful petals, as in tulips; the collective name for these parts is tepals. Flowers may have radial symmetry (with the parts arranged around the flower's center, like the spokes of a wheel) or bilateral symmetry, meaning the flower can be divided in half and will appear the same on both sides.

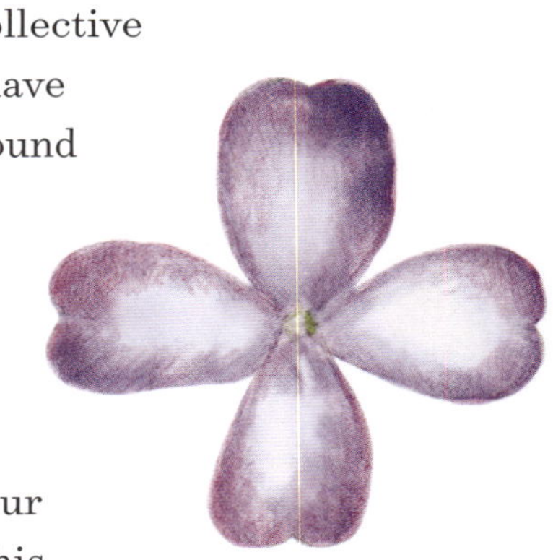

Dame's Rocket (*Hesperis matronalis*), from the mustard family

All flowers in the mustard family (Brassicaceae) have four petals in a cross pattern, four tall stamens, and two short stamens. Learn this one pattern, and you can identify all plants in this family—over 3,700 individual plant species. This is a great way to differentiate what may seem like sea of green color in the landscape. Soon you will be able to pick out individual plants and put them in a plant family! Even when you do not know a plant's name, you will be able to know what family it is in.

# List of Plant Families by Number of Species

*The scale of the illustrations corresponds to the number of species in each family.*

The plant families illustrated in this book together encompass more than 140,000 species, or almost 37 percent of the approximately 380,000 accepted species of land plants identified by scientists. About 2,000 new plant species are identified each year.

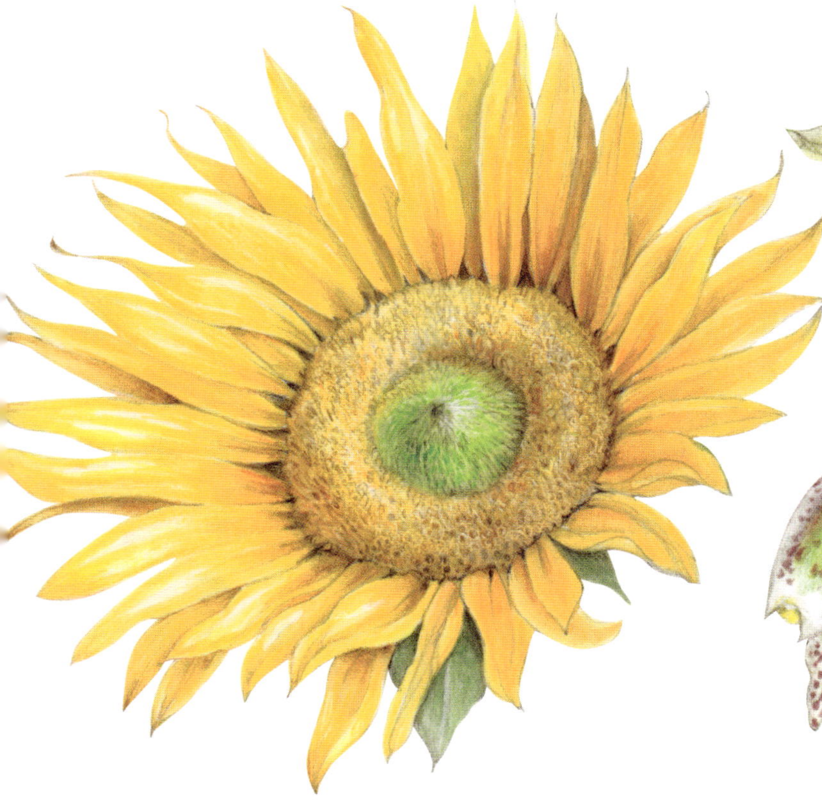

ASTERACEAE
Sunflower Family
30,000+ species

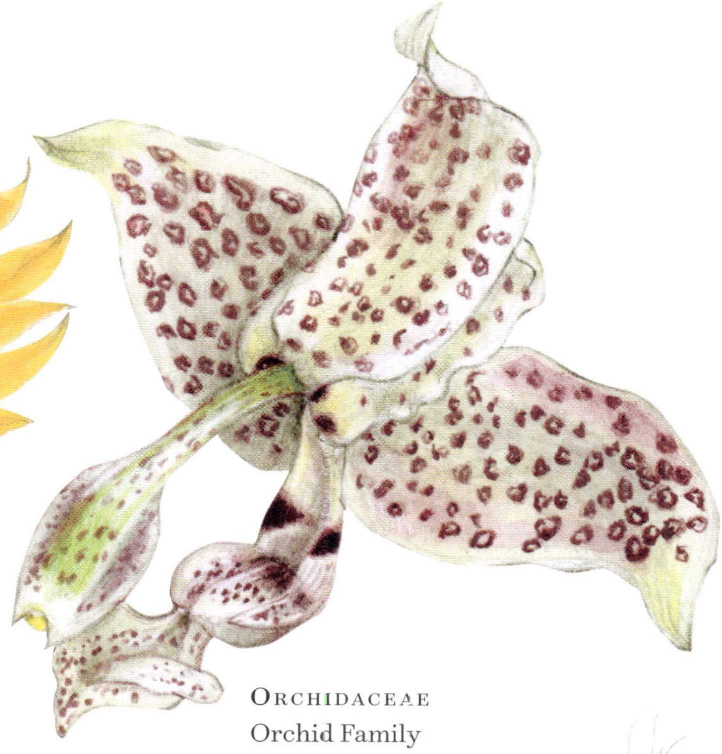

ORCHIDACEAE
Orchid Family
28,000+ species

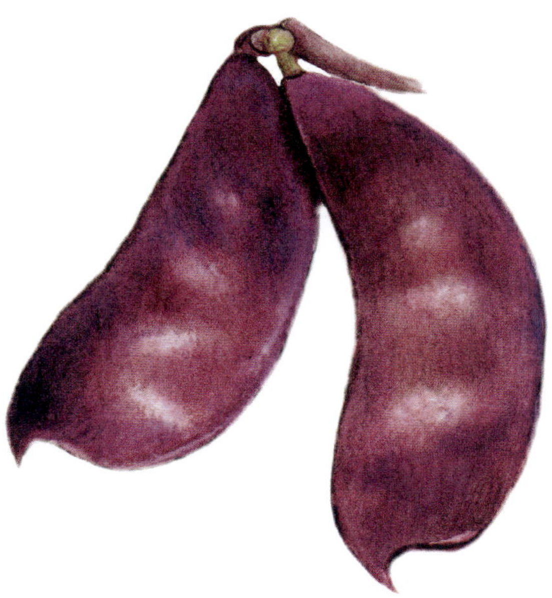

FABACEAE
Pea Family
19,000+ species

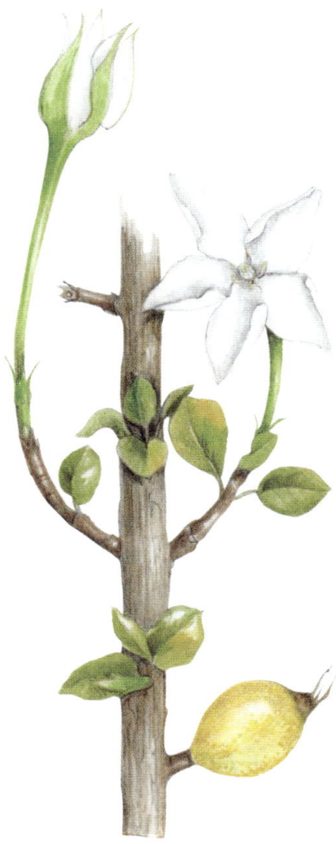

RUBIACEAE
Coffee Family
14,000+ species

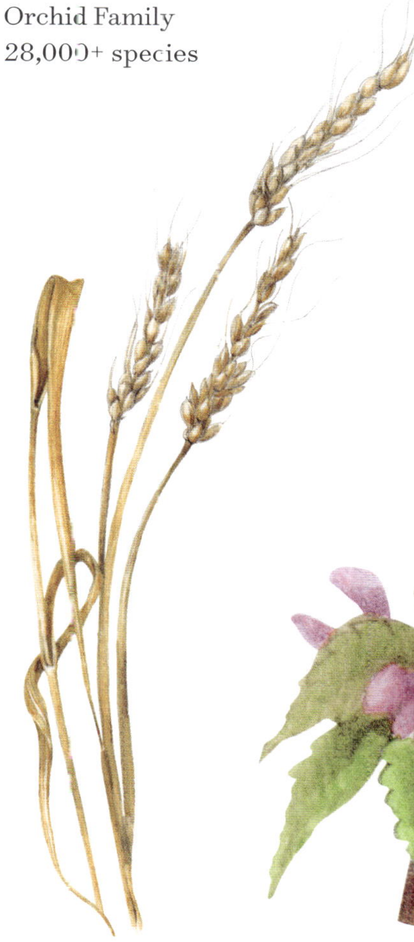

POACEAE
Grass Family
11,000+ species

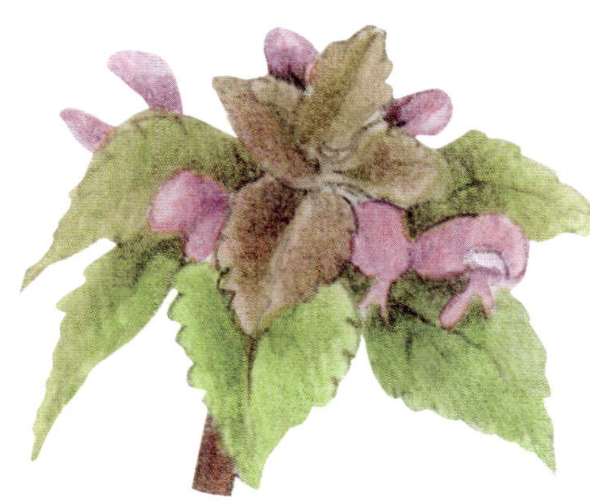

LAMIACEAE
Mint Family
7,000+ species

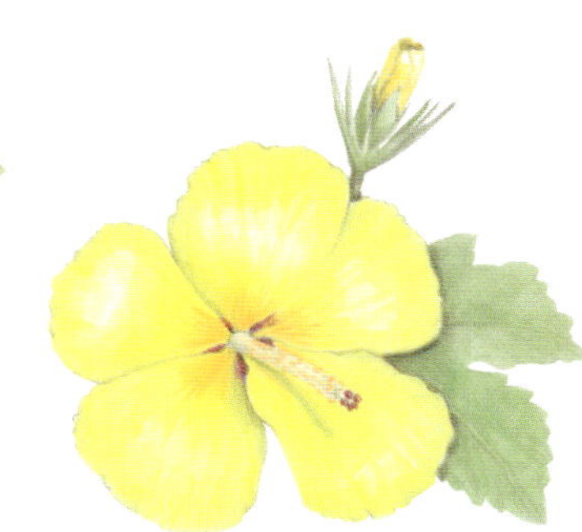

MALVACEAE
Mallow Family
4,000+ species

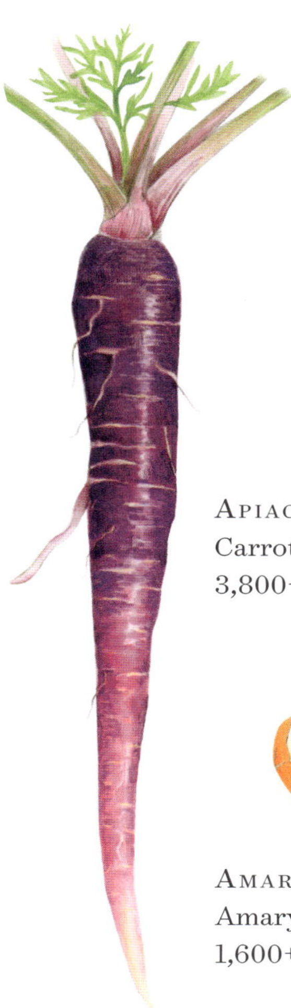

APIACEAE
Carrot Family
3,800+ species

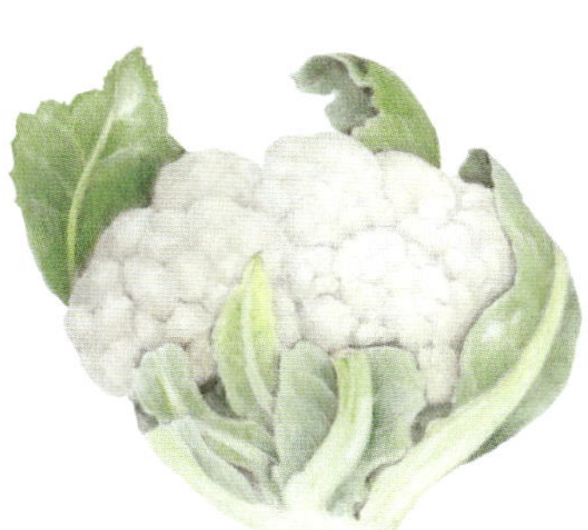

BRASSICACEAE
Mustard Family
3,700+ species

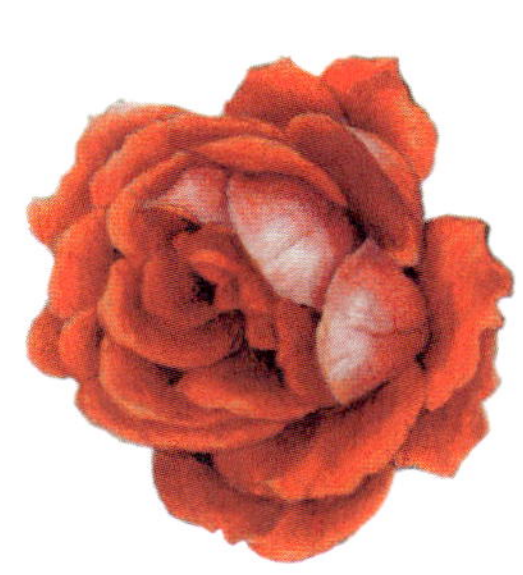

ROSACEAE
Rose Family
3,000+ species

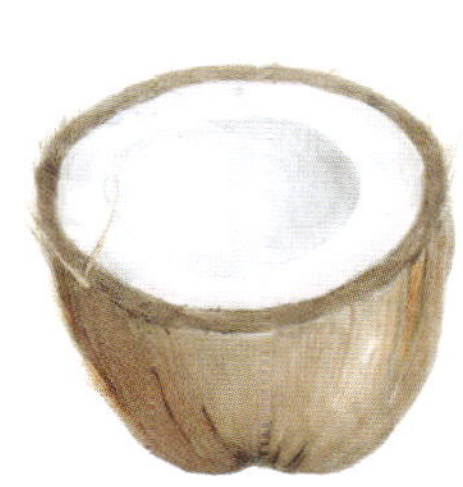

ARECACEAE
Palm Family
2,600+ species

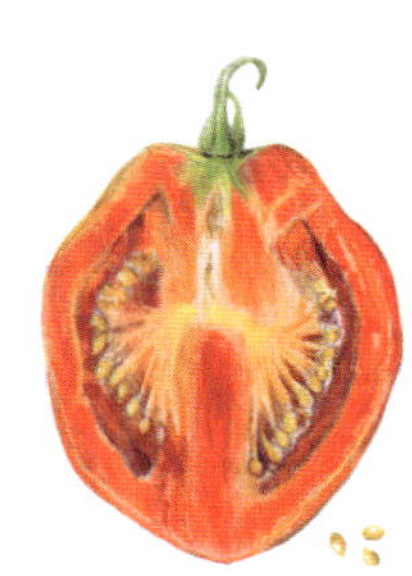

SOLANACEAE
Nightshade Family
2,500+ species

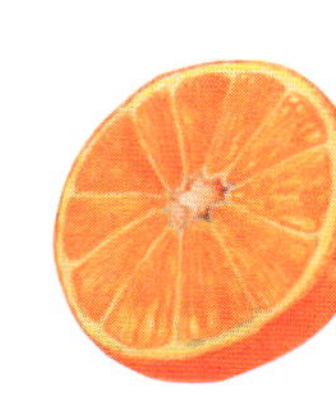

RUTACEAE
Citrus Family
2,000+ species

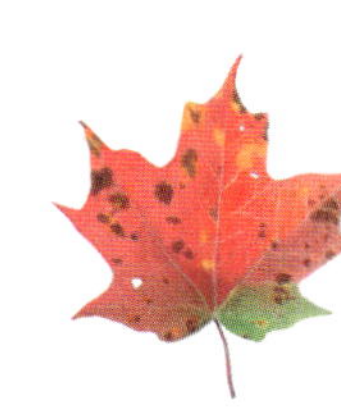

SAPINDACEAE
Soapberry Family
1,700+ species

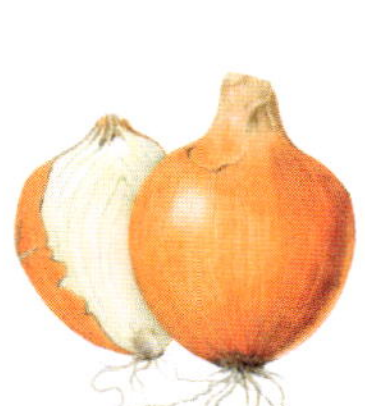

AMARYLLIDACEAE
Amaryllis Family
1,600+ species

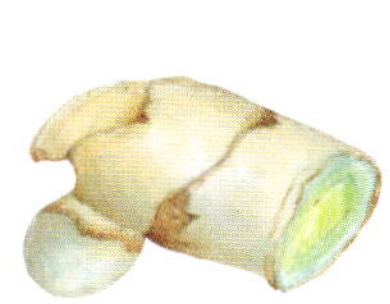

ZINGIBERACEAE
Ginger Family
1,500+ species

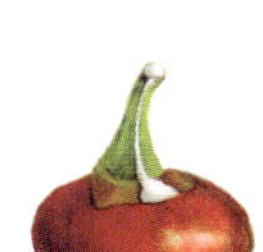

MORACEAE
Fig Family
1,000+ species

CUCURBITACEAE
Gourd Family
940+ species

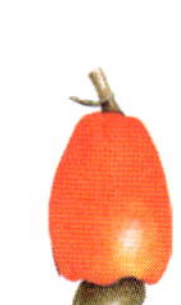

ANACARDIACEAE
Cashew Family
800+ species

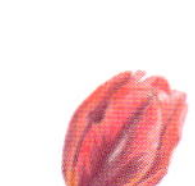

LILIACEAE
Lily Family
600+ species

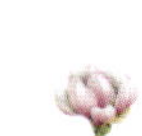

MAGNOLIACEAE
Magnolia Family
220+ species

# How I Selected and Illustrated the Plant Families

My selection of plant families is guided by the importance of plants to humans, their familiarity in our lives, the beauty we delight in, and the often-unnoticed connections they share as members of the same family. I considered plants' importance to humans for these purposes: food and medicinal uses; historical context and beauty; economic importance (timber and food, for example); environmental impact (either positive or detrimental).

I established basic criteria to guide my process in creating these paintings:

- Plants featured in a painting must belong to the same family.

- I worked from real plants, not photographs, though I studied photos to supplement my knowledge if the plant was no longer available.

- Paintings, measuring 22 inches by 30 inches, encompass all subjects on a single page, with the composition and interaction determined by the family's unique dynamics.

- Each painting includes text for identification, along with dates, locations, and other notes, including thoughts, connections, and discoveries.

- The margins are an integral part of the paintings, featuring color samples, test swatches, and notes. My medium is a combination of colored pencils and watercolor on hot pressed watercolor paper.

As these paintings have evolved, my process has been refined. I knew right away that I wanted each painting to be guided by the characteristics of the plant family. I consider myself in partnership with the plants, and I leave time and space for the plants to help me decide on the composition for the painting.

Nature is a built-in classroom for anyone to use. All you have to do is go outside and see what is happening. Take time to notice, and you will be amazed that every day there is a new lesson to explore with built-in questions and answers.

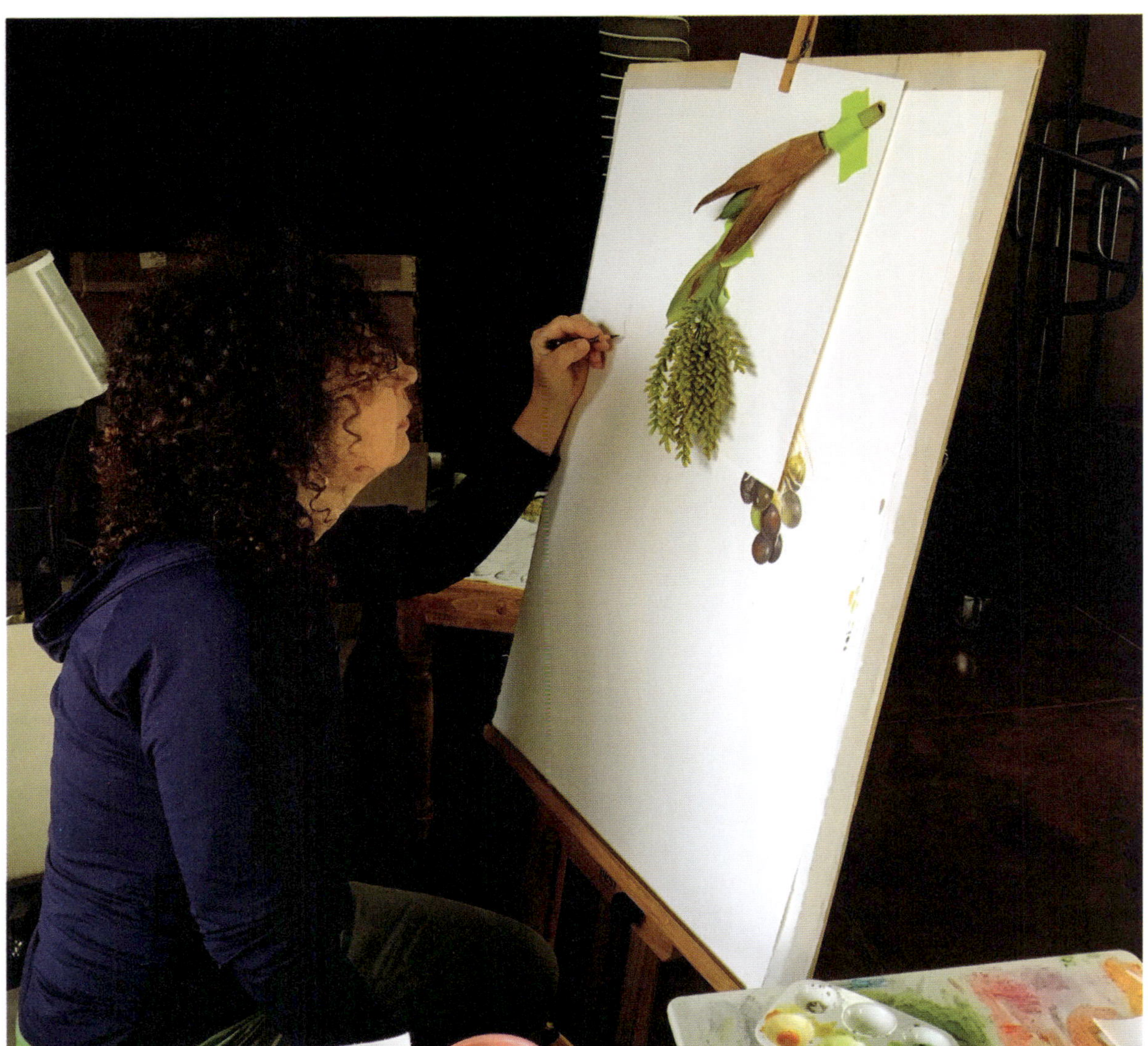

As I sketch the plants for my painting, I attach the live specimens to the paper in order to capture all of the details.

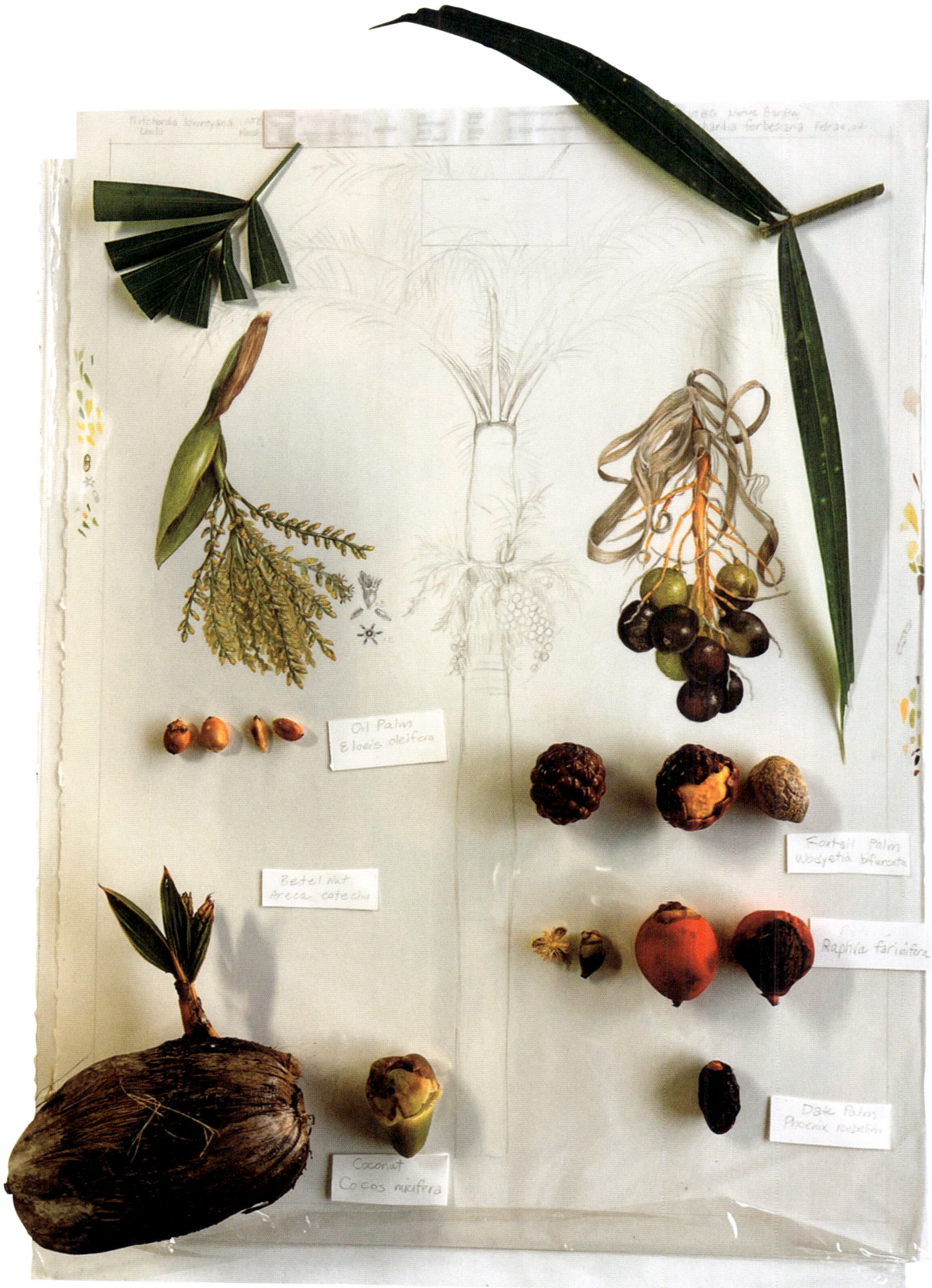

In the early stages of a plant family painting, I experiment with different compositions by placing a large sheet of clear plastic film or plain paper over the paper. This allows me to see how the plants speak to one another on a page even while my painting progresses, and protects the paper.

I placed these color samples in the margins of the rose family painting.

PROCESS STEP-BY-STEP

1. Research the plant family; note the characteristics that make it unique and the individual plants I might want to include.
2. Create a quick thumbnail sketch for composition.
3. Figure out where I can get the plants and when they will be ready to draw.
4. Introduce the plants to each other on the page to see how they might interact. I put the plants on large sheets of clear plastic film and then lay this over my real painting to consider new plants next to already painted plants on the page. I considered focal points and where text labeling would go.
5. Compare the color or color combinations of plants that might be used in a painting.
6. Draw the plants when they are ready, sometimes leaving details to finish later once the composition was complete.

# 1

# Amaryllis Family
## Onions (*Allium*)

### Amaryllidaceae
### 1,600+ species

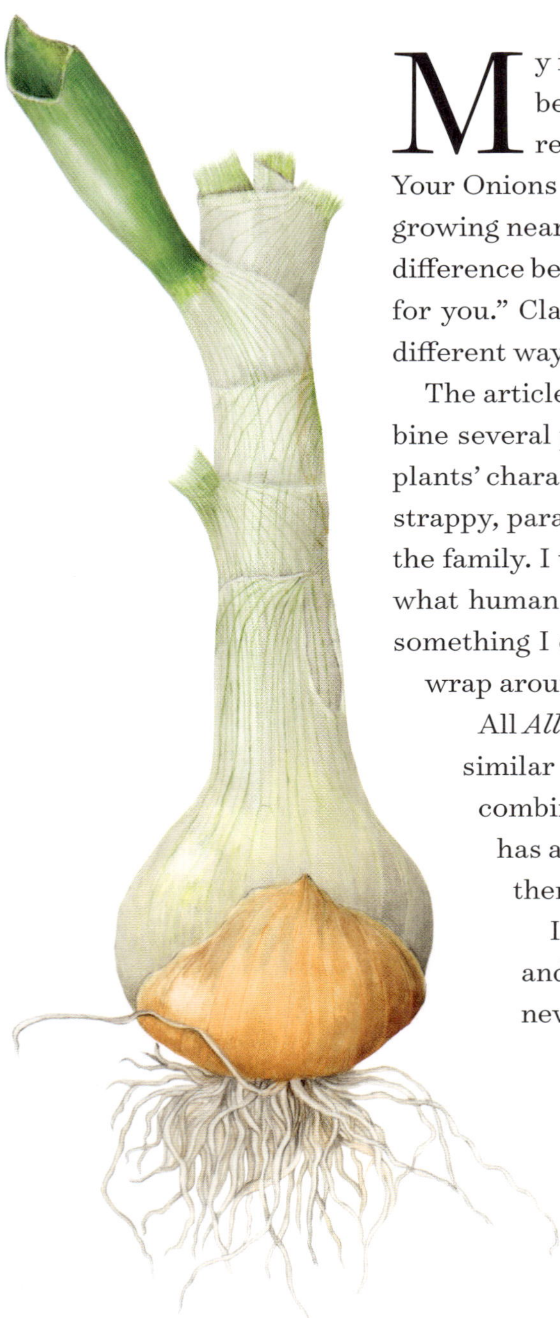

My first large painting composition, which inspired this project, was of onions (*Allium*), members of the amaryllis family (Amaryllidaceae). I was inspired to start this painting after reading an article by Melissa Clark in the *New York Times* on May 17, 2021, entitled "Know Your Onions (and Shallots and Leeks and Ramps)." I had just discovered my first large patch of ramps growing nearby and had created a painting of them. The article's subtitle asks: "Ever wonder about the difference between a spring onion and a scallion? A ramp and a scape? Melissa Clark has a crash course for you." Clark goes on to describe many of the popular alliums (members of the genus *Allium*) and different ways to use them.

The article inspired me to try a new idea. Instead of creating individual plant portraits, I would combine several plants from the same family on one large painting, weaving together some aspects of the plants' characteristics to make the composition cohesive. Because alliums are monocots, they have long, strappy, parallel-veined leaves, so I utilized their elegant movement to connect the different species in the family. I wanted the painting to be about connection, the plants relating to each other—not unlike what human family members do when they gather. In my family we hug and sit close to each other, something I especially love to do these days now that I am a grandma. I allowed the plants' leaves to wrap around one another and the roots to reach out to other plants on the page.

All *Allium* species are edible, so I ate these plants as part of my research, noticing how they were similar and different from each other. It was exciting to search out the subjects for this painting, combining cultivated and foraged plants. I am fortunate to live in a rural area and have a son who has an organic vegetable farm. This allowed me to track and study these plants as they grew and then add them to my painting.

I started the painting in May, after reading the article, and added to it through the summer and fall until it was finished. It was a profound moment for me. I felt as if I had discovered a new way of working. I decided to create more of these plant family paintings.

## Plants Illustrated

*Allium cepa*, Onion, Patterson
*Allium cepa*, Red Onion, Red Wing
*Allium cepa*, Spring Onion, Forum
*Allium cepa* var. *aggregatum*, Shallot, Conservor
*Allium porrum*, Leek, King Richard
*Allium sativum*, Garlic, German White
*Allium schoenoprasum*, Chive
*Allium tricoccum*, Ramp or Wild Leek
*Allium vineale*, Field Garlic or Onion Grass

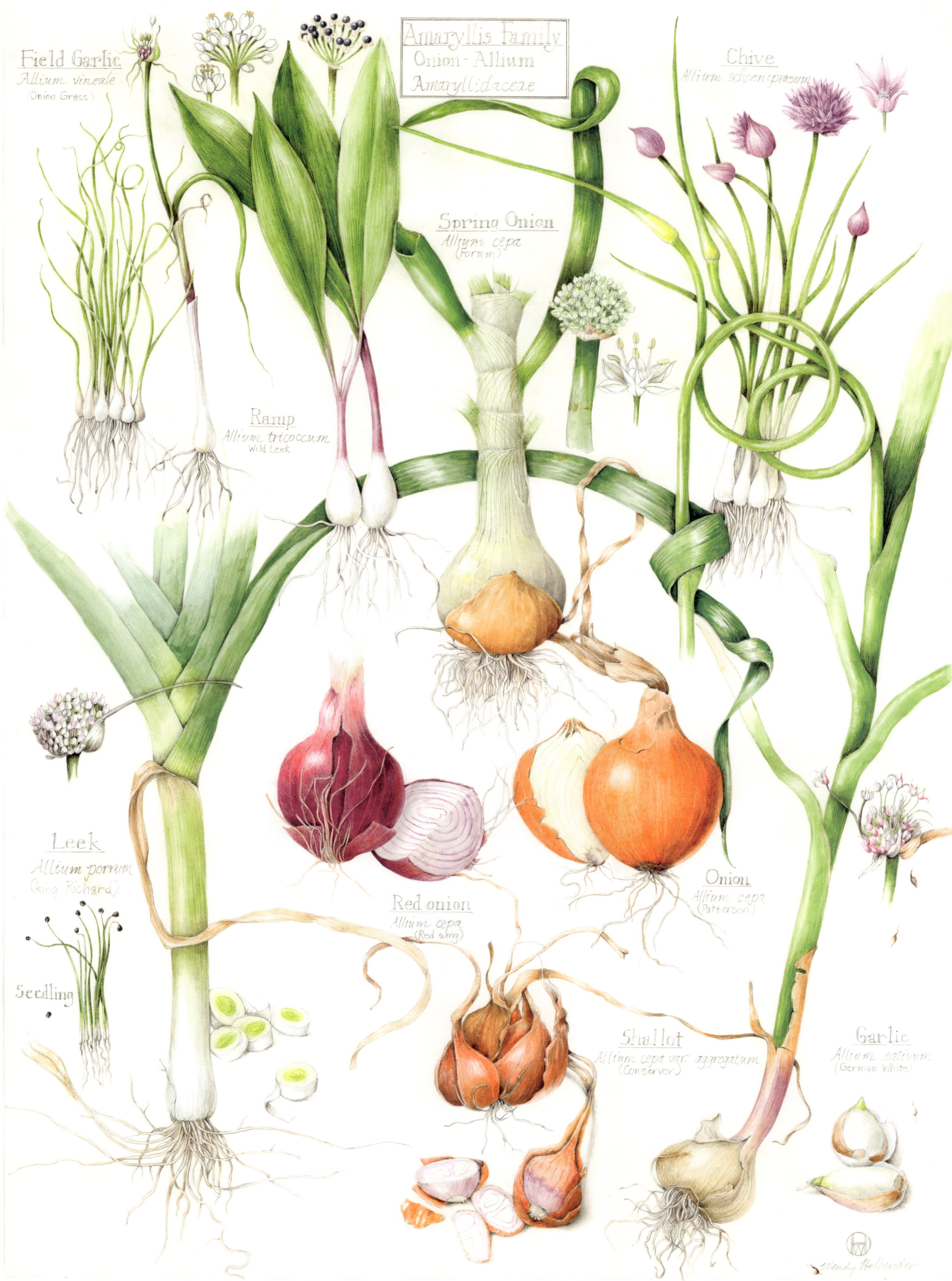
Amaryllis Family
Onion - Allium
Amaryllidaceae
Field Garlic
Allium vineale
(Onion Grass)
Ramp
Allium tricoccum
Wild Leek
Spring Onion
Allium cepa
(Forum)
Chive
Allium schoenoprasum
Leek
Allium porrum
(King Richard)
Seedling
Red onion
Allium cepa
(Red wing)
Onion
Allium cepa
(Patterson)
Shallot
Allium cepa var. aggregatum
(Conservor)
Garlic
Allium sativum
(German White)

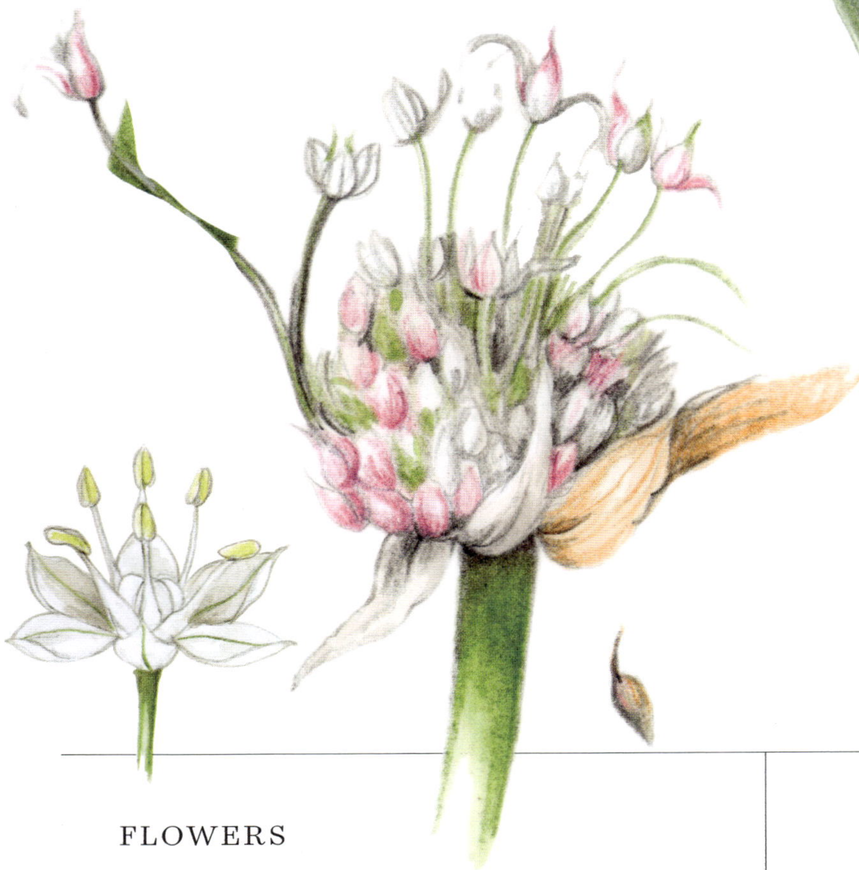

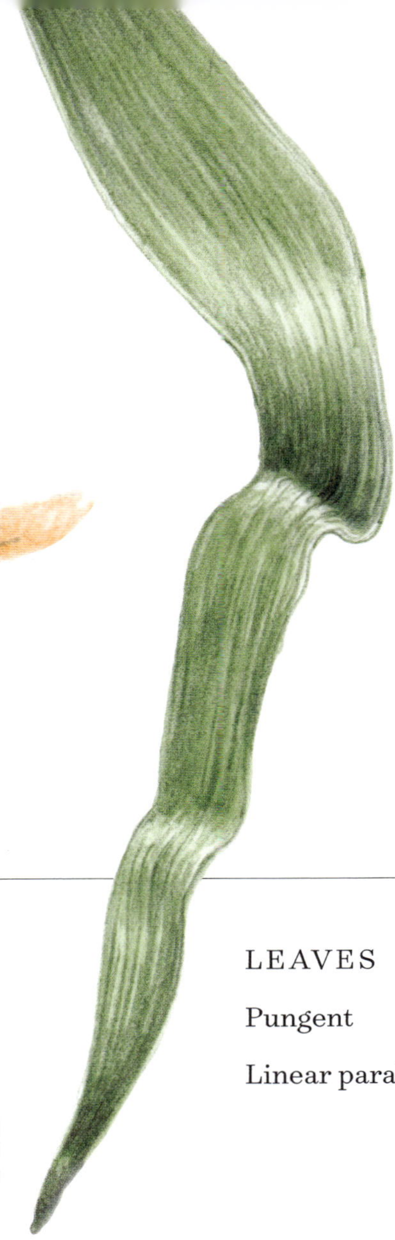

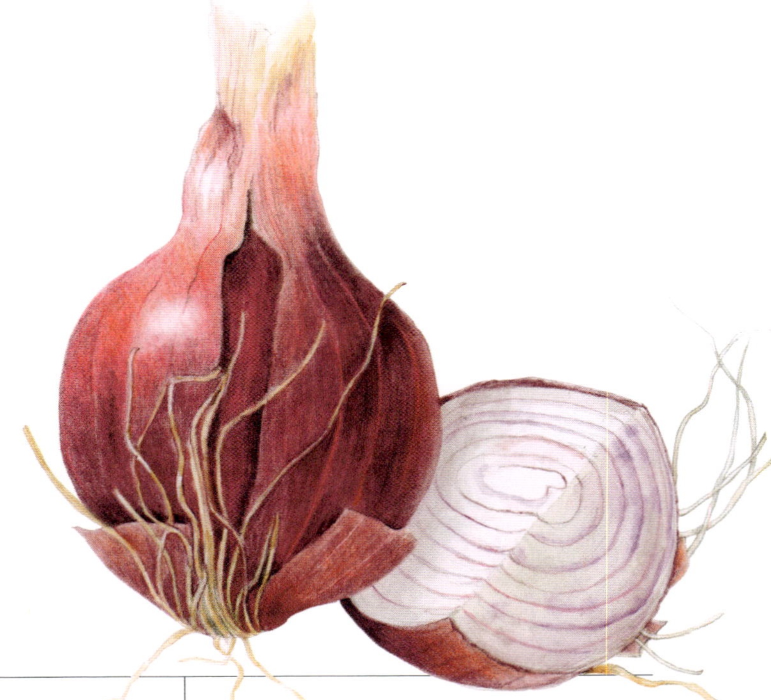

FLOWERS

Produced on scapes (leafless flower stalk)

6 tepals produced in 2 whorls

6 stamens, 1 pistil

LEAVES

Pungent

Linear parallel veining

BULBS

Reform annually from the base of the old bulb (most are perennials)

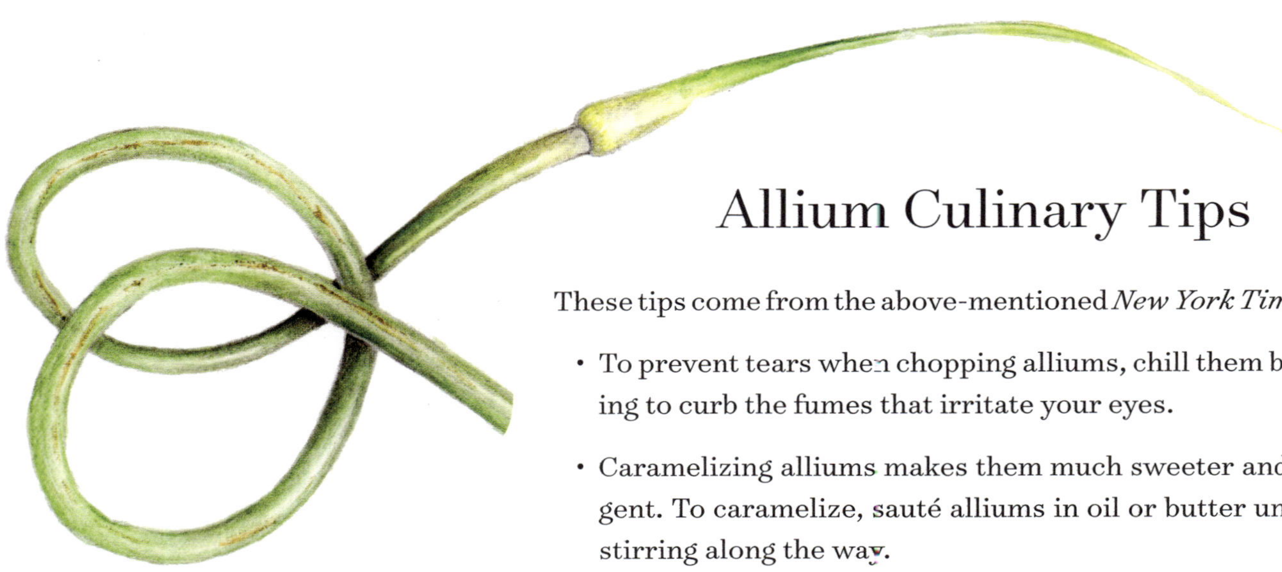

## Allium Culinary Tips

These tips come from the above-mentioned *New York Times* article:

- To prevent tears when chopping alliums, chill them before slicing to curb the fumes that irritate your eyes.
- Caramelizing alliums makes them much sweeter and less pungent. To caramelize, sauté alliums in oil or butter until brown, stirring along the way.
- Garlic scapes are the plant's developing flower heads. Clip them in the spring to encourage the bulbs to grow bigger. The scapes are delicious; they have a mild but flavorful taste with just a hint of garlic. To prepare, sauté or grill like string beans, or make into a pesto.

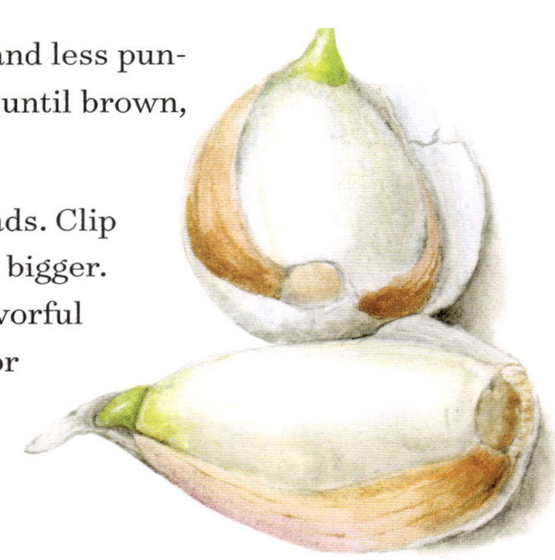

Garlic, German White (*Allium sativum*)

# More about Alliums

Onions are one of the most basic vegetables that are also a kitchen staple. They last a long time and add flavor to almost any recipe. Spring onions, as their name implies, are harvested in spring because they are usually planted in fall and overwinter. This delicious vegetable is harvested when its stems are still fresh. The stems and bulbs are both edible. Other varieties of onions are planted in spring and then harvested in the summer when they can be cured for storage. Onions are usually harvested before they flower, since once they flower, the plant puts its energy into growing seed, and thus the bulb won't grow as much.

I also grow ornamental onions in my garden and kids love to eat the flowers after they learn that they're edible.

Field garlic is another favorite allium of mine. My grandson Leo especially loves to pick and munch on them every time he runs by them in the grass.

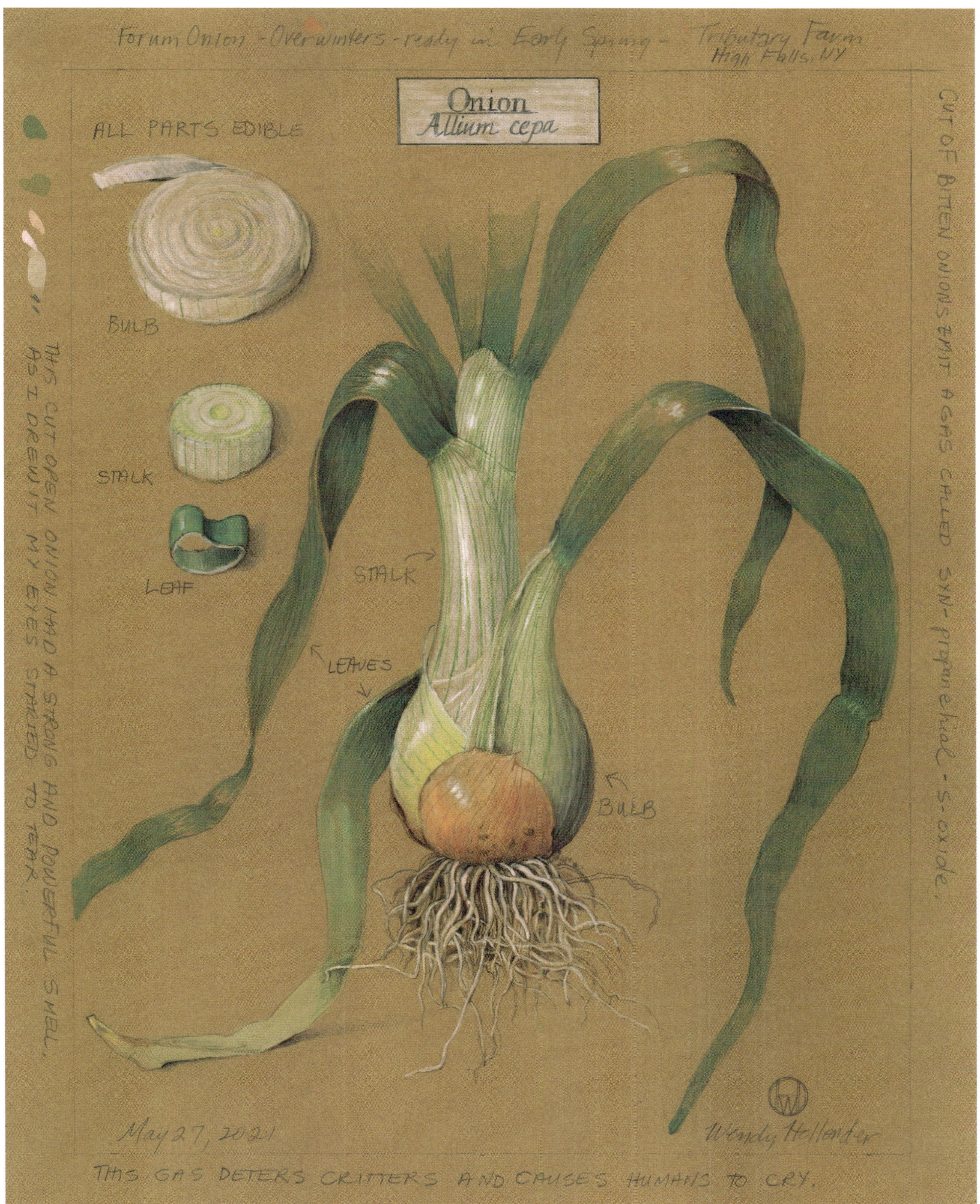


I created this painting on Stonehenge Kraft paper right after reading the Melissa Clark article. It was the inspiration to start the amaryllis family painting, and I knew I could have so much fun creating ribbonlike leaves that would weave throughout the different alliums. Kraft paper provides a way to explore drawing on a midtone paper, allowing the artist to add in whites for vibrant contrast.

# 2

# Amaryllis Family
# (*cont.*)

## Amaryllidaceae
## 1,600+ species

I created a second painting of the Amaryllidaceae after my first, of the onion genus (Allium), which represented only part of this family. This painting shows some of the flowering plants in this family alongside some of the edible allium or onion plants.

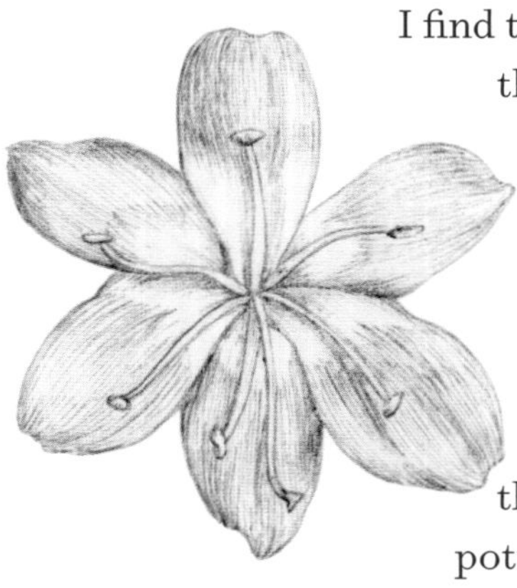

I find the flower structure of this family exciting to look at and to draw. When you get to see some of the showstoppers, such as amaryllis, blooming in pots in the winter, and daffodils, one of the first flowers of spring, it is always memorable. I love drawing flowers that have a distinctive three-dimensional structure because it is easier to create enticing depth with their forms. Flowers with a trumpet shape, such as daffodils, are especially gratifying to draw. I also love to draw bulbs, showing the three-dimensional form on round objects with vibrant highlights along with the twisting and turning of overlapping roots and long strappy leaves.

The amaryllis family allowed me to combine some plants that grow in temperate zones with those found in a more tropical setting. Many people, especially in colder zones, get amaryllis potted bulbs during the holiday season in order to enjoy their beautiful blooming flowers in the winter. These same plants grow outside in tropical landscapes. Daffodils are an early bulb flower in temperate zones, welcoming spring in the garden.

I started this painting in January 2024 after I had finished creating the first nine paintings for this book. I felt I could catch my breath and start a new painting. I was able to get Natal lily, paper whites, and amaryllis. I forced the paper whites to flower and watched them grow by my sunniest windowsill during the winter. A friend gifted me an Apple Blossom amaryllis slightly nibbled by her cat after she learned that these plants are poisonous to cats and can make them sick. Since amaryllises are so spectacular, I started the painting with it right in the center. I used the idea of entwining leaves and roots again. I had winter-storage red onions and garlic from my son's farm. I saved lots of room for tropical plants to be added when I went on my annual trip to Hawai'i in February.

## Plants Illustrated

*Allium cepa*, Red Onion
*Allium sativum*, Garlic
*Clivia miniata*, Natal Lily
*Crinum pedunculatum*, Swamp Lily
*Hippeastrum*, Amaryllis Apple Blossom
*Hippeastrum striatum*, Striped Barbados Lily
*Hymenocallis littoralis*, Beach Spider Lily
*Narcissus*, Daffodil
*Narcissus papyraceus*, Paper White

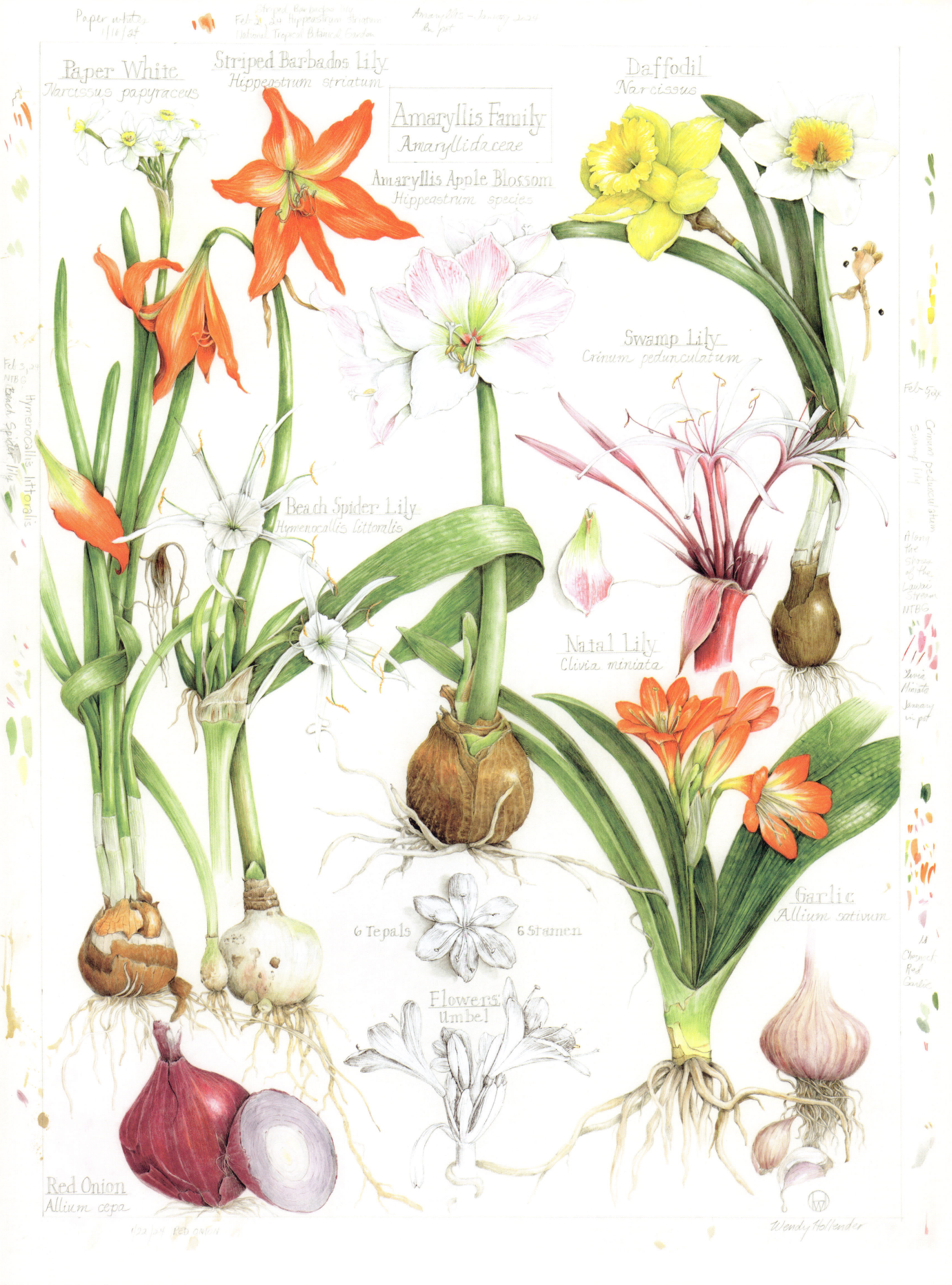
Paper White
Narcissus papyraceus
Striped Barbados Lily
Hippeastrum striatum
Amaryllis Family
Amaryllidaceae
Amaryllis Apple Blossom
Hippeastrum species
Daffodil
Narcissus
Swamp Lily
Crinum pedunculatum
Beach Spider Lily
Hymenocallis littoralis
Natal Lily
Clivia miniata
6 Tepals
6 Stamen
Flowers
Umbel
Garlic
Allium sativum
Red Onion
Allium cepa
Wendy Hollender

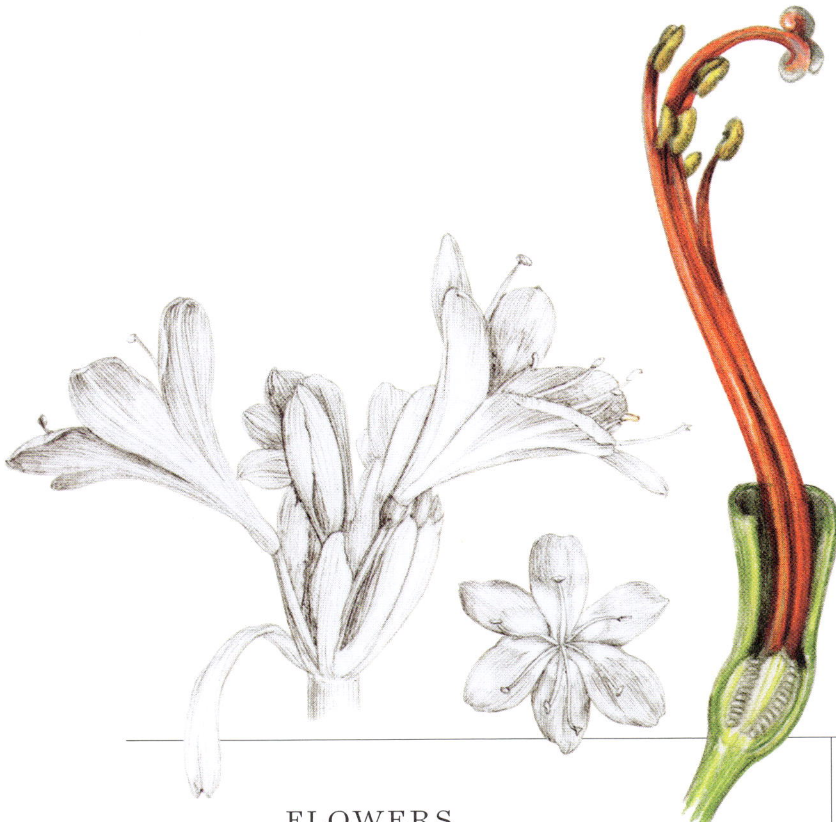

FLOWERS

Produced on scapes (leafless flower stalks)

6 tepals produced in 2 whorls

6 stamens

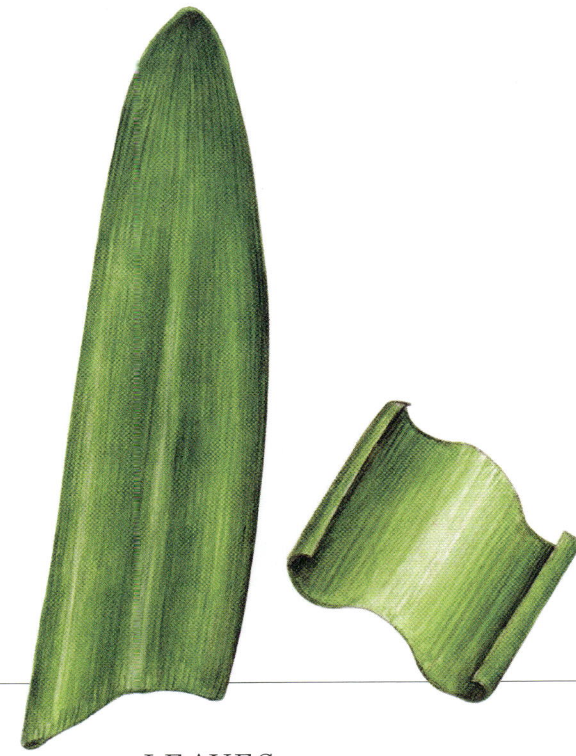

LEAVES

Pungent or not

Parallel linear veining

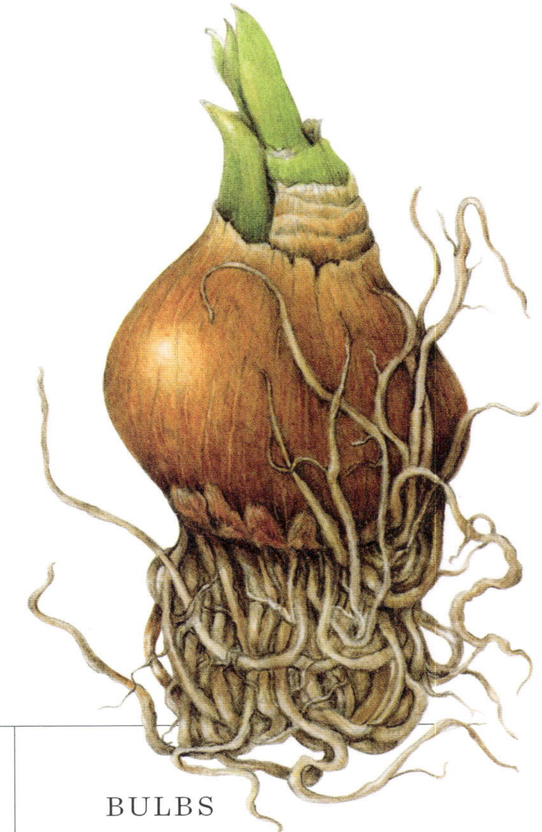

BULBS

Re-form annually from base of old bulb (most are perennials)

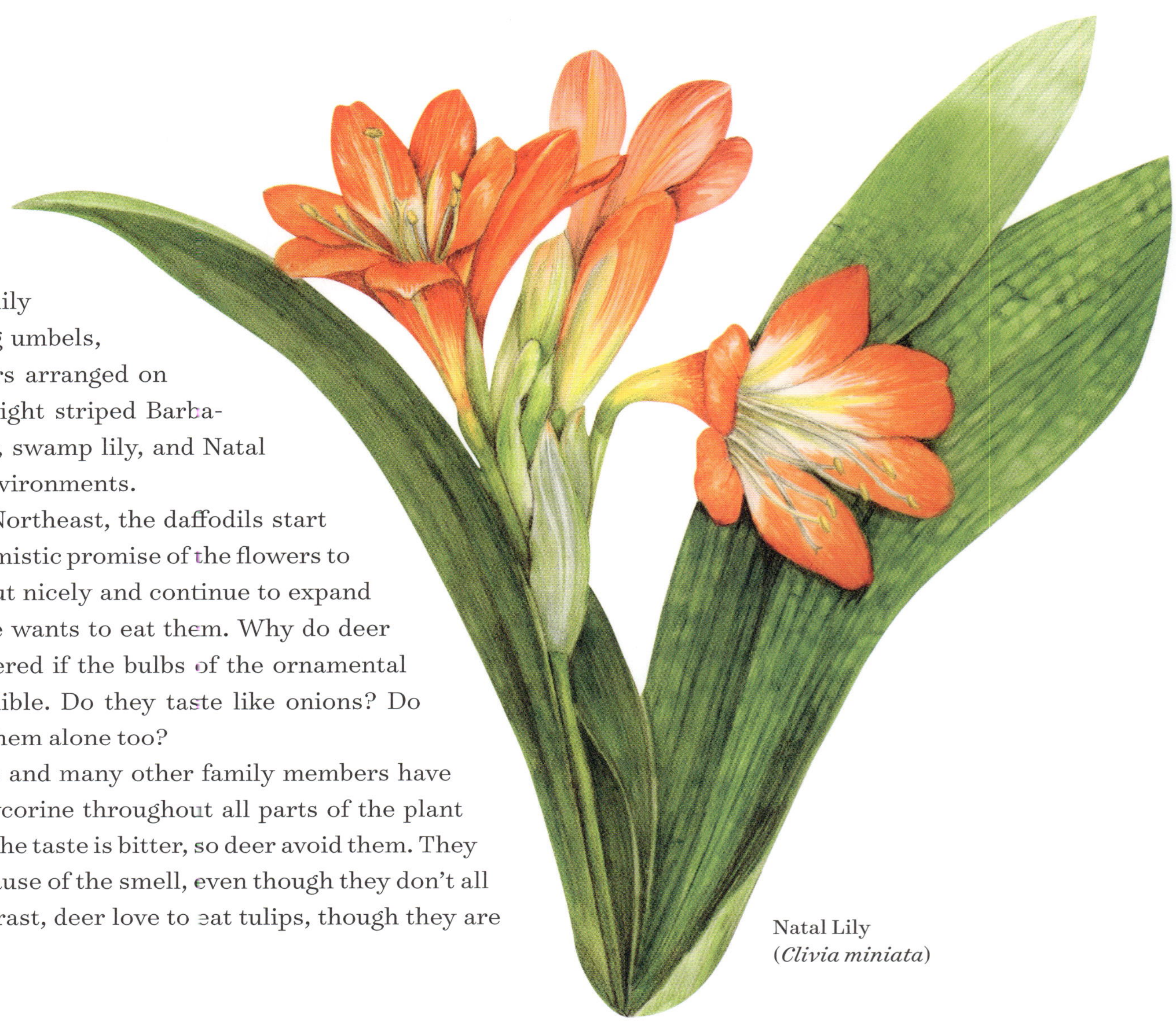

Natal Lily
(*Clivia miniata*)

The tropical flowering plants of the amaryllis family are full of large flowering umbels, clusters of several flowers arranged on one central stem. The bright striped Barbados lily, beach spider lily, swamp lily, and Natal lily all grow in tropical environments.

In early spring in the Northeast, the daffodils start to bloom, offering an optimistic promise of the flowers to come. Daffodils spread out nicely and continue to expand in the garden, and no one wants to eat them. Why do deer ignore daffodils? I wondered if the bulbs of the ornamental flowering plants were edible. Do they taste like onions? Do deer eat onions or leave them alone too?

I learned that daffodils and many other family members have a toxic chemical called lycorine throughout all parts of the plant that can make deer sick. The taste is bitter, so deer avoid them. They also avoid all alliums because of the smell, even though they don't all contain lycorine. By contrast, deer love to eat tulips, though they are poisonous to humans.

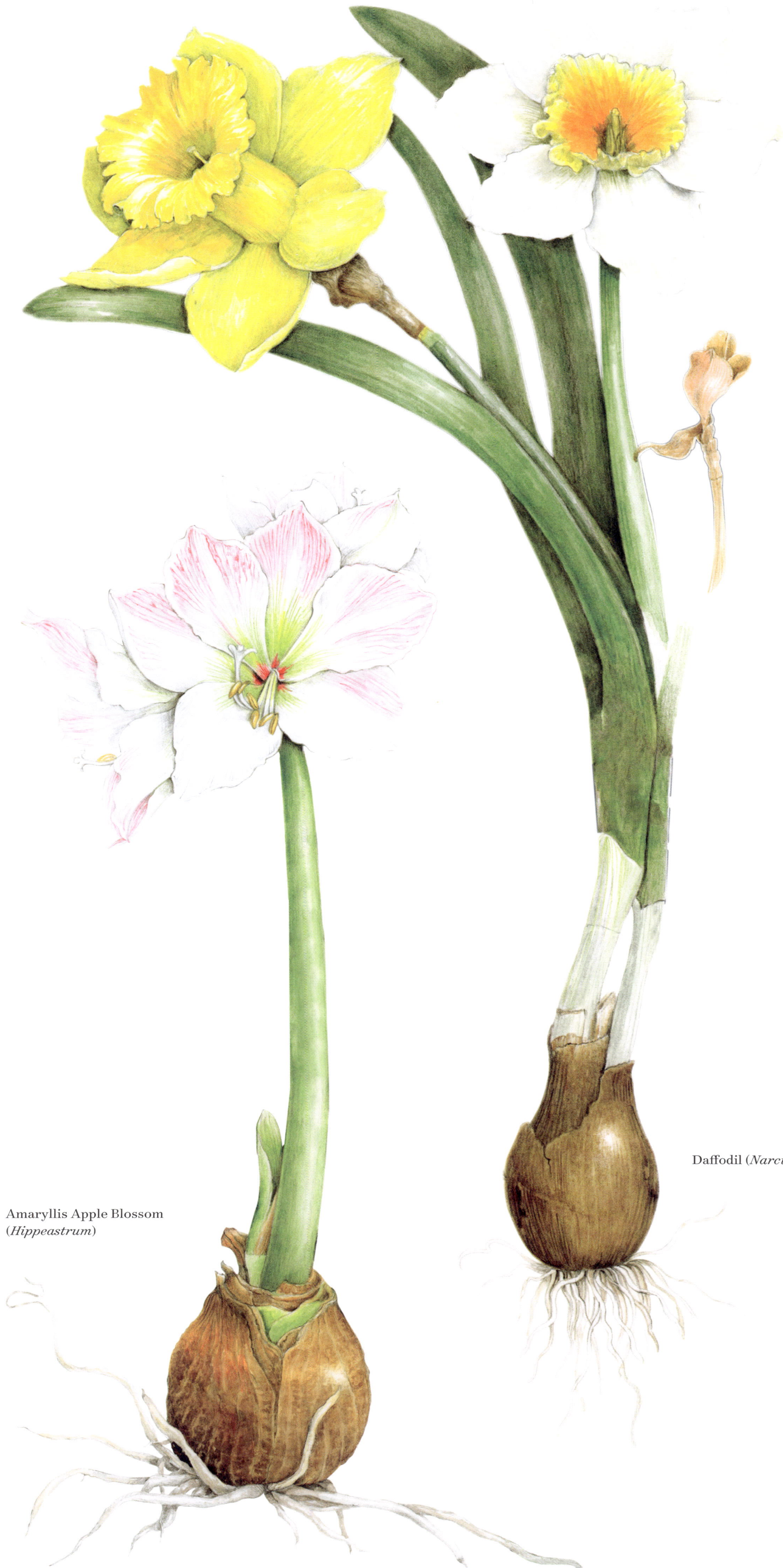

Amaryllis Apple Blossom
(*Hippeastrum*)

Daffodil (*Narcissus*)

3

# Carrot Family

## Apiaceae
## 3,800+ species

I remember as a child my mom encouraging me to eat carrots because they would improve my eyesight. When I would be "crangry"—a combination of hungry and cranky—while waiting for dinner, she would say, "eat a carrot stick," so I did. At the Jewish holiday Passover, at the beginning of spring each year, part of the celebration at the seder dinner included a ritual using bitter herbs such as parsley to remember the hard times of the past. Chicken soup, a family recipe made by my grandmother and then passed on to my mother and to me, has a base of carrots, celery, parsnips, parsley, and dill, all members of the carrot family, in addition to onion and a whole chicken. My grandmother even included the chicken's feet in her stockpot, a tradition that, I will admit, I haven't tried. Nevertheless, this is still the broth of the gods in my opinion and a wonderful cure for everything from the common cold to flu and anything else that ails you. Before serving, fresh chopped dill is sprinkled on top. Sadly, dill did not make it into my painting. I ran out of room!

Drawing the umbel (umbrella-like flower cluster) indicative of this family is a great way to learn about this unusual but fascinating pattern that nature uses and repeats over and over. I spent a lot of time looking under the microscope in the hopes of understanding the tiny flowers, which are mostly bisexual, meaning that each flower has both male and female reproductive parts. Because there are so many individual flowers—first on an individual stem and then grouped on a cluster of stems, and then these clusters are on an even bigger stem—made this quite confusing and also challenging. This is a pattern in nature called fractals, where part of the whole is repeated and resembles the larger structure as well.

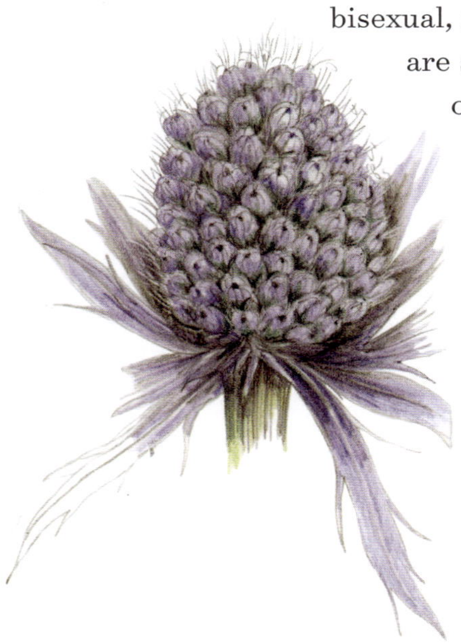

I noticed that about half of Queen Anne's lace flowers have one individual purple flower in the very center of the cluster. At first I thought it was a small bug, and I kept trying to brush it away. But under magnification I could see it was a purple flower,

### Plants Illustrated

*Apium graveolens*, Celery
*Apium graveolens* var. *rapaceum*, Celeriac
*Cicuta maculata*, Water Hemlock
*Coriandrum sativum*, Cilantro/Coriander
*Daucus carota* subsp. *sativus*, Carrot
*Daucus carota*, Queen Anne's Lace
*Eryngium planum*, Blue Sea Holly
*Foeniculum vulgare*, Fennel
*Heracleum sphondylium*, Hogweed
*Pastinaca sativa*, Parsnip
*Petroselinum crispum*, Parsley
*Zizia aurea*, Golden Alexander

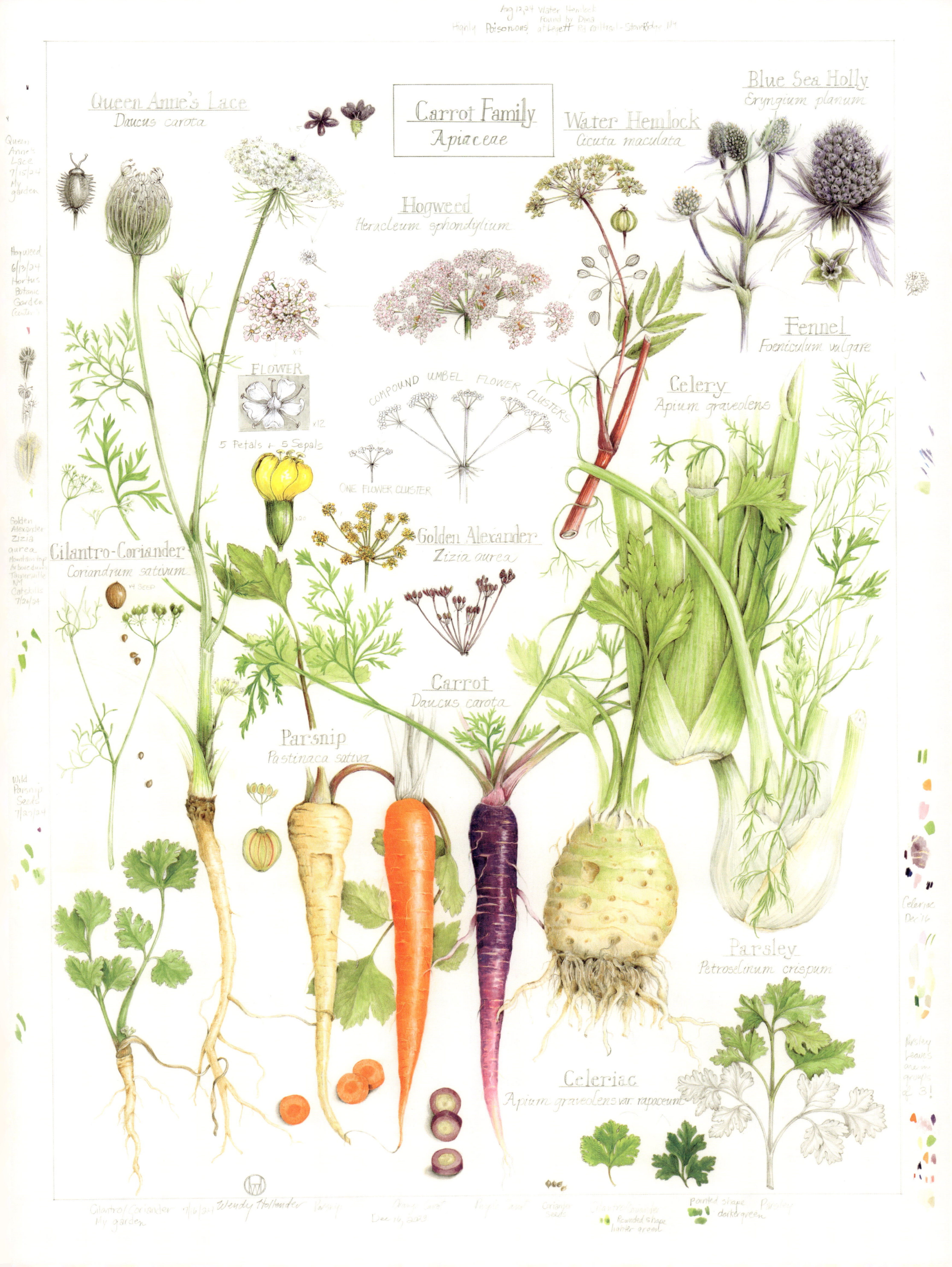

Carrot Family
Apiaceae
Queen Anne's Lace
Daucus carota
Water Hemlock
Cicuta maculata
Blue Sea Holly
Eryngium planum
Hogweed
Heracleum sphondylium
Fennel
Foeniculum vulgare
FLOWER
5 Petals + 5 Sepals
COMPOUND UMBEL FLOWER CLUSTERS
ONE FLOWER CLUSTER
Celery
Apium graveolens
Golden Alexander
Zizia aurea
Cilantro-Coriander
Coriandrum sativum
Carrot
Daucus carota
Parsnip
Pastinaca sativa
Parsley
Petroselinum crispum
Celeriac
Apium graveolens var. rapaceum
Wendy Hollender

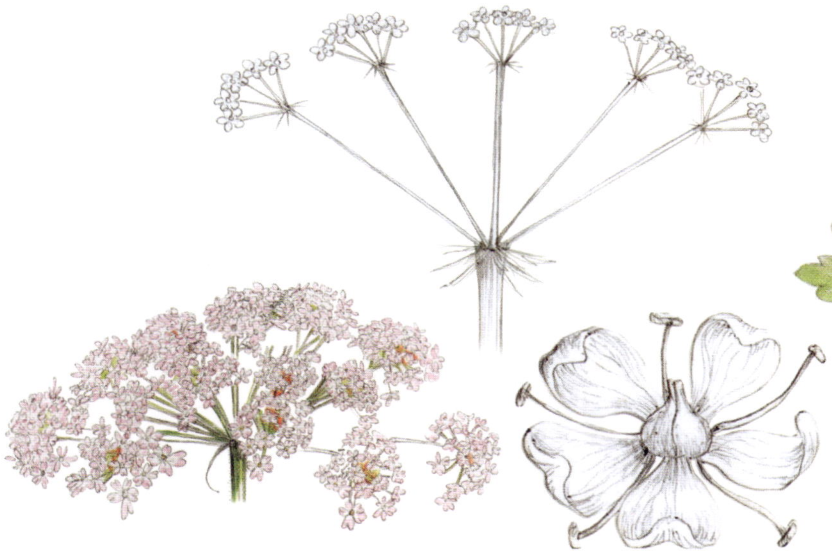

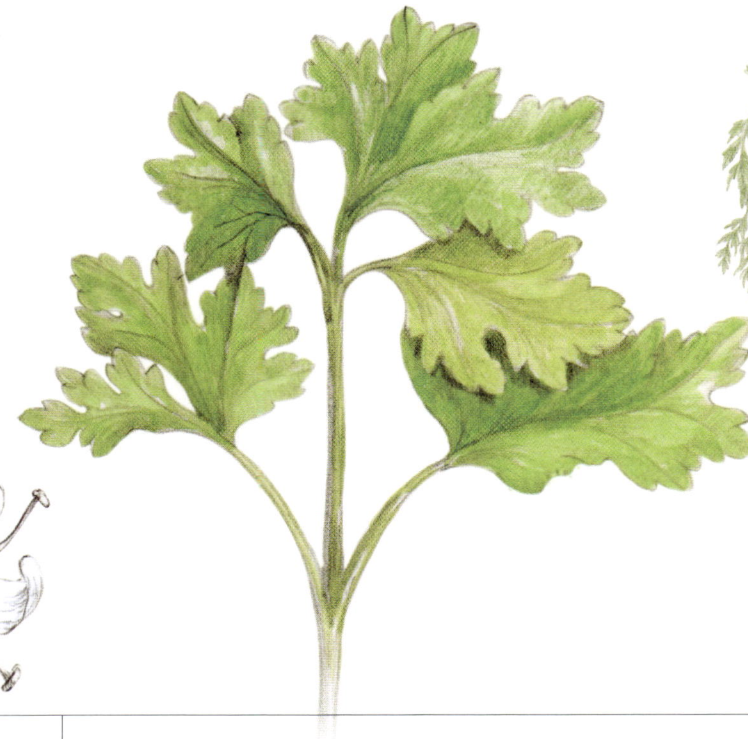

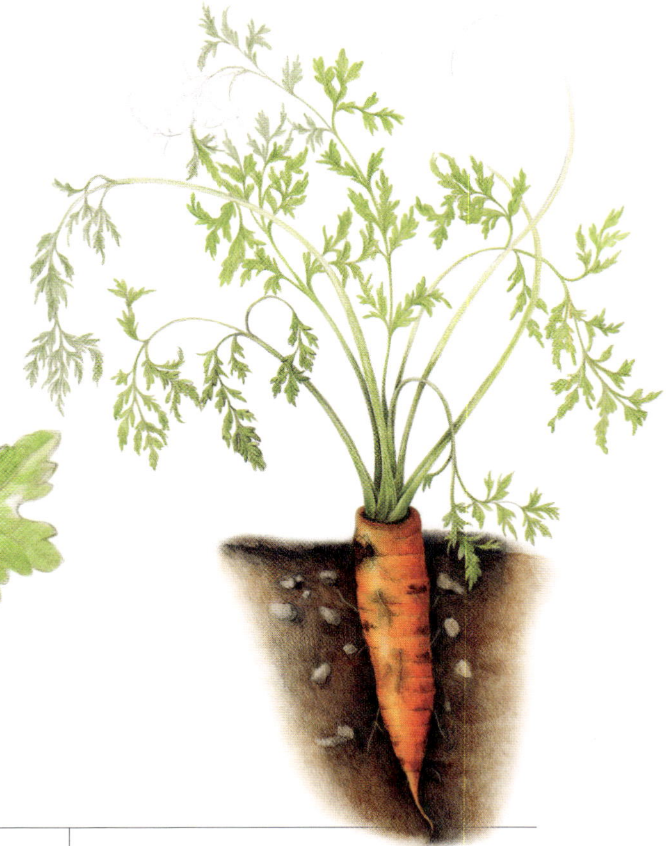

FLOWERS

In compound umbels

5 sepals, 5 petals

Usually bisexual

Flower stalks usually hollow

LEAVES

Pinnately compound

ROOT

Taproot (large central root) in some species

Plants aromatic

and my curiosity was raised. Why did nature do this? Nature usually seems to have a reason for doing things, in my experience, especially when it is repeated over and over again. I did some research, but the purpose of this one purple flower is a mystery. Perhaps it is meant to attract a pollinator, but no one knows for sure.

Apiaceae plants often have aromatic properties. Some members of the family can be toxic, so proper identification is crucial before consumption or use. When I collected water hemlock I was sure to wear gloves as it is highly poisonous. I notice when harvesting parsnip that it is important to protect my arms as the leaves can cause an uncomfortable rash. We often eat the taproots of this family, though stems and leaves of some plants can be consumed as well. I love to make a carrot pesto out of the leafy tops of carrots, for example. Originally, carrots were purple. Queen Anne's lace, a prolific wildflower in the meadows around my home and in temperate zones all over, is known as wild carrot.

In my research I found out that sea holly is part of this family. This confused me to no end because in my opinion, the sea holly resembles a thistle plant, and thistles are in the sunflower family. This led to more and more time under the microscope with the tiny flowers, and eventually I could see why sea holly is in the carrot family rather than the sunflower family—because each flower is bisexual (also known as a complete flower). I did have one nightmare or two where I accidentally put the sea holly into the sunflower painting, but thankfully this did not actually happen.

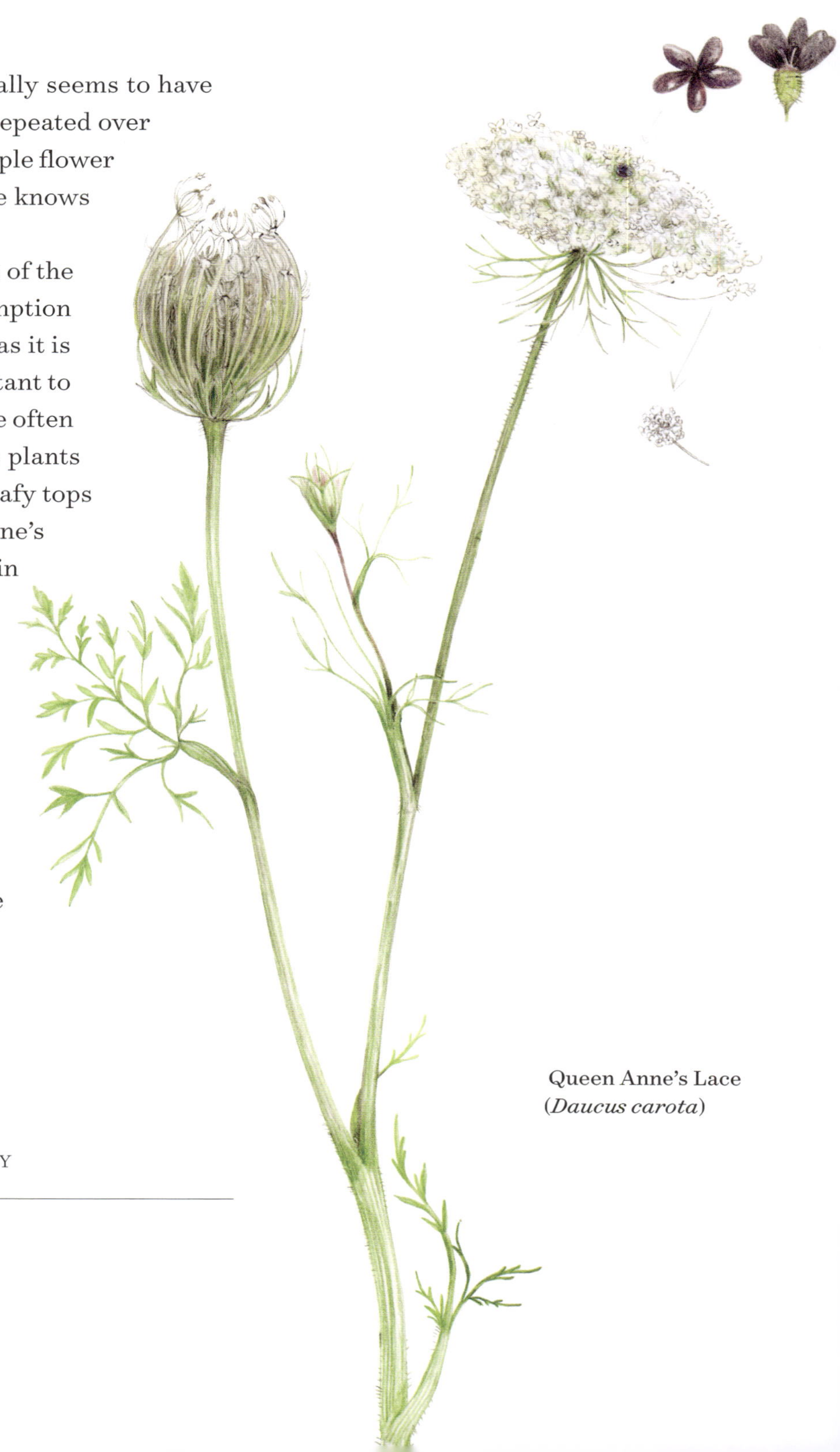

Queen Anne's Lace
(*Daucus carota*)

# A Fragrant and Familiar Family

### BY MARTA McDOWELL

In the world of flora, there is one family that, for me, has special appeal, in part because of its associations with a cartoon character and a poem. The former is a cartoon rabbit, a long-eared rascal with the signature phrase "What's up, Doc?" For decades beginning in the 1940s, Bugs Bunny headlined Warner Bros. Looney Tunes and Merrie Melodies cartoons while munching on a member of this plant family: the carrot.

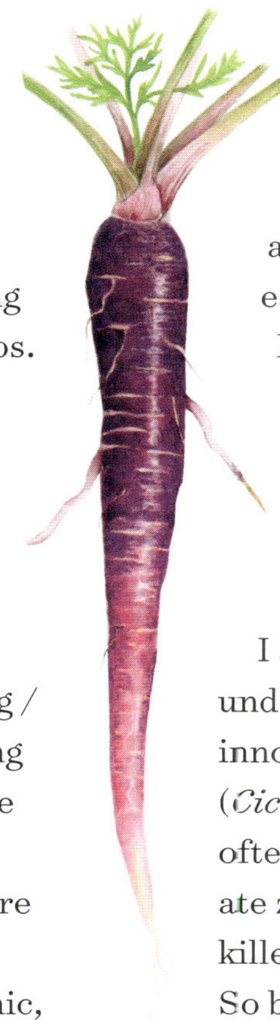

Carrot (*Daucus carota* subsp. *sativus*)

American poet William Carlos Williams wrote "Queen-Anne's Lace" in 1921. In this exploration of nature, desire, and diversity, he described its flower with "the fibres of her being / stem one by one, each to its end." This branching umbel shape is a feature common to many of the Apiaceae. I confess a nostalgia for the family's earlier name, Umbelliferae, as more lyrical, more descriptive of its skeletal form. Like parasols, umbels are radially symmetrical—actinomorphic, if you want the precise term. Most Apiaceae flowers look like the ribs of an umbrella shorn of fabric by gusts of wind and left abandoned on a city street corner.

Nostalgia is also likely at play in my fondness for this plant family. Saturday mornings of my childhood were spent on a nubby turquoise living room couch, watching TV cartoons with my older brother, Jerry. Outrageous, irrepressible Bugs Bunny was a favorite. I can still hear Jerry's whoops of laughter—bittersweet, as he died over twenty years ago.

Family vacations were summer driving trips from New Jersey to Illinois to visit my mother's parents, her siblings, and our twenty-six first cousins. The youngest of four, I sat between my two sisters, Pat and Kay, in the back seat or napping in the "way back" of the Rambler station wagon. These were the years before seat belts or personal electronic devices. Interminable highway hours were occupied with playing car games like I Spy, arguing, or staring out the window at miles of Queen Anne's lace nodding along roadside verges.

These memories seem to have transplanted themselves into my present-day garden, populated with many Apiaceae kin. Dill and coriander have their place, along with the decorative but still edible bronze fennel. Petroselinum crispum, or parsley, fills my November raised beds. Seedlings self-sown from last year's flowering plants have grown in the shade of larger vegetables for months. They have outlasted the tomatoes, basil, and cucumbers, ballooning in size with cool autumn temperatures and shrugging off the early frosts. Now they are ready for transformation into parsley pesto and Mediterranean-inspired recipes.

My little patch of carrots is still going strong after a period of benign neglect. Those I harvested in early summer were less baby carrot than gestational. Bugs Bunny would have scoffed. They were hardly noticeable chopped into a salad, though the carrot tops were tasty. Now they have bulked up significantly. I noticed my carrot leaves, like most leaves in Apiaceae genera, are sheathed where they are attached to the starchy orange roots.

I am reminded by Wendy's plant selection to underscore that not all Apiaceae are edible. Some innocent-looking umbels, including water hemlock (*Cicuta maculata*), are deadly. This plant of ill repute is often called the most poisonous of the northern temperate zone. While less well known than the hemlock that killed Socrates (*Conium maculatum*), it is equally lethal. So be safe. Don't ingest any wild plant unless confident of your foraging creds.

While I don't grow Queen Anne's lace—invasive in the mid-Atlantic where I garden—my flower beds are full of two self-sowing Apiaceae kin: the annual White Laceflower (*Orlaya grandiflora*) and the perennial Golden Alexander (*Zizia aurea*). Their sturdy stems stand tall without staking and spread into gaps in my borders. But even I edit their offspring, as their ability to propagate is pronounced.

As a gardener, I appreciate that Apiaceae species are generally easy to grow and offer accessible pollen and nectar to a variety of beneficial insects. Many are host plants for the black swallowtail butterfly. To me, each caterpillar I find decimating one of my garden plants is a tiny, significant victory for nature.

MARTA McDOWELL lives and gardens in Chatham, New Jersey. She teaches landscape history and horticulture at the New York Botanical Garden and is a popular lecturer and writer. Her latest book is *Gardening Can Be Murder* (Timber Press), about the horticultural connections to crime fiction. Timber Press has also published *Unearthing "The Secret Garden"*; *Emily Dickinson's Gardening Life*; *The World of Laura Ingalls Wilder*; the *New York Times* best-selling *All the Presidents' Gardens*; and *Beatrix Potter's Gardening Life*, now in its ninth printing. She was the 2019 recipient of the Garden Club of America's Sarah Chapman Francis Medal for outstanding literary achievement.

# 4

# Cashew Family

## Anacardiaceae
## 800+ species

On a walk one day with my colleague and friend, herbalist Dina Falconi, we naturally started talking about my favorite topic, plant families. Dina mentioned that the cashew family was interesting because its plants grow in both temperate and tropical landscapes. Cashew, mango, pistachio, sumac, and poison ivy are all in the same family. I had never seen a cashew fruit in the flesh but really wanted to because it is visually unusual and sumptuous looking. I could not resist. I knew it would be interesting to introduce the family's temperate and tropical plants to each other on the page. I imagined these plants from different climate regions would enjoy meeting each other inside my painting, like long lost cousins at a family reunion. I wondered if they had originated together at one time and then through dispersion and travel adapted and changed all over the world.

I started my painting in Hawai'i with mangoes growing at the National Tropical Botanical Garden (NTBG) in Kaua'i. I spent a week on this plant, gathering flowers and fruits that were growing side by side on the same tree. I started with mango, which can be an irritant, allergen, or even quite poisonous to some people. I had a student coming to a workshop the second week who I knew could literally not be in the same room with mango, so I planned to do all my drawing of this subject in week one.

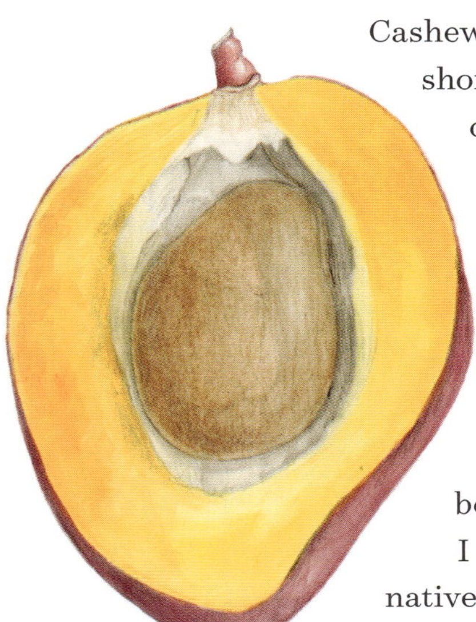

Cashew is one of my favorite nuts. I found a grower in Hawai'i, the Neu Mana Hui Farm on the north shore of Kaua'i, where I could see the plant growing. I was able to see it flowering in the farm's cashew orchard. It was an adventure driving down the long, bumpy dirt road that leads to the farm. When I arrived, I saw large cashew trees everywhere. They were flowering beautifully, but no fruit would develop until much later in the season. NTBG also had one cashew tree, but it was not yet flowering. Sadly, the cashew did not develop fruit for the six weeks I was in Hawai'i, but have no fear—I had connected with a source that I emailed for months to check on the cashew fruit. In late June, they alerted me to a package arriving the next day. I had to cancel all my plans and run home to meet my cashew fruits. It was a real joy to spend the next few days drawing them, with the nuts hanging out of the bottom of the fruit.

I am in awe of poison ivy, partly because it is such a powerful and mysterious plant. It is native to the United States and can grow in many environments. It even knows how to change its appearance so it can be difficult to detect. Even though poison ivy is dangerous to humans, many

## Plants Illustrated

*Anacardium occidentale*, Cashew
*Koordersiodendron pinnatum*
*Mangifera indica*, Mango
*Pistacia vera*, Pistachio
*Rhus typhina*, Staghorn Sumac
*Toxicodendron radicans*, Poison Ivy

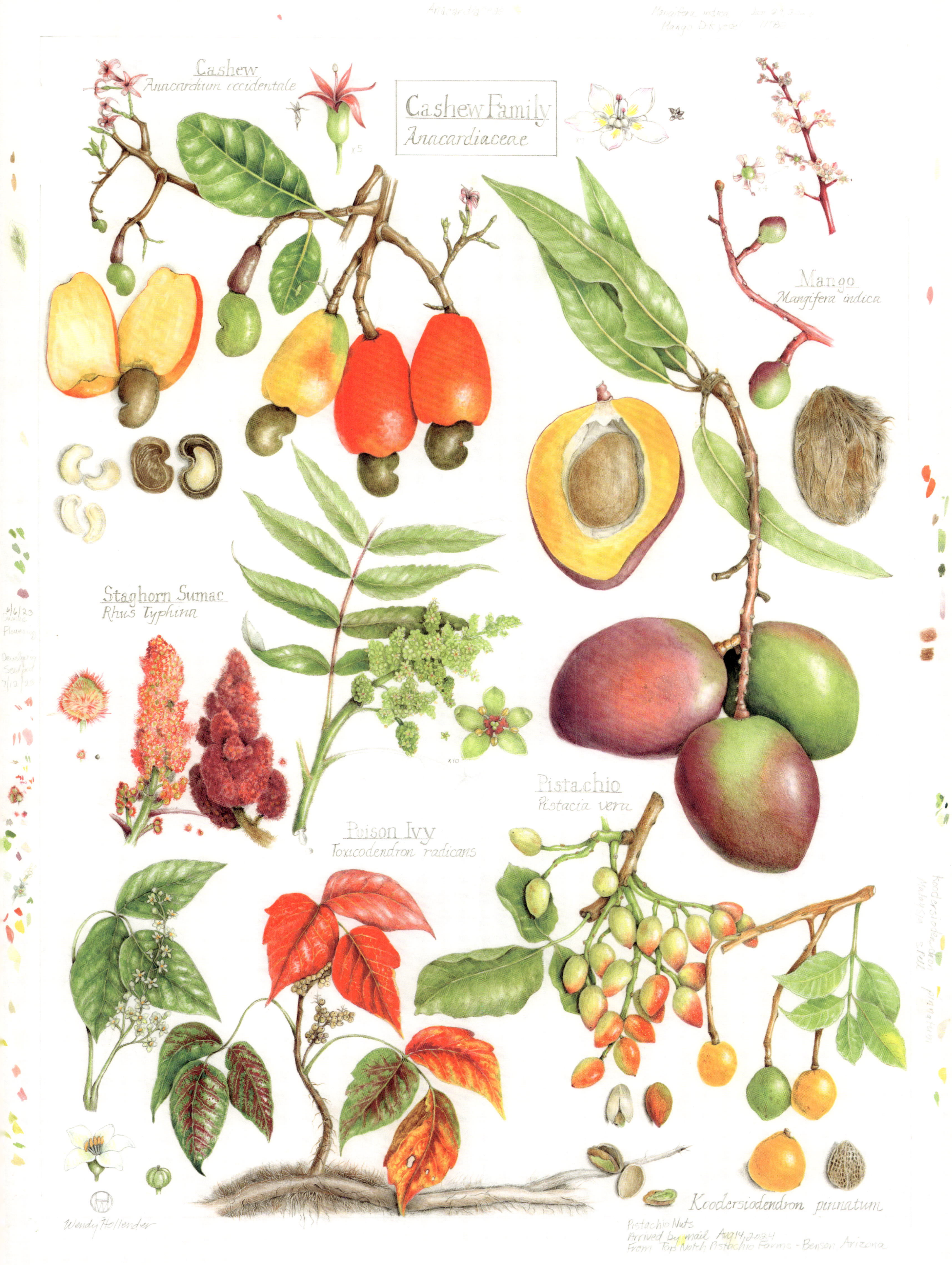

Cashew
Anacardium occidentale
Cashew Family
Anacardiaceae
Mango
Mangifera indica
Staghorn Sumac
Rhus Typhina
Poison Ivy
Toxicodendron radicans
Pistachio
Pistacia vera
Koodersiodendron pinnatum
Wendy Hollender
Pistachio Nuts
Arrived by mail Aug 14, 2024
From Top Notch Pistachio Farms - Benson, Arizona

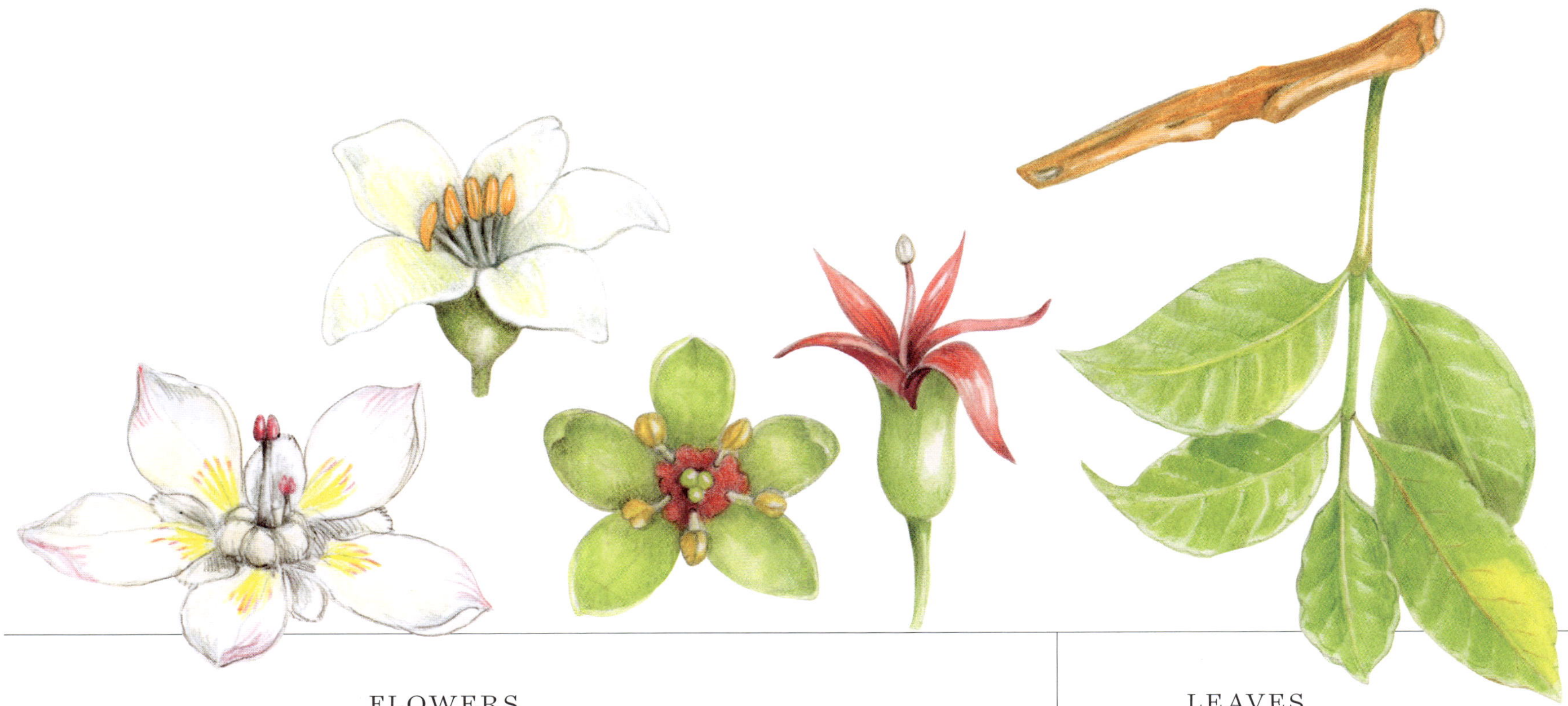

FLOWERS

Often with 5 petals and 5 sepals

LEAVES

Alternate

Simple or pinnately compound with leaflets in opposite pairs

Irritating oils and resins

other species consume it. It is an important source of valuable nutrition for certain animals in the winter months, when very little else is available to eat. Poison ivy was a challenge to draw. I am quite allergic and did not want to collect the plants. Instead, I took a stool out into my garden and found a sugar maple tree that had a poison ivy vine growing on it. I got on the stool with my magnifier and studied and photographed the flowers, which I was then able to draw. I waited for the white berries to ripen and the beautiful fall color to permeate the leaves and added them. The root system of poison ivy anchors it to trees and is really intricate, and I loved the opportunity to study and add it to the painting.

Sumac leaves reveal one of autumn's most vibrant red colors.

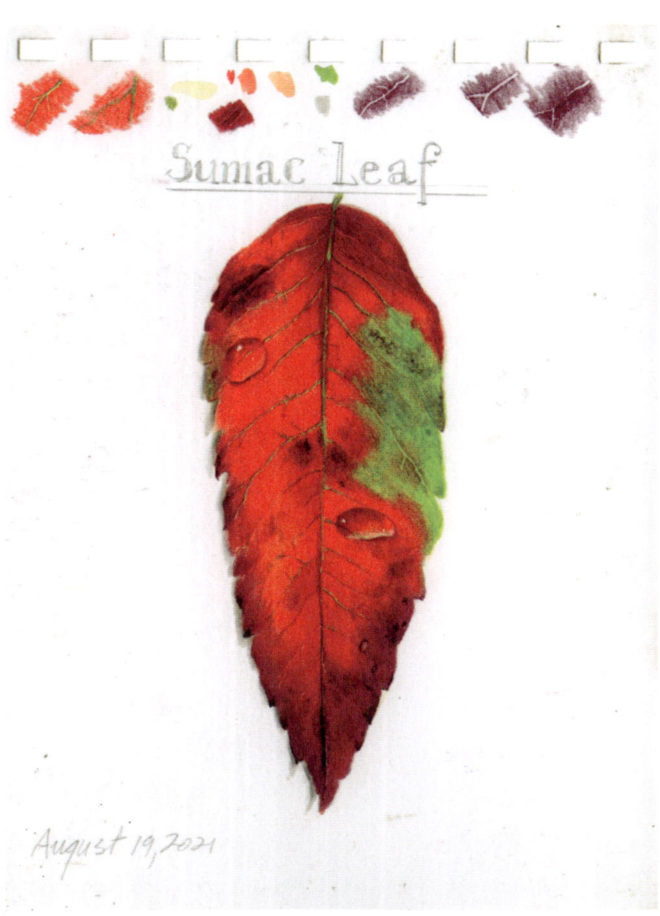


Sumacs were flowering in the spring, and I collected and studied the minute flowers under the microscope to see the similarities to the mango and cashew flowers that I had drawn in Hawai‘i. As the season developed, I anticipated the fuzzy, deep red fruits, which are profuse in my area, followed by the vibrant colors that develop on fall leaves. On October 31, I realized I had not added a colorful red leaf, so I went on a neighborhood bike ride to find one. They all seemed to be gone, though I knew I had seen many just days before. Finally, I found a couple of stragglers on some sumac trees and collected these leaves, but alas, my completed painting didn't seem to yield any good spots to add in the leaf, so I didn't.

The final challenge in this painting was pistachio nuts. I needed to find a grower in the western United States. I contacted several in California, but no one got back to me. Finally the botanist Susan Pell, who studies and writes about this plant family, suggested I contact growers in Arizona, saying, “they will get back to you.” And wouldn't you know it—the first grower I contacted did. Sarah Ogilvie of Top-Notch Pistachios in Benson, Arizona, was kind enough to send me exactly what I needed: specimens of the gorgeous pistachios fruiting on her trees.

# Anacardiaceae: A Diverse Family

BY SUSAN K. PELL &
JOHN D. MITCHELL

The cashew family (Anacardiaceae) is a cosmopolitan family of approximately 800 species in over eighty genera, with the greatest diversity in tropical regions. It is best known for its edible, ornamental, and dermatitis-causing plants. Mango (*Mangifera*), pistachio (*Pistacia*), cashew (*Anacardium*), and pink peppercorns (*Schinus*) are of greatest economic importance in the family, but dermatitis caused primarily by poison ivy and poison oak (*Toxicodendron*) in the United States is also economically impactful in that it is expensive to treat and causes substantial loss of work. Many other Anacardiaceae members are regionally important such as the edible *Spondias* (e.g., hog plum, jocote), *Rhus* (sumac), and *Sclerocarya* (marula), and plants whose sap is used to create lacquerware (*Toxicodendron*, *Gluta*) in Asia. Plants of horticultural value include trees and shrubs used in garden landscapes (e.g., *Cotinus*, *Rhus*, *Searsia*) and along city streets (*Harpephyllum*, *Pistacia*, *Schinus*).

Wendy's beautiful illustration of the cashew family reveals the family's morphological diversity, highlighting identifying characteristics of selected species. The fleshy red and yellow parts of the cashew fruit (hypocarps) are known as false fruits because they develop from the stem of the flower instead of the flower itself, but these juicy structures attract animal dispersers to the fruit, within which is the seed we eat. The white fruits of poison ivy stand in contrast to its red fall foliage, perched above a stem covered in adventitious roots—all diagnostic characters for this beautiful and toxic liana (vine).

DR. SUSAN K. PELL is the Executive Director at the United States Botanic Garden. As a botanist and an educator, she has spent her career at botanic gardens and has traveled the world to collect plants for scientific study. During her tenure at the New York Botanical Garden and Brooklyn Botanic Garden, Susan had the opportunity to teach a diversity of plant science topics to students from toddlers to retirees, to develop interpretation for the public, and to direct molecular research and plant taxonomic studies using samples she collected on her expeditions.

JOHN D. MITCHELL is an Adjunct Scientist at the New York Botanical Garden and Research Associate at the Smithsonian Institution–National Museum of Natural History. He is the former chairman of the Beneficia Foundation and now serves on the board of directors of Bat Conservation International and Turtle Survival Alliance.

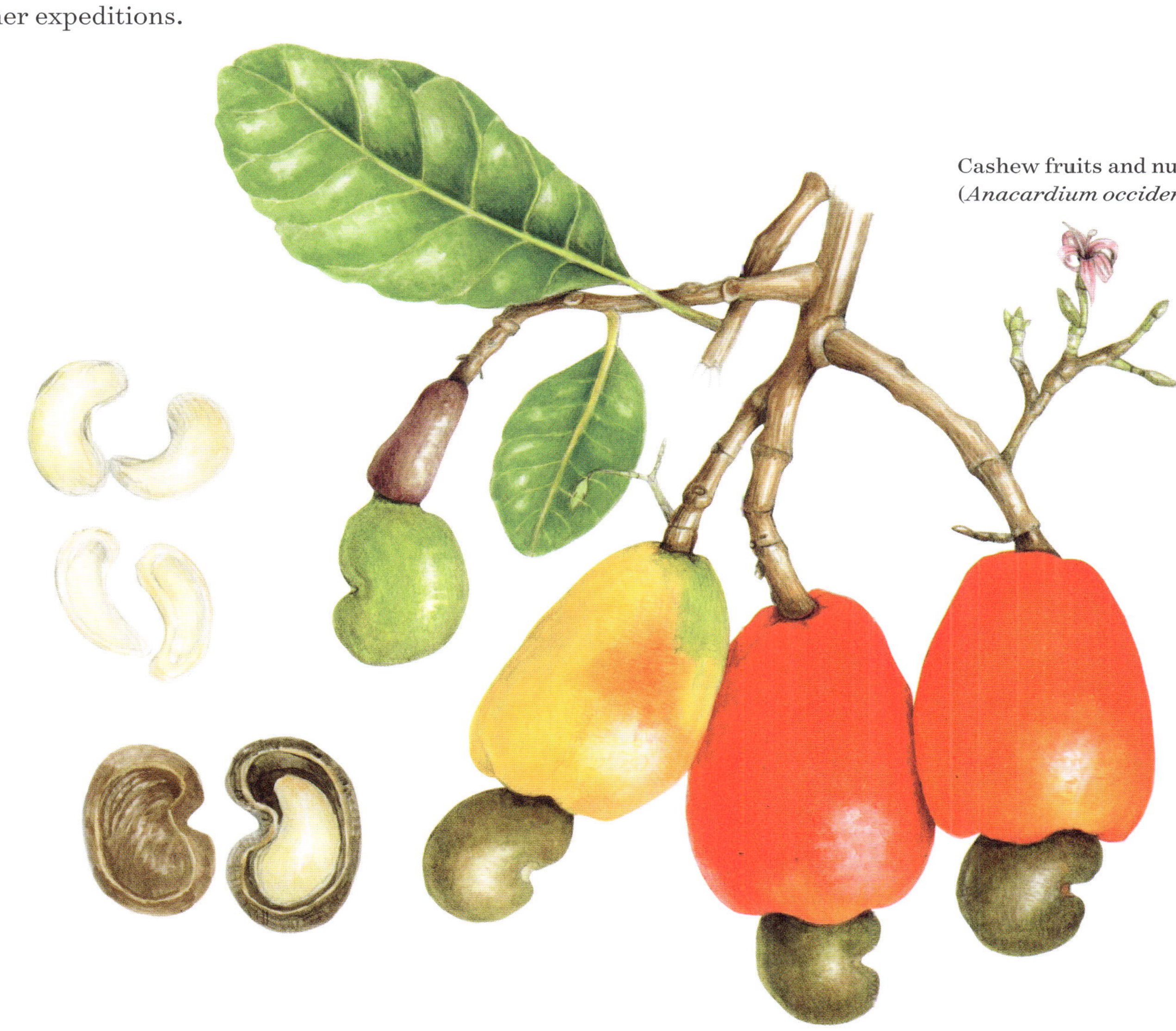

Cashew fruits and nuts
(*Anacardium occidentale*)

# 5

# Citrus Family

## Rutaceae
## 2,000+ species

The citrus family reminds me of sunshine. It starts with the color of the fruits, which are mostly in the orange and yellow range. When a fruit is cut in half, the pattern created by the cross section looks like a sunburst, with rays that radiate out of the center just like the sun. This type of fruit is known as a hesperidium. The flesh is divided into segments packed with tiny juice-filled vesicles and covered with a thick, leathery rind full of fragrant oil glands. Citrus fruits are the only fruits that have this structure.

I have been going to Hawai'i for eighteen years. Each winter I organize a botanical drawing workshop at the National Tropical Botanical Garden on the island of Kaua'i. A cornucopia of various citruses grows at the botanical garden and elsewhere on Kaua'i. Almost every Hawaiian home has citrus trees growing in the yard.

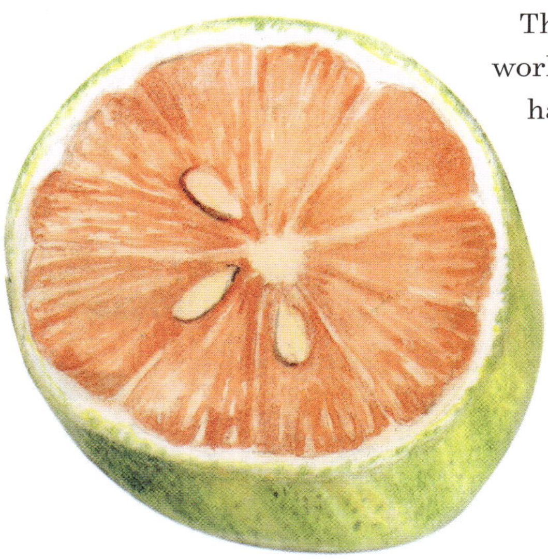

There are many varieties of citruses, and they are among the most popular fruits in the world. I decided to start this painting with some of my favorite specific fruit trees that have the juiciest and sweetest fruits. There is a pink grapefruit growing in Chipper and Hau'oli Wigmore's garden, former president and executive assistant, respectively, at NTBG, that I visit each year. The fruits and flowers are always abundant. Pink grapefruit is tart and sweet at the same time. And I love the color of the flesh. I chose the tangelo orange from my student and good friend Jane's garden. The top of the rind on this fruit pulls off easily. Then all you need to do is tip the opened end into your mouth and squeeze: a complete serving of orange juice is delivered. Kumquat, another citrus, requires no peeling. Just pop the whole small fruit into your mouth.

## Plants Illustrated

*Citrus aurantifolia*, Lime
*Citrus australasica*, Finger Lime
*Citrus japonica*, Kumquat
*Citrus limon variegata*, Zebra Lemon
*Citrus medica* var. *sarcodactylis*, Buddha's Hand
*Citrus × meyeri*, Meyer Lemon
*Citrus × paradisi*, Grapefruit
*Citrus reticulata × paradisi*, Tangelo
*Citrus tangerina*, Tangerine
*Clausena lansium*, Wampi
*Murraya paniculata*, Mock Orange
*Poncirus trifoliata*, Hardy Sour Orange

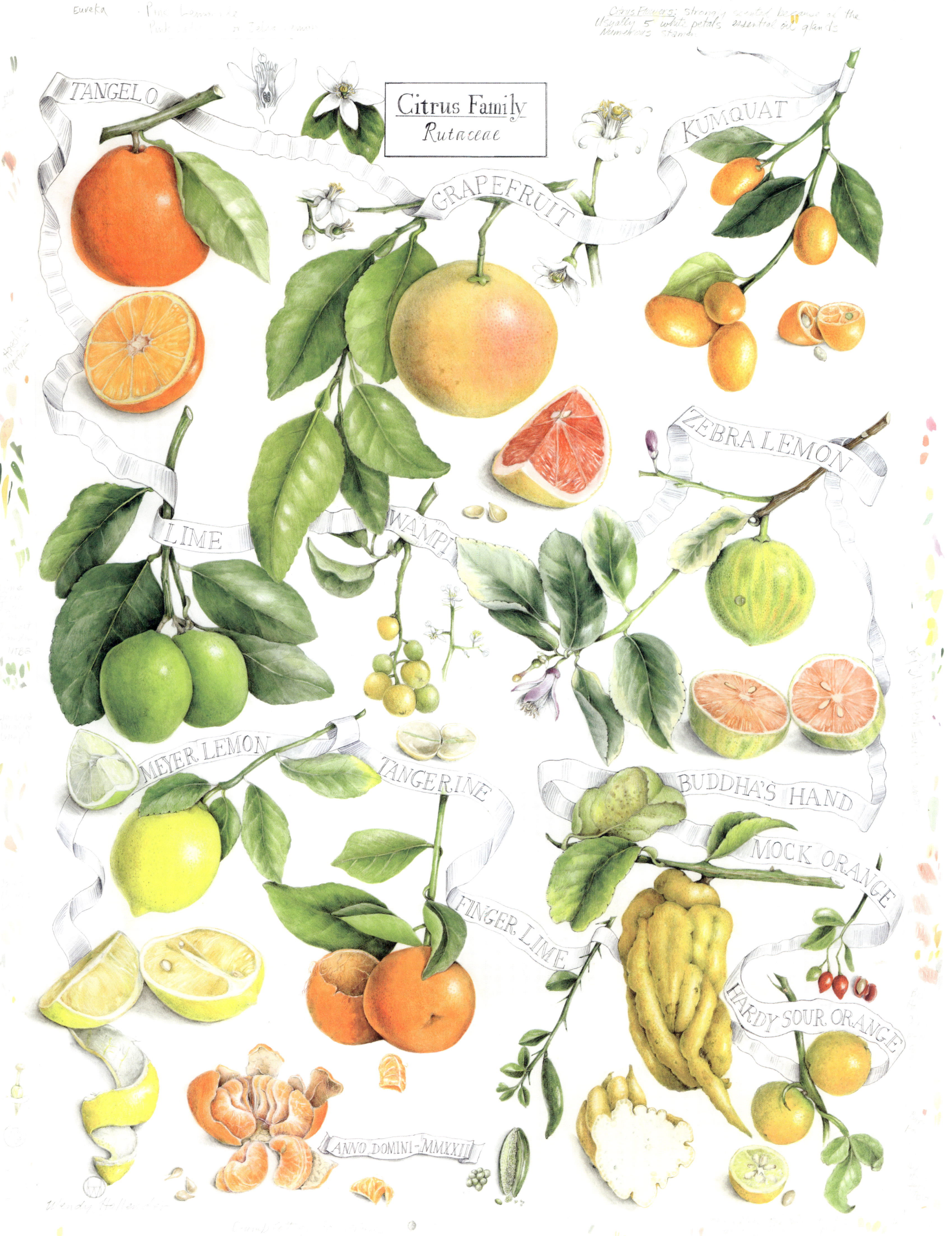

Citrus Family
Rutaceae
TANGELO
GRAPEFRUIT
KUMQUAT
ZEBRA LEMON
LIME
WAMPI
MEYER LEMON
TANGERINE
BUDDHA'S HAND
MOCK ORANGE
FINGER LIME
HARDY SOUR ORANGE
ANNO DOMINI-MMXXII

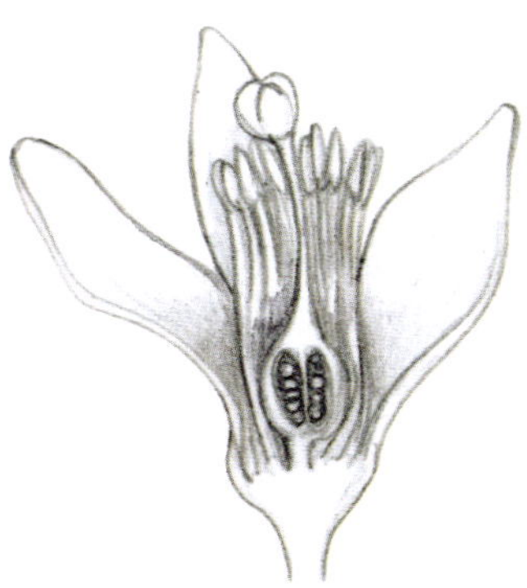

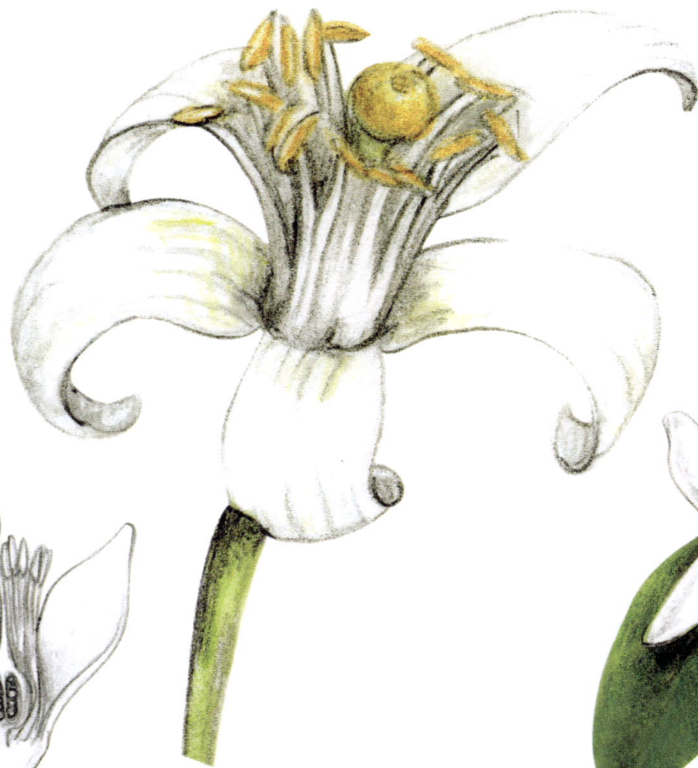

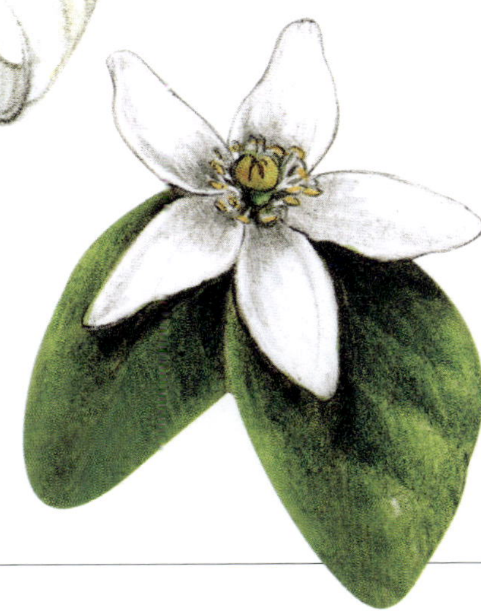

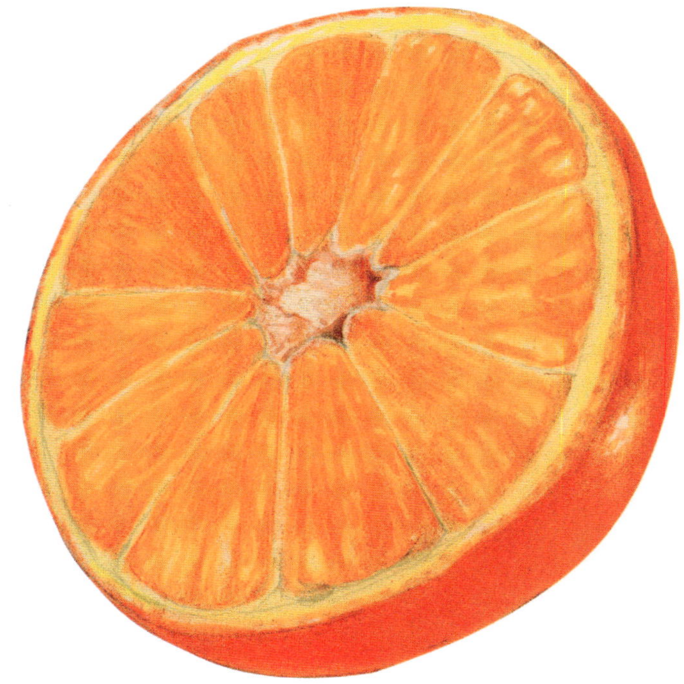

FLOWERS

Radially symmetrical
4–10 petals
4–10 stamens
Scented

LEAVES

Glossy
Oval-shaped

FRUITS

Segmented, juice-filled
Covered in leathery rind

There are about 2,000 species in approximately 160 genera in this family. The *Citrus* genus encompasses over 1,000 different plants, mostly cultivars or hybrids. This means that most citruses are descendants of the same original parental species and have been bred over time for various desirable characteristics.

One of the earliest original citruses was the Buddha's hand, named for its fruit, which looks like a many-fingered hand. It was originally used for its edible rind and has very little flesh inside. It is also used as a temple offering and for its aroma.

Most citruses are sensitive to frost and cannot grow in temperate climates. The one exception is the hardy sour orange, which is grown near me in New York's Hudson Valley at Hortus Arboretum and Botanical Gardens. Of course I saved a space for it in my painting. I was also gifted a delicious marmalade that Allyson Levy and Scott Serrano make each year with the fruit from their prolific sour orange tree. The fruit is very seedy, but the rinds make a great marmalade or can be steeped in honey. These two unusual delicacies are jam-packed with citrus flavor!

There is a long and interesting history of citrus fruits in art. The genre of still-life painting has often relied on citrus as a subject. Sometimes the rind is carefully peeled in one piece, creating a swirling spiral design. I knew I would use this idea somewhere in my painting.

In 1646, Giovanni Battista Ferrari, a Jesuit priest and professor of Hebrew at Collegio Romano, published a book called *Hesperides; sive, De Malorum aureorum cultura et usu* (Hesperides: The cultivation and use of golden apples), replete with magnificent illustrations by many of the finest artists of the time. Most of the citrus illustrations in the book include twisting ribbons with text throughout. I have always admired these beautifully rendered ribbons, their twists and turns creating complex shadows and highlights. If you practice drawing ribbons, you will be well on your way to drawing convincing botanicals too, especially plant families that are monocots, with their twisty, strappy leaves. I decided to use this technique for plant labels in my citrus family painting.

The flowers of citrus plants are quite fragrant. They bloom profusely as the fruits develop. As a result, you can create an illustration with both fruit and flower, showing all the stages of development together on a branch. And if this isn't enough, it is fascinating to know that citrus flower nectar contains caffeine, enticing pollinators to come back again and again.

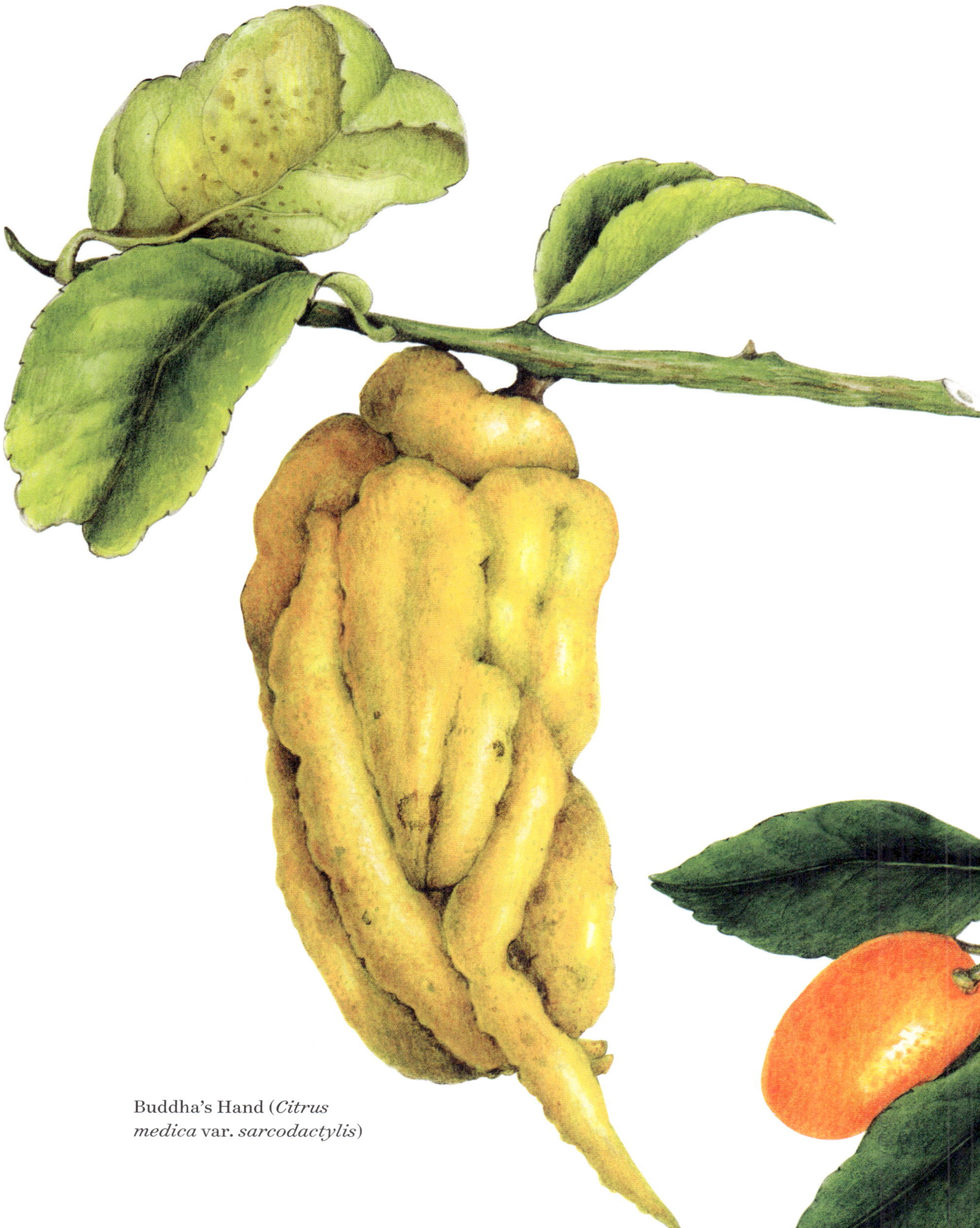

Buddha's Hand (*Citrus medica* var. *sarcodactylis*)

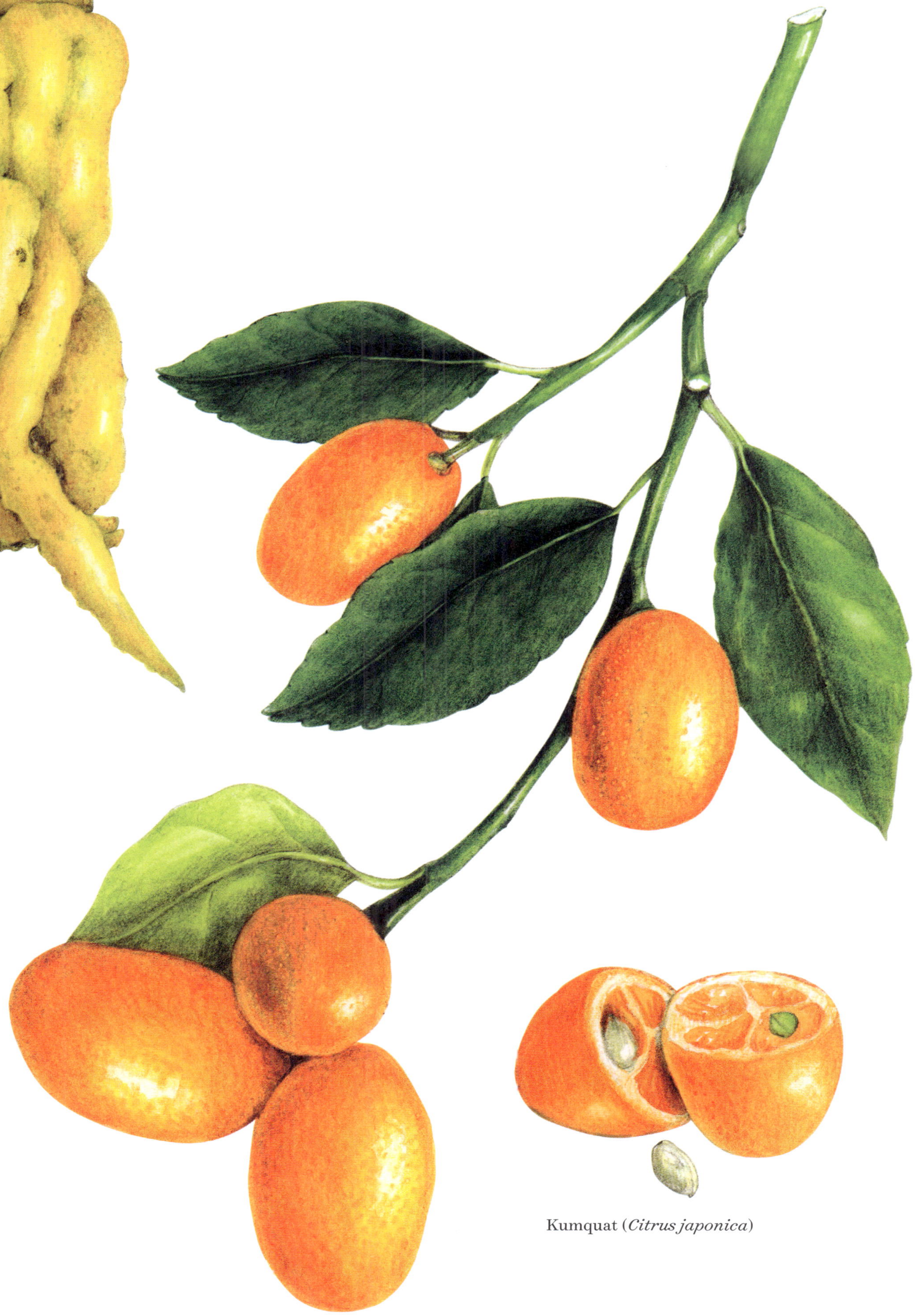

Kumquat (*Citrus japonica*)

# 6

# Coffee Family

## Rubiaceae

## 14,000+ species

In the United States, over 60 percent of adults start their day with coffee. For this reason alone I included this family. It doesn't hurt that I stay each winter in Hawai'i right next door to the largest coffee grower in the United States. I often drive by Kaua'i Coffee's miles of coffee shrubs, stretching from the ocean to the mountains in Kalāheo.

I start each day with a cup of hot coffee. When I was a child, my mom would pour a tiny bit of coffee in my morning glass of milk and called it coffee milk. It is no wonder I always have morning coffee—and my favorite ice cream flavor is coffee—and a cappuccino or latte is still my favorite way to drink it. I love the ritual of making the morning brew and sipping it slowly. It helps me to wake up and ease into the day. I stopped drinking coffee last year, just to see if I could. I thought it would be difficult, but it turned out to be easy. I still kept up the ritual of a hot morning beverage, but I switched to a toasted barley brew from the grass family, which looks like coffee but has no caffeine. Once I knew I could control my coffee drinking, it seemed OK to continue to drink it.

The coffee family (Rubiaceae) is quite large and varied, and I was fortunate to collaborate with David Lorence, the National Tropical Botanical Garden's foremost botanist on the family. The NTBG has a large collection of Rubiaceae plants, so there were many to choose from.

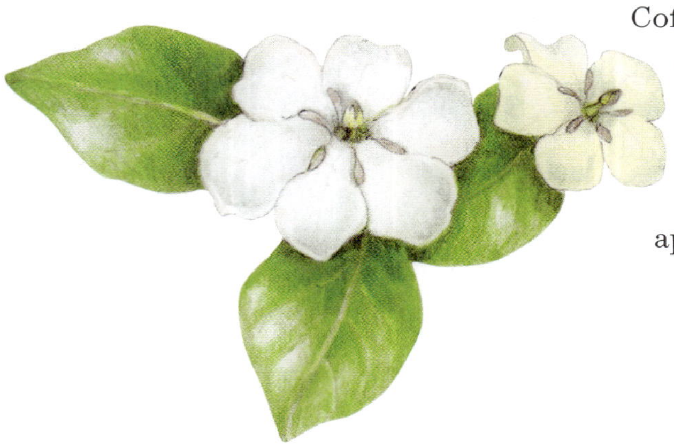

Coffee beans come from coffee berries, also called "cherries." The berries start out green and then turn yellow or red, depending on the variety. Ripe berries are cleaned of pulp, roasted, and then ground to be made into a beverage rich in aroma, flavor, and of course caffeine. The coffee plant is very interesting to see growing. The flowers cover the bushes in spring, giving the appearance of snow-covered plants, and the smell is very enticing. The white

## Plants Illustrated

*Catunaregam spinosa*, Mountain Pomegranate
*Coffea arabica*, Coffee
*Gardenia cornuta*, Horned Gardenia
*Gardenia latifolia*, Indian Boxwood
*Gardenia taitensis*, Tahitian Gardenia
*Morinda citrifolia*, Noni
*Morinda latibractea*, Kesengelengel
*Mussaenda raiateensis*, Pacific Mussaenda
*Portlandia grandiflora*, Bell Flower
*Pseudomussaenda flava*, Dwarf Mussaenda
*Rosenbergiodendron formosum*, Blackberry Jam Fruit

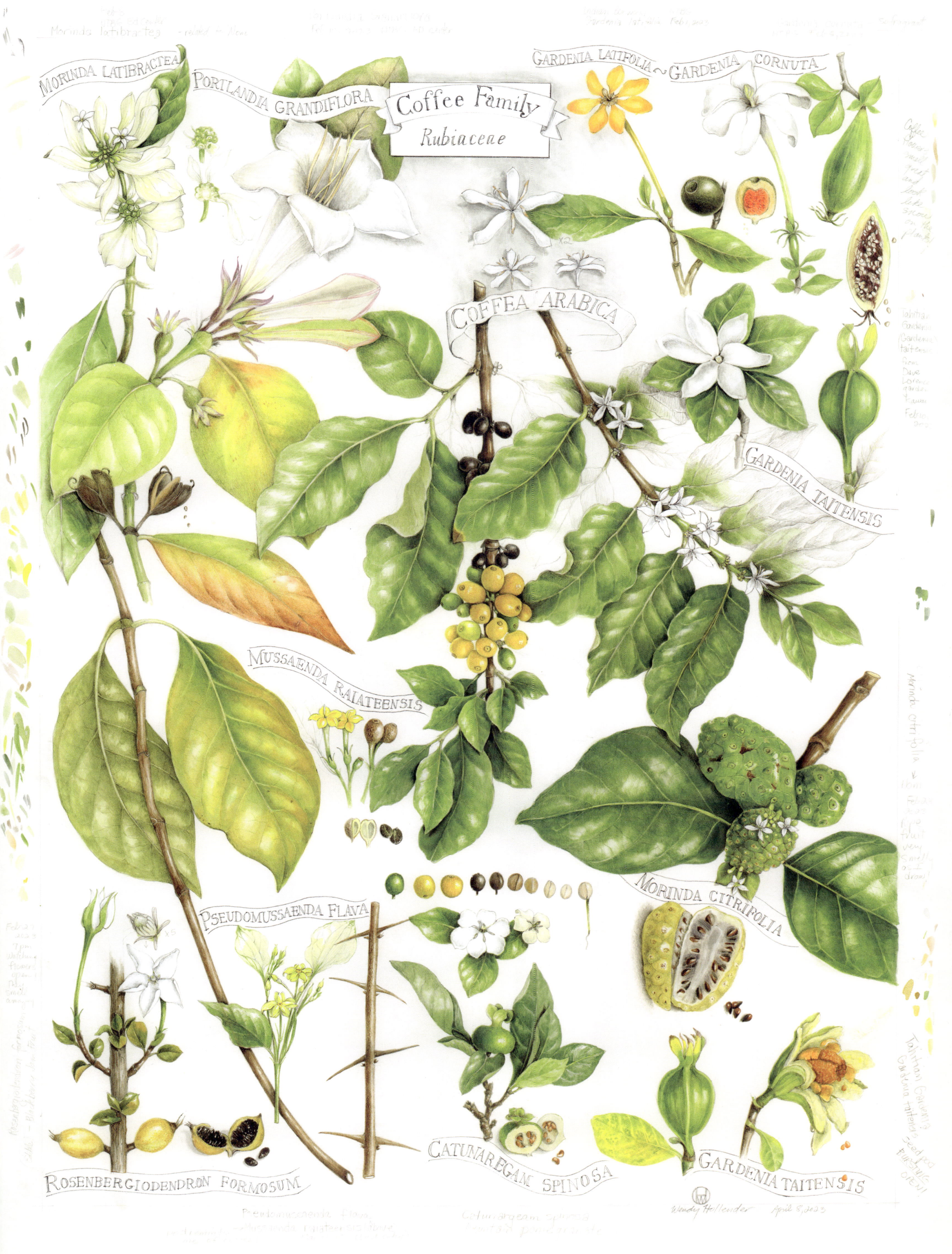
Coffee Family
Rubiaceae
Morinda latibractea
Portlandia grandiflora
Gardenia latifolia ~ Gardenia cornuta
Coffea arabica
Gardenia taitensis
Mussaenda raiateensis
Morinda citrifolia
Pseudomussaenda flava
Rosenbergiodendron formosum
Catunaregam spinosa
Gardenia taitensis
Wendy Hollender April 8, 2023

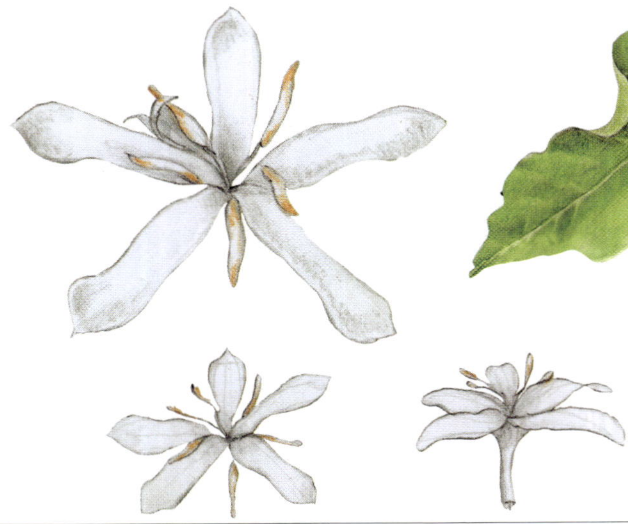

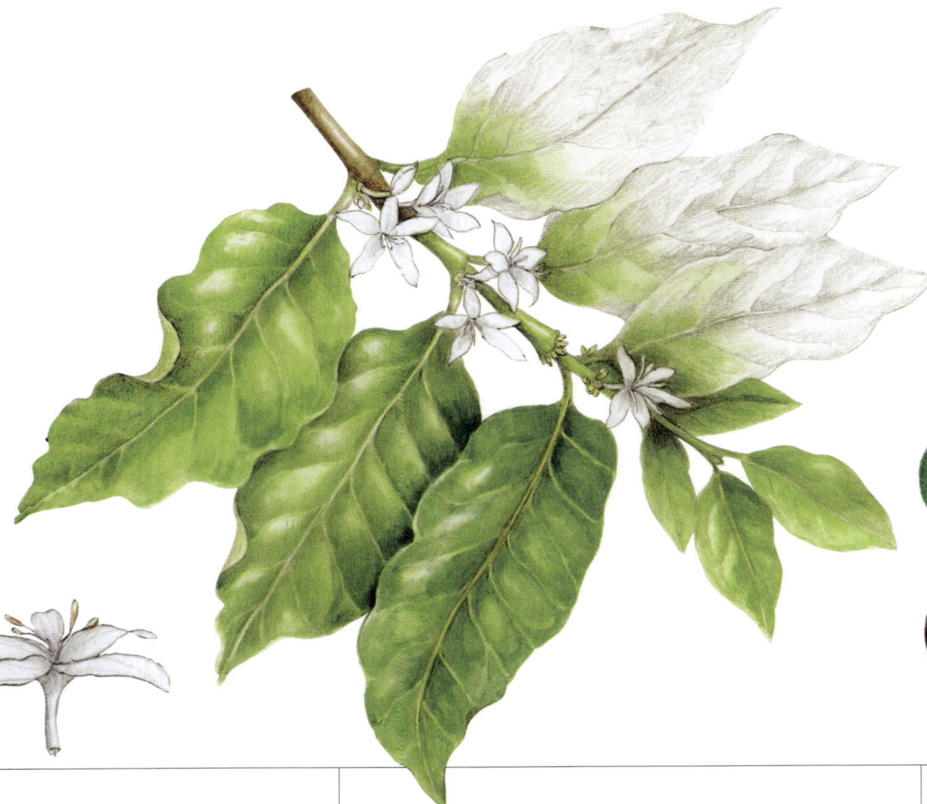

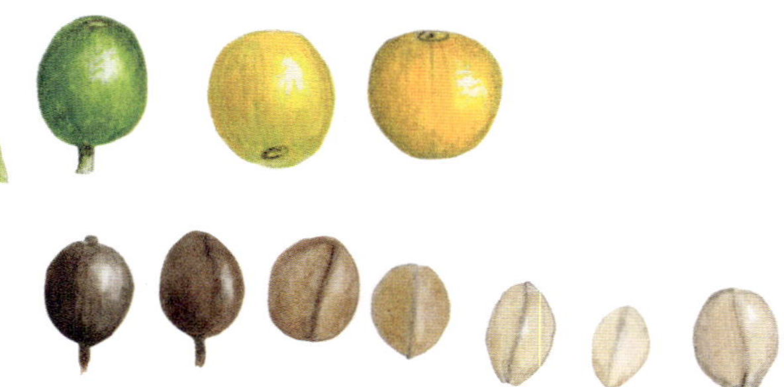

| | FLOWERS | LEAVES | FRUITS |
|---|---|---|---|
| Evergreen shrubs or small trees | Radially symmetrical<br>4–10 petals<br>4–10 stamens<br>Scented | Elliptical and waxy<br>Oppositely arranged along the stem<br>Often prominently veined | Usually a capsule or berry with numerous seeds |

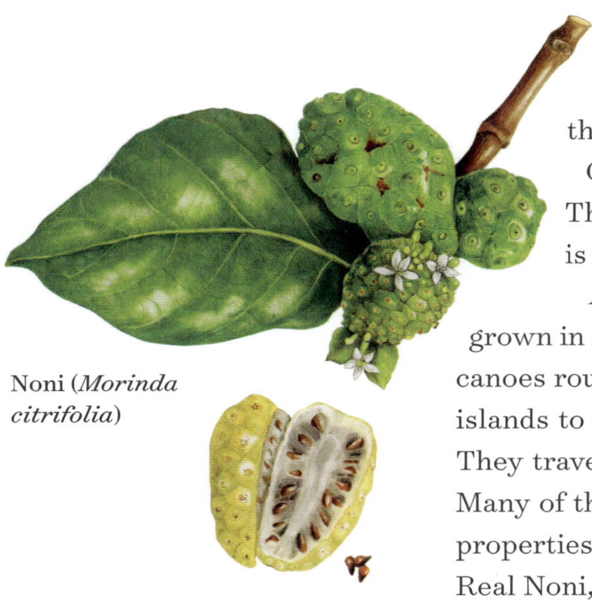

Noni (*Morinda citrifolia*)

flowers are five-petaled with a tubular base. I discovered that they are related to gardenias, the flowers of which come in many sizes and shapes, have four to ten petals, and smell divine.

Caffeine is in the nectar of coffee and citrus flowers. Bees like it and will come back for more. The bee remembers the scent, and it comes back when the scent is present again. Perhaps this is why caffeine is known to help our memory?

Another interesting species in this family is the noni, or Indian mulberry, a "canoe plant" grown in Hawai'i. Canoe plants are two dozen–plus plant species the Polynesians transported in their canoes roughly 1,000 years ago when they traveled across the southern Pacific Ocean to discover new islands to settle. They carried with them plants important for food, building, clothing, and medicine. They traveled 2,000 miles to reach Hawai'i, where they grew these plants, which are now abundant. Many of the canoe plants are as crucial to life today as they were then. The noni has many medicinal properties. Right before I started this project, I illustrated the noni plant for a Hawaiian company called Real Noni, which sells noni-based products such as pain relief lotion and edible fruit leather. The ripe noni fruit is very smelly, and I became acutely aware of this as I cut a ripe fruit open to draw it. I drew it very quickly for this reason.

The flowers of many gardenias and other Rubiaceae species open at dusk and exude a beautiful scent. One afternoon I picked a branch of *Rosenbergiodendron formosum*, the common name of which is blackberry jam fruit. The branch had buds and fruits on it. I hurried back to my desk to paint it. As the hours ticked by and the sun set, the flowers got more and more fragrant as they opened. I was indeed in paradise. I worked late that night because I didn't want to miss out on the intoxicating smell. The fruit is very seedy but edible and sweet.

I started my painting with the coffee plant right in the center. I chose a variety with yellow berries and decided not to include plants with red berries. I wanted to keep this painting focused on yellow, green, and white and didn't want the red to start to pull the painting in a different direction. There are plants with red in them in this family. *Rubia tinctorum*, or herb madder, is grown for its red pigment, which is used in dyes and paints. Honestly, there were so many plants to choose from, I cut out red in part to make it easier to choose my plants for this painting; though, ironically, the Latin root of Rubiaceae (*ruber*) means red. It was not a very scientific decision, and perhaps I disappointed Dave Lorence when he suggested a red-flowering plant one day. But as an artist, I let my intuition be my guide. I also did not have room for both common and scientific name labels on this painting, so I opted for the scientific names, which I did not do in any other painting. I was inspired by the idea of vintage coffee labels for the header of this painting.

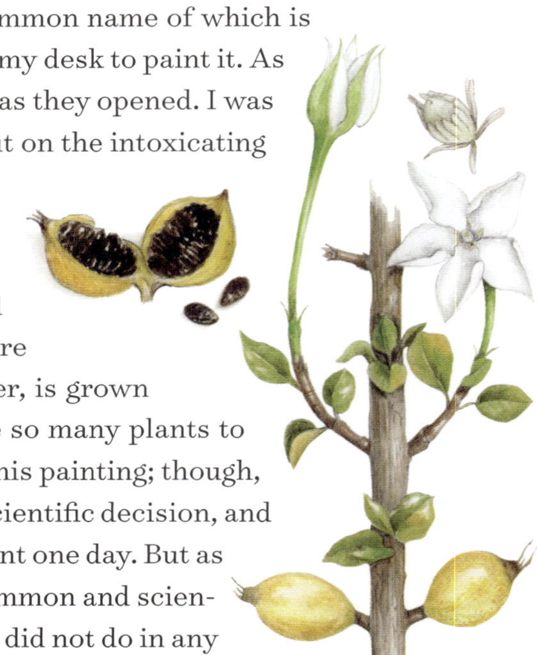

Blackberry Jam Fruit (*Rosenbergiodendron formosum*)

# From Morning Brew to Medicine

BY DAVID LORENCE

My first position as a newly minted PhD was at the National Autonomous University of Mexico in Mexico City, where I was hired as a research botanist/plant taxonomist at the National Herbarium (MEXU). The head of the herbarium and my boss, Dr. Mario Sousa, said, "Dave, you should choose a large family to work on. How about the Rubiaceae? Nobody here is working on that family." Mario and his students all studied the Fabaceae, or legume family. Although my previous research focused on ferns and the Monimiaceae family, I agreed, and that's how I began studying the coffee family. This is an interesting family, I thought, as I began digging into the literature. It's extremely diverse in Mexico, and many of the herbarium specimens at MEXU were misidentified or unnamed, so I had my work cut out for me. Now, after studying the family for almost forty-five years, I've named and published over eighty new species myself or with collaborators. During the first day of my very first field trip in Mexico, I discovered a new species in Oaxaca that I later named *Rondeletia ginetteae*, in honor of my wife, Ginette, in partial atonement for leaving her behind with our baby daughter in Mexico City, newly arrived and knowing only a few words of Spanish.

With approximately 600 genera and 14,000 species, Rubiaceae is the world's fourth-largest flowering plant family, after the asters (Asteraceae), orchids (Orchidaceae), and legumes (Fabaceae). Rubiaceae is most diverse in the humid tropics and subtropics, although a few, usually herbaceous species grow in cold, temperate areas and deserts, including species of *Galium* (madder) with small ball-shaped fruits that hitch a ride by sticking to clothing and fur. Rubiaceae species, including herbs, shrubs, vines, epiphytes, and even large canopy trees, are extremely important floristic components of tropical plant communities. Indeed, they are often the dominant woody plants in the understory and midstratum of many wet and moist tropical forests.

Some Rubiaceae species are familiar to us because they produce alkaloids that afford them economic or medicinal utility. By far the best known is coffee, the family's most important commercial crop, derived largely from the roasted seeds of *Coffea arabica*. So many of us depend on a caffeine boost from our morning cup of java. Other useful Rubiaceae include species of *Cinchona*, whose bark is the source of quinine and other alkaloids that have saved millions of lives from deadly malaria. Roots of *Carapichea ipecacuanha* yield ipecac syrup, used medicinally as an expectorant in cough syrups and as an emetic in cases of poisoning (remember the little brown bottle in the medicine chest?). The dye alizarin crimson originally was extracted from the threadlike roots of the herb madder (*Rubia tinctorum*) and used to dye clothes in ancient times. However, synthetic pigments used by artists today have replaced the original madder dye. The noni or Indian mulberry (*Morinda citrifolia*) from Hawai'i and other Pacific islands is widely used as a medicinal plant and a famine food and yields a yellowish dye. A related species, *Morinda latibractea*, with showy white bracts, is also featured in Wendy's drawing.

Several tropical Rubiaceae species, such as *Genipa americana*, *Borojoa patinoi*, and *Vangueria madagascariensis*, furnish edible fruits, but these are generally little known or appreciated outside of their native ranges. A number of genera, such as *Gardenia*, *Ixora*, *Mussaenda*, *Pentas*, and *Warszewiczia*, are often cultivated for ornamental or horticultural value because of their showy or fragrant flowers. Most *Gardenia* species, including *G. latibractea*, featured here, have flowers with a delicious heady fragrance used to scent perfumes and cosmetics. I hope you will enjoy experiencing the diversity of this large and important family.

DR. DAVID LORENCE is Senior Research Botanist at the National Tropical Botanical Garden (NTBG) on Kaua'i, where he has worked since 1987. He serves on the board of directors of the Heliconia Society International and curates NTBG's living collections of Zingiberales and Rubiaceae. His research specialties are systematics and floristics of tropical vascular plants including Pacific and Neotropical members of the Rubiaceae family as well as other tropical plant families and ferns.

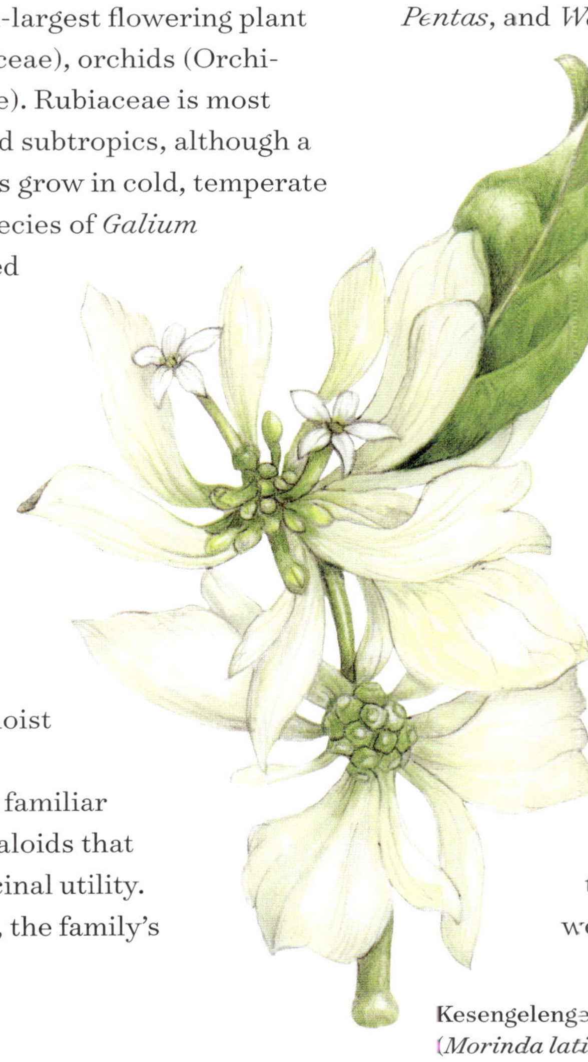

Kesengelengel
(*Morinda latibractea*)

# 7

# Fig Family

## Moraceae
## 1,000+ species

Figs and mulberries are some of my favorite fruits, making this a sumptuous family to draw. The mulberry tree has a long history in the creation of silk. Silkworms, the larval stage of the silkworm moth, eat mulberry leaves. When ready to pupate, they spin cocoons with a fiber made from their saliva. The cocoon is harvested and unspun to create silk thread. In May 2024, I went to Andalusia, Spain, to hold a botanical drawing workshop at the Garden of Lemons. Andalusia is in the south, in a hilly and mountainous region close to the historic city of Granada and also to the coast of the Mediterranean Sea. It is primarily an agricultural region and is very warm in the summer. The white mulberry tree, *Morus alba,* was introduced to Andalusia in the eighth century from Syria to further the silk industry. The silkworm requires a highly nutritious leaf as its primary diet, and it is found on mulberry trees but also on the Osage orange tree, another plant in this family. Apparently, silkworms prefer the leaves of the white mulberry, and these trees were planted all over the region. Even though the Andalusian silk industry is long gone, the trees are still growing there.

Several mulberry trees also grow in my garden. I have been comparing the fruits of these trees each year to experience the different flavors. By far my favorite mulberry comes from the Illinois Everbearing variety. The fruit cluster is large and deep purple in color when ripe. The white mulberry is a weedy tree that sprouts up and grows quickly in my area. Its ripe fruit may be white, red, or deep purple. The berries are rounder and smaller than those of Illinois Everbearing. The taste is mild but refreshing, like drinking flavored water. The really sweet mulberries are longer and turn dark purple. I have to compete with the birds for these fruits, as everyone wants them because they have a rich, delicious, deeply sweet flavor and are juicy. I included a Baltimore oriole that was eating these berries when I was working on the painting.

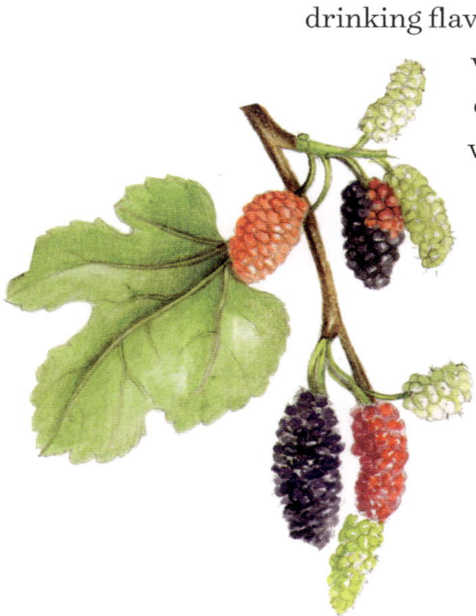

The berries grow underneath the leaves, almost hidden from anything looking from above. They grow in clusters but ripen only one fruit at a time, tricky for collecting, as you can't grab an entire bunch but must pick one from each cluster each day. The ripest, sweetest fruit falls to the ground when ready. Perhaps putting a cloth under a tree to collect them would work, and you could just shake the tree until the ripe ones fall to the ground.

## Plants Illustrated

*Artocarpus altilis,* Breadfruit
*Ficus auriculata,* Roxburgh Fig
*Ficus carica,* Common Fig
*Ficus religiosa,* Bodhi Tree
*Maclura pomifera,* Osage Orange
*Maclura tricuspidata,* Che
*Morus alba,* White Mulberry
*Morus alba × rubra,* Red Mulberry, Illinois Everbearing

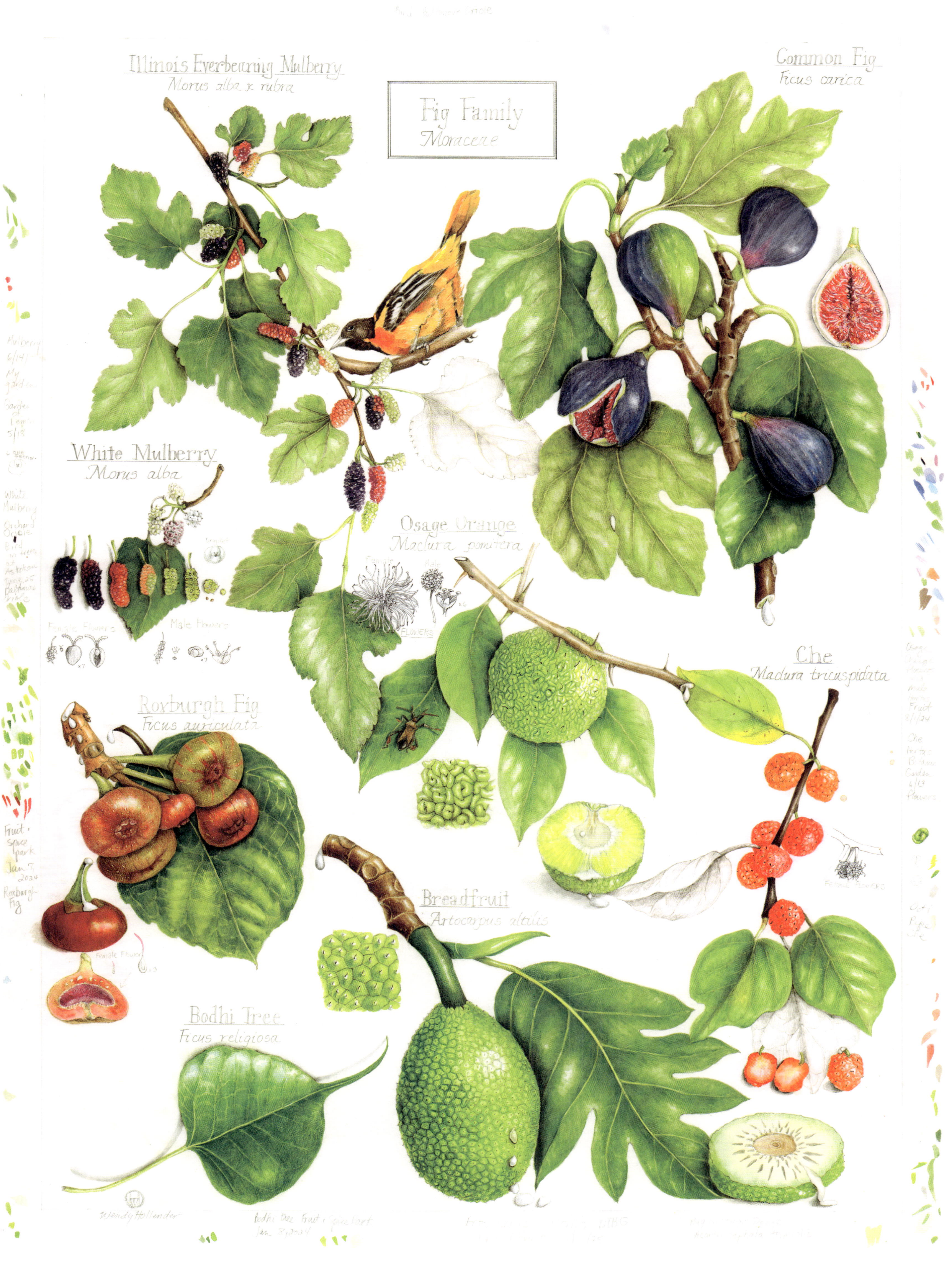
Fig Family
Moraceae
Illinois Everbearing Mulberry
Morus alba x rubra
Common Fig
Ficus carica
White Mulberry
Morus alba
Female Flowers
Male Flowers
Osage Orange
Maclura pomifera
FLOWERS
Che
Maclura tricuspidata
Roxburgh Fig
Ficus auriculata
Breadfruit
Artocarpus altilis
Bodhi Tree
Ficus religiosa
Wendy Hollender

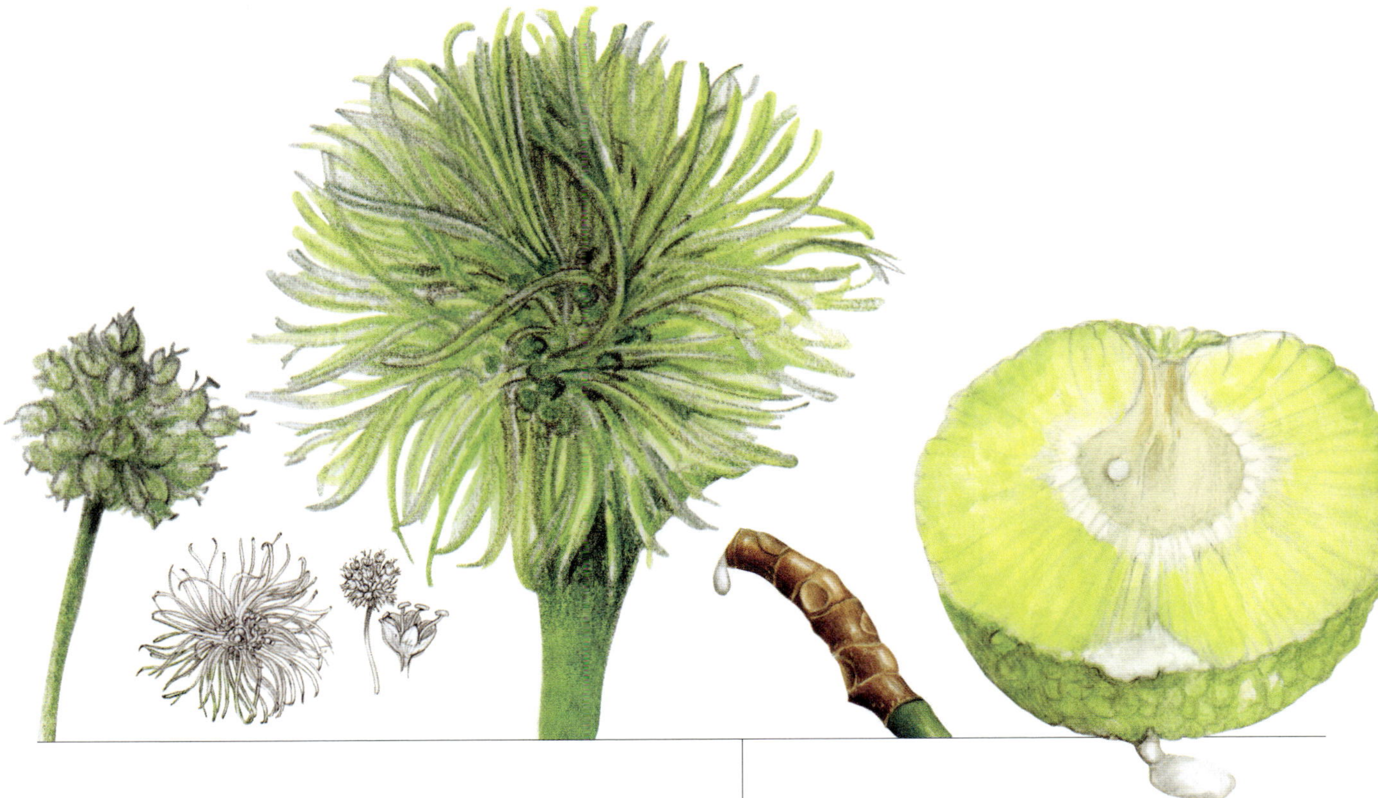

FLOWERS
Small, often clustered

Milky latex

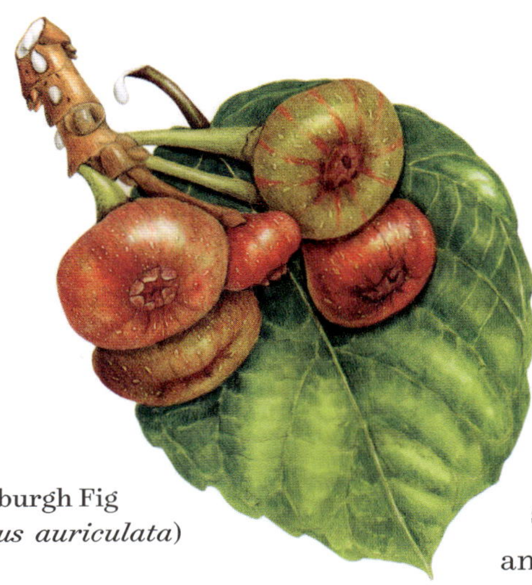

Roxburgh Fig
(*Ficus auriculata*)

I researched and started drawing figs at the Garden of Lemons too. Fig trees are weeds all over the region. There are green figs and purple figs. I love the color of the purple figs and waited to include them in my painting, as they ripen later in the season. I had the garden's proprietors, Ingo and Henrik, document their ripening so I could get the color right. I mostly work from real plants but occasionally I use photographs when that is the best option. I drew the plant first and waited to add the color from the photographs. If only I could have eaten the fruits too . . .

In January 2025, I went to the Fruit and Spice Park in Florida to draw some of the tropical fruits growing there. The Roxburgh fig was fruiting, and I added it to my painting. The fig has a mild taste and is refreshing to eat when ripe.

I enjoyed studying the bumpy texture of the Osage orange and enlarged the form so I could draw it in detail. This is such a strange-looking fruit, and I was lucky to find a local tree growing nearby, in a neighbor's garden. A related species, the che, grows at the nearby Hortus Arboretum and Botanical Gardens, and its fruit has a mild but pleasant taste.

The bodhi tree, also known as the sacred fig, is from India. The Buddha was said to have reached enlightenment under the original bodhi tree circa 500 BC. That tree is gone, but a replacement has been planted on the same spot and is a place of pilgrimage for many people. These trees are large and can live for a very long time, sometimes over 1,000 years.

I finished my painting with the addition of breadfruit, which grows at the National Tropical Botanical Garden in Hawaiʻi. The Breadfruit Institute of the NTBG promotes conservation, study, and use of breadfruit for food and reforestation, and over 150 varieties grow there. The pattern on the breadfruit is a hexagon shape repeated over and over again. Visualize the pattern on a soccer ball, with one difference: instead of equal hexagon shapes all over, the shapes and sizes are irregular, as in nature things grow at different, uneven rates. Breadfruit is considered a nutritious and quick-growing tree that can provide a readily available food source in the tropics all over the world. At the garden, they are always trying different recipes for breadfruit. I love to sample these dishes. My favorite way is to prepare breadfruit like a thin-sliced, crispy French fry.

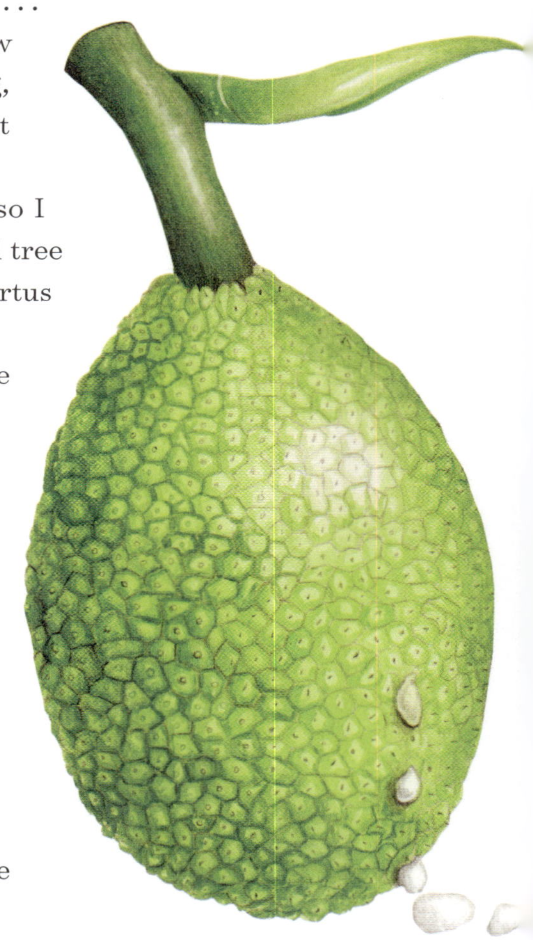

Breadfruit
(*Artocarpus altilis*)

# Moraceae: Abundant, Fruitful, and Unusual

BY LEE REICH

I met the Moraceae family decades ago, when I first dove into gardening in earnest, my hands in the dirt and my head in graduate studies in agriculture. I planted a fig tree (*Ficus carina*) although I'm not sure why *that* plant, considering I was living in winter-frigid Madison, Wisconsin. Then again, the plant was very inexpensive, and I did grow it in a pot. And it's a plant with roots in ancient history. Needless to say, I never plucked a ripe fruit from the plant, although in the years since, with experience and education under my belt, I've harvested plenty of ripe figs here in the less frigid, but still very cold, Hudson Valley.

Shortly after my introduction to Moraceae, I met and planted another family member, a mulberry tree. This tree has had close associations with humanity, black mulberry (*Morus nigra*) for its fruits, among the most delicious of all fruits, and white mulberry (*Morus alba*) for its leaves, which provide fodder for silkworm larvae.

Over the years, my personal associations with the Moraceae continued. I planted creeping fig (*Ficus pumila*), che (*Maclura tricuspidata*), which bears a fruit something like a mix of mulberry and fig, and Osage orange (*Maclura pomifera*). The last fruit, not at all edible, has a convoluted green skin in folds like a human brain. Its wood is extremely rot resistant.

Oh, I forgot weeping fig (*Ficus benjamina*), a large tropical tree that I have maintained in a pot, after fourteen years of growth, as a four-inch-high bonsai; and the tropical breadfruit, which I've seen but never tasted. I met this last member of the family in high school in reading *Mutiny on the Bounty*; Captain Bligh's mission was to collect breadfruit on Tahiti for feeding enslaved people on British sugar plantations.

Moraceae family members don't broadcast their kinship with a commonality of prominent, showy flowers. Their flowers are often small and aggregated into a larger structure. And you'd never even see fig flowers unless you cut the fruit open. The fruit is, botanically, a syconium, which is essentially inside-out stem tissue lined with flowers on the inside. Pollination, which is required for some kinds of figs, could obviously be problematic; enter *Blastophaga psenes*, a special wasp that enters the "eye" of the syconium, inadvertently carrying and spreading pollen from male fig plants. Most fig varieties and many members of the Moraceae do not require pollination to set fruit.

If there's one obvious family characteristic that unites Moraceae, it is the milky latex exuded by cut portions of the plant. One tropical family member, the rubber tree, a common houseplant, has been harvested as a source of rubber. That unifying field characteristic of Moraceae is not, however, unique to this family; members of the spurge family (Euphorbiaceae), for instance, also share this characteristic.

Moraceae is a cosmopolitan family, most widespread in tropical and subtropical regions but also extending into temperate regions. Mulberries, mostly wild ones—either white mulberry, red mulberry, or their hybrids—are among the twenty most common trees in New York City. To further confuse things, don't try to identify mulberry species by the color of their fruits: white mulberry fruits range in color from black to purple to, of course, white.

LEE REICH, PhD, dove into gardening decades ago, with one foot in academia, as an agricultural scientist with the USDA and Cornell University, and one foot in the field, the organic field. He eventually expanded his field to a farmden (a farm-garden hybrid) and left academia to lecture, consult, and write. He has authored nine books (most recently, *Growing Figs in Cold Climates*) and was a syndicated columnist for Associated Press. Besides a year-round supply of fruits and vegetables, his farmden provides a testing ground for innovative techniques in soil care, pruning, and growing fruits and vegetables, and serves as an educational site for workshops and training. Science and an appreciation of natural systems underpin his work.

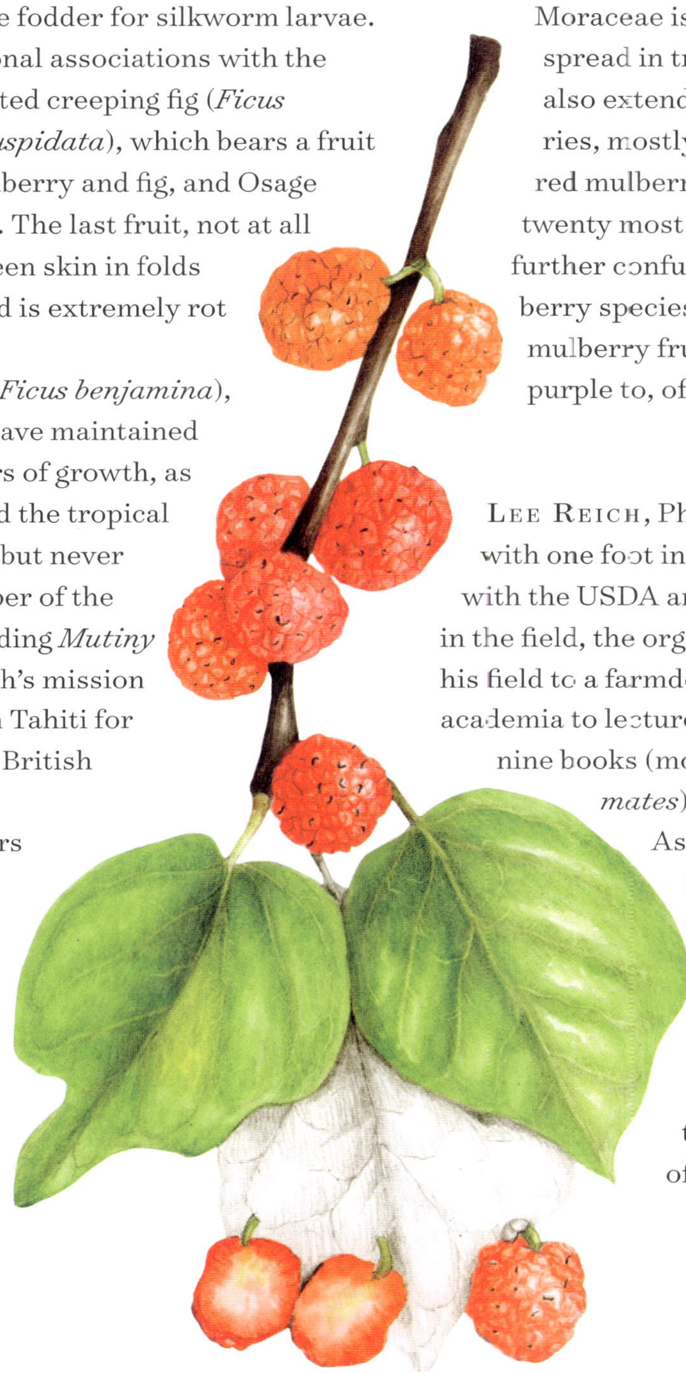

Che (*Maclura tricuspidata*)

# 8
# Ginger Family

## Zingiberaceae
## 1,500+ species

I started the ginger family (Zingiberaceae) painting at home in the fall, working with local growers to acquire edible ginger and turmeric. Starling Yards, a farm in Red Hook, New York, grows both these nutritious and edible plants in greenhouses for local consumption. Most people are used to seeing ginger and turmeric as dried rhizomes (underground stems) or in powder form that they can purchase in grocery stores. When used fresh culinarily, these rhizomes provide a different experience from the dried rhizomes we normally use. I got several fresh plants to use in my paintings, but also enough to happily use them in tea and food preparation. They tasted so fresh. I highly recommend getting fresh rhizomes if you can.

Blue turmeric is especially unusual visually and has a strong flavor. Starling Yards grows it exclusively for one of the first farm-to-table restaurants, Blue Hill at Stone Barns, in Tarrytown, New York. There they pickle the blue turmeric, retaining the unusual color for use. Myoga ginger is also grown locally, at Hortus Arboretum in Stone Ridge. The edible flowers and buds taste like ginger and can be used raw or in stir-fries.

The bottom of my painting was started first, and then I moved on to two plants growing in the glasshouses at the New York Botanical Garden. Here, blooming in October, I found white ginger and shampoo ginger. White ginger is grown for its lovely scent and used extensively in beauty products. Shampoo ginger was the original inspiration for shampoo and has been used as such by Hawaiians for over 1,000 years by squeezing the pods and rubbing the gooey substance into the hair. It is one of the original "canoe plants" (known as " 'awapuhi") that I mentioned in the coffee family (Rubiaceae) account. I was happy to find it flowering at the NYBG, when Marc Hachadourian brought me some to draw, so I did not have to wait to see if I could find it in Hawai'i, which would be my next and last stop for this painting. There it is usually dormant in the winter, growing in the spring and summer.

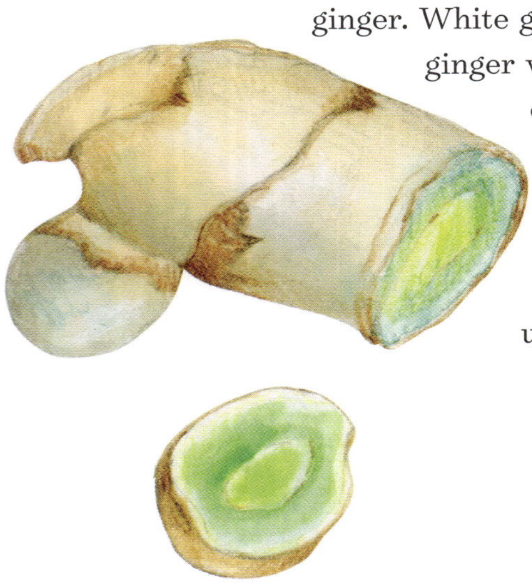

## Plants Illustrated

*Alpinia purpurata*, Red Ginger
*Alpinia zerumbet*, Shell Ginger
*Curcuma aeruginosa*, Blue Turmeric
*Curcuma longa*, Red Turmeric
*Curcuma longa*, Orange Turmeric or 'Ōlena
*Elettaria cardamomum*, Cardamom
*Etlingera elatior*, Torch Ginger
*Hedychium coronarium*, White Ginger
*Zingiber mioga*, Myoga Ginger
*Zingiber officinale* Roscoe, Yellow Peruvian Ginger
*Zingiber zerumbet*, Shampoo Ginger

# A Lifetime of Healing, Flavor, and Ceremony

BY LEI WANN

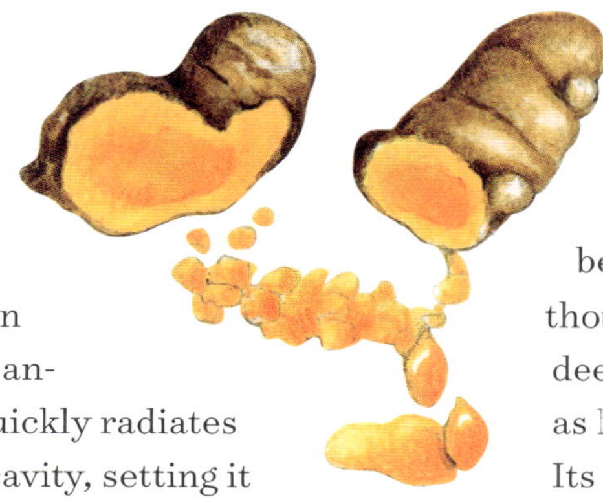

Orange Turmeric or ʻŌlena (*Curcuma longa*)

The smell of *ʻōlena* (*Curcuma longa*) quickly takes me back to my childhood. I hear the sound of it being squeezed into my ear by my grandmother and feel the sensation of my earache battling the natural antibacterial remedies of ʻōlena. It quickly radiates down into every part of my nasal cavity, setting it aflame and reminding me of the interconnectiveness of my body while serving a lesson on the physiology of the human face. Even my eyes start to tear, and I swear my tears are yellow, like the color of ʻōlena. Every sense is awakened, and it's no wonder I have never forgotten this plant and its medicine. The flames are hot and vibrant but a comforting feeling—just like the memory. I know it won't be long until my ears feel better and the infection goes away. Tūtū always knew how to fix my ear infections.

In the years following my youth, I would learn how to make dyes with ʻōlena to color my *pāʻū*, or hula skirts. I would learn how to make medicines for internal and topical infections. ʻŌlena would often be a part of cleansing and purification ceremonies. This cultivar now grows in abundance in our botanical garden. The strains we have come from old wild populations left by our ancestors. The genetics of our strains are from the cultivars that came in canoes with our ancestors generations ago. In our garden, living collections are also living connections. When I use our ʻōlena in Limahuli for medicine, dyeing, or ceremonial practices, I can't help but feel so deeply connected. Some say our genetic memory awakens—I think there is truth in that because if there weren't, traditional medicine would not be alive in my daily life today.

When it is in bloom, *ʻawapuhi* (*Zingiber zerumbet*) takes me back to summers in Hāʻena. In the evening, after long days in the sun, often at the beach, we would go to our freshwater stream and bathe in a deep pond where generations of my ancestors bathed with the same plant. We would search for red bulbs popping out of the bushes—the plant's inflorescences. The bright red ones with fresh flowers emerging or dried on the side had the most soap. We would collect as many as we needed and usually a few extra, but not all, to ensure there would be enough for tomorrow's bath. We would wash our hair and clean our bodies, sometimes using the extras to hit or throw at each other. The round bulb would splatter on my sister or cousins like a water balloon, sending soap and laughter throughout the pond. Sometimes as kids we used the ʻawapuhi for perfume, rubbing the root or the inflorescence on ourselves. We also use the ginger root in medicine. It's an antiseptic and used in some internal medicines for inflammation and infections. Most commonly, I have seen medicine people in our culture use it for ringworm or eczema—you just rub the root on the infected skin. ʻAwapuhi is much more subtle than ʻōlena when used in medicine. ʻŌlena is potent and strong from beginning to end, no matter what you use it for.

There are a number of introduced gingers in Hawaiʻi that add to our biodiversity. They have become iconic plants in the imagery of the islands, though they are not actually from here. Some have been deeply adopted into our culture through traditions such as lei making. The yellow *Kāhili* ginger is a great example. Its scent is mesmerizing, and its flowers can be made into gorgeous leis in many different styles. When I make these leis, I am reminded of how common these plants have become along many trails in Hawaiʻi and how they have become quite problematic for our native forests. When we were kids, they were less pronounced. We would gather them and suck on the peduncle as a sweet treat while hiking.

The Hawaiian red ginger (*Alpinia purpurata*) is another ginger I use frequently in lei making. Like the Kāhili, it is introduced, but it is excellent for making leis in the traditional *kāmoe* style. When completed, this lei resembles our traditional *lei hulu*, or bird-feather leis, which are often a sign of regal status, high rank, or prestige. I gift this lei to people who present themselves with such genealogy or hold strong leadership in my life. The same can be said about the lei made with the hanging red and yellow gingers. The bright yellows and reds signify our *aliʻi*, or royalty colors, in Hawaiʻi.

As a kid, I don't think I realized what all these connections and memories with plants would one day mean. I am thankful for the burn and the healing I experienced as a child, those experiences and practices that guide me today. Gingers are among many plant species with which I feel a very strong relationship in Limahuli, and they continue to be part of my work in biocultural conservation.

Since January 2020, LEI WANN has been the Director of Limahuli Garden and Preserve. Her ancestral ties to Limahuli and passion to care for Hāʻena have created a deep connection that shines strongly in her leadership. She is able to share knowledge of the past while bringing in modern research. Lei Wann went to the University of Waikato, the University of Hawaiʻi at Hilo, and the University of Hawaiʻi at Mānoa, earning degrees in anthropology, Pacific Island studies, and elementary education. She began her career in conservation, then spent many years in education leading and developing *ʻāina*-based curriculum for *keiki* at Kawaikini NCPCS (New Century Public Charter School) and Kamehameha Schools. Currently, she is carrying on her *kuleana* at Limahuli Garden and Preserve by teaching and perpetuating the biocultural traditions of Hāʻena.

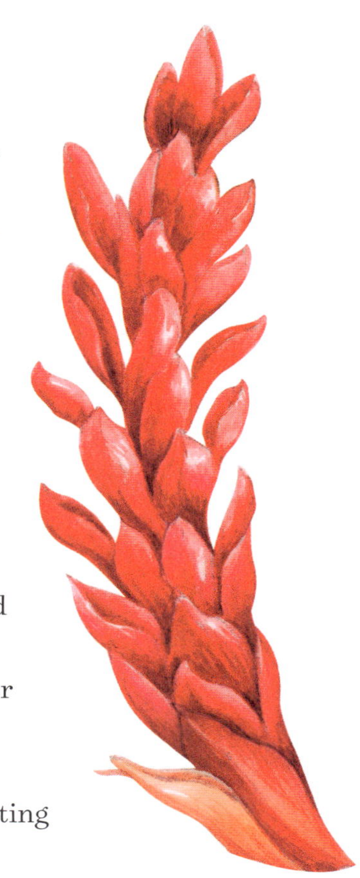

Red Ginger (*Alpinia purpurata*)

# 9

# Gourd Family

## CUCURBITACEAE
## 940+ species

I needed to go wild with the gourd family. I wanted my painting to be a celebration of the beautiful and abundant fruits and flowers, the vining stems and tendrils, and the explosion of seeds.

This is an unusual family. Cucumbers, zucchini, melons, and pumpkins are all closely related, as is apparent once you have studied their fruits and flowers. All these plants grow as vines with tendrils and unisexual flowers. The first time I ever saw a flower in this family was on a trip to Italy soon after I had started to study botanical illustration. In a farmers' market, I was struck by a zucchini with a flower still attached at the top. Of course, I had to draw it that way.

Many of these plants are exciting to draw because they have gorgeous yellow to peachy-orange flowers with five attached petals that form a funnel shape. In many species, the flowers are large and showy with ruffled edges. Learning to create three-dimensional-looking flowers starts with identifying a geometric shape that closely resembles the flower; in this case I used a funnel shape. This aids in drawing but also in creating a convincing three-dimensional light source. Squash flowers are good subjects for this. Since the flowers are either male or female, they display clear differences in their reproductive parts, providing an easy opportunity to learn some botany. The female flower's ovary, said to be "inferior" because it sits below the flower, already resembles a small zucchini. When the female flower has been fertilized with pollen from a male flower, the ovary develops quickly into a fruit. (In botany, we refer to all seed-bearing vessels as "fruits," even those we might call vegetables in culinary use, as in the case of cucumbers and zucchinis.) It is a good idea to be diligent about collecting your fruits in this family as they grow fast. The taste is best when they are still small, but anyone who grows zucchini knows that sometimes it can rapidly develop into a huge fruit the size of a baseball bat! These are truly prolific fruits. When cut in cross section, the fruits reveal the three-chambered ovary with seeds.

I planted cucumbers and summer squash in my garden so I could have numerous flowers to choose from for drawing. At first the plants produced lots of male flowers, but no female

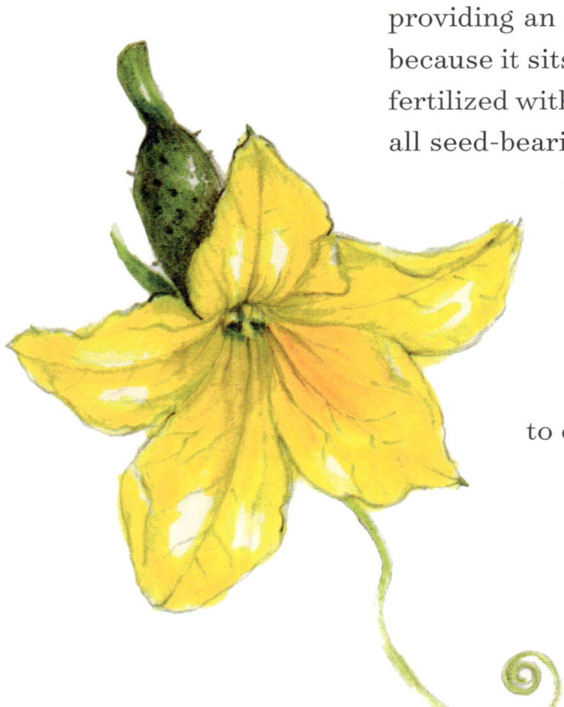

## Plants Illustrated

*Citrullus lanatus*, Watermelon, Sugar Baby
*Cucumis melo*, Cantaloupe, Sarah's Choice
*Cucumis sativus*, Cucumber
*Cucurbita maxima*, Hubbard Squash
*Cucurbita maxima*, Kabocha Squash, Sunshine
*Cucurbita moschata*, Honeynut Squash
*Cucurbita pepo*, Ornamental Gourd
*Cucurbita pepo*, Pumpkin
*Cucurbita pepo*, Zucchini
*Luffa aegyptiaca*, Luffa Gourd

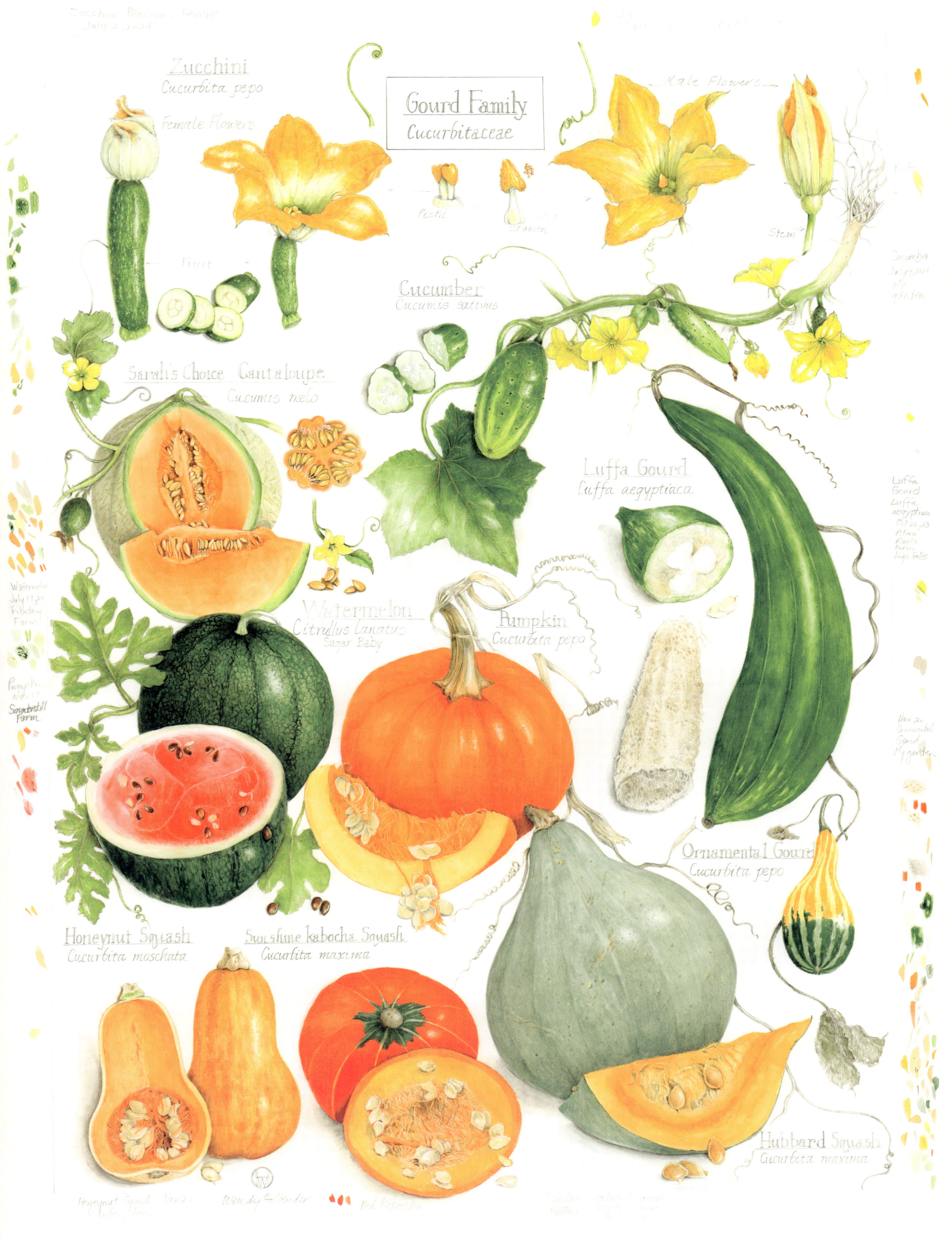
Gourd Family
Cucurbitaceae
Zucchini
Cucurbita pepo
Female Flowers
Fruit
Pestil
Stamen
Male Flowers
Stem
Cucumber
Cucumis sativus
Sarah's Choice Cantaloupe
Cucumis melo
Luffa Gourd
Luffa aegyptiaca
Watermelon
Citrullus lanatus
Sugar Baby
Pumpkin
Cucurbita pepo
Ornamental Gourd
Cucurbita pepo
Honeynut Squash
Cucurbita moschata
Sunshine kabocha Squash
Cucurbita maxima
Hubbard Squash
Cucurbita maxima
Wendy Hollender

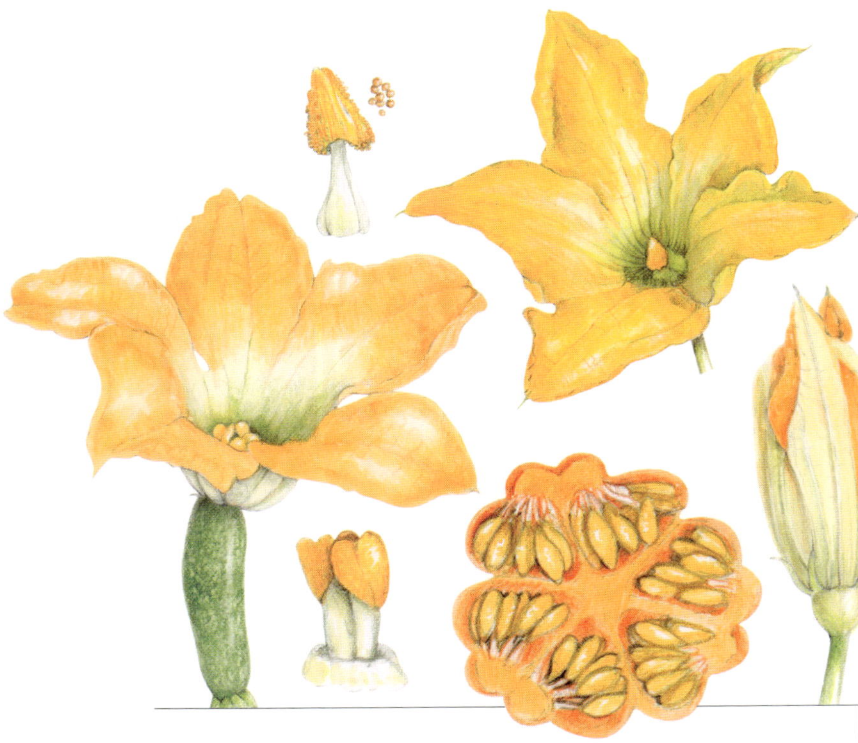

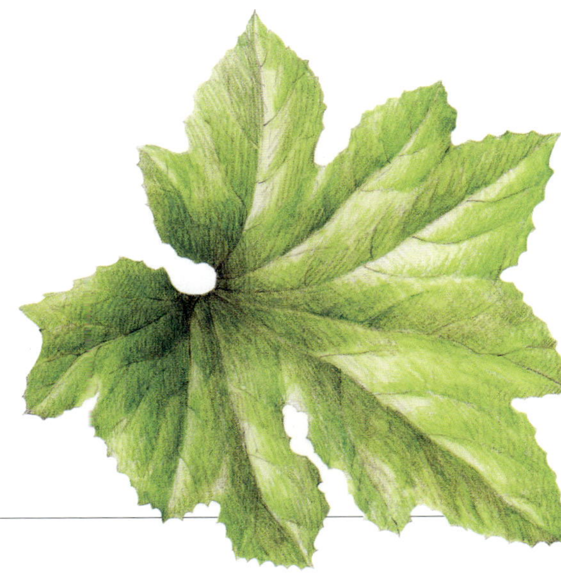

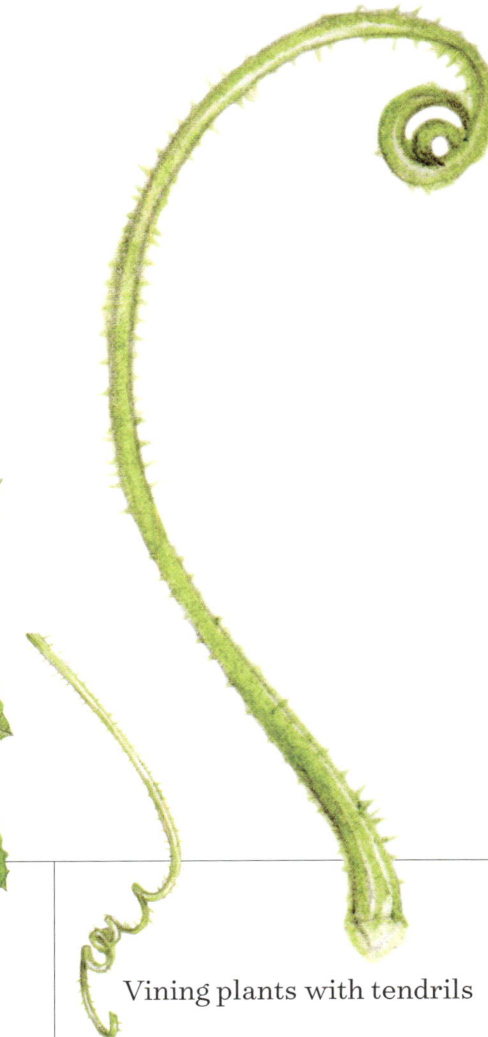

FLOWERS

Unisexual (either male or female)
Funnel-shaped with 5 petals united at the base
5 separate sepals
5 stamens
Ovary inferior (see text) with 3 united carpels

LEAVES

Palmately veined

Vining plants with tendrils

flowers were showing up for me to draw. Finally, the female flowers started to arrive, and the fruits started to develop soon after. It was amazing to see how quickly they grew. I had so many cucumbers from just a couple of plants that I started to make cucumber lemonade and cold cucumber soup. I recommend both of these to anyone with an abundance of cucumbers.

My compost pile is a source of wonder in my garden because so many plants grow in it from seeds thrown in from the food I eat. Squash seeds love my compost, and year after year they produce amazing gourds, mostly ornamental. An interesting thing about this family is that the seeds will grow into plants that often are not the same as what was originally planted. This is because either they are hybrid seeds that will not produce the same fruit, or they are open-pollinated (because of those separate male and female flowers), and the pollen came from a different squash, which created a new variety with shared characteristics.

I started my Cucurbitaceae painting in the fall with the varieties of winter squashes available from Tributary Farm. The challenge for this painting was how to include the subjects life-size on the page with room for all of them. I knew that a pumpkin, for example, would have to be a small one, as would the watermelon I chose. I found a grower of the luffa (or loofah) gourd, used to make sponges—which I never would have guessed was a squash similar to a cucumber.

My compost in fall was in full glory with several squashes, my favorite ornamental gourds, and some other newly formed squashes that had developed on their own. I had to include the ornamental gourd in my painting. Its vines spread out like strings of bright Christmas lights with dangling colorful striped gourds and the most voluptuous tendrils attaching and dangling everywhere. I love how nature's magic just shows up to make my day. At least fifty of these gourds were produced this year. I keep them in baskets all around to enjoy the orange-green color combination, my grandson Toby's two favorite colors. Next year I will put some of these gourds out by the road for my neighbors to enjoy!

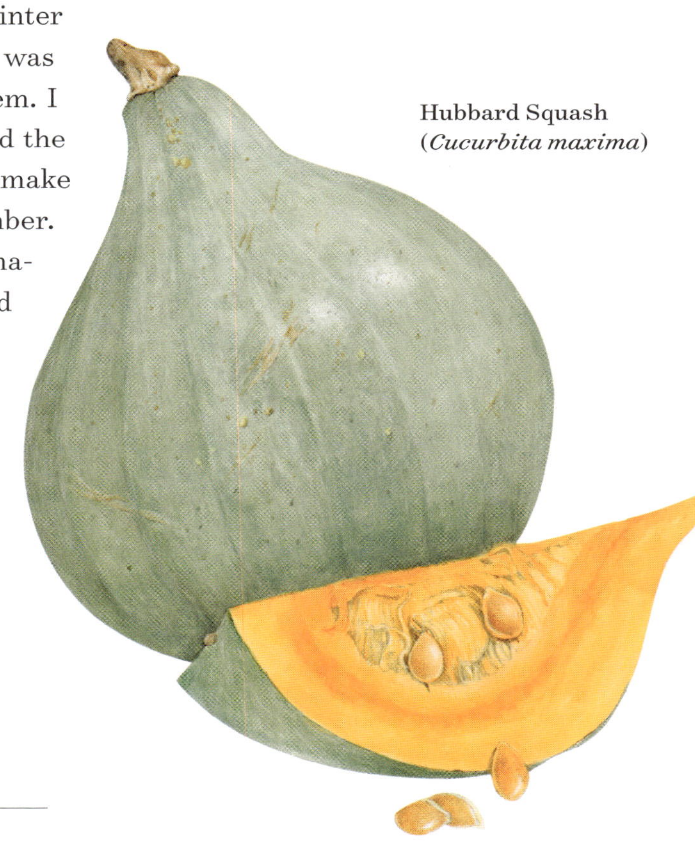

Hubbard Squash
(*Cucurbita maxima*)

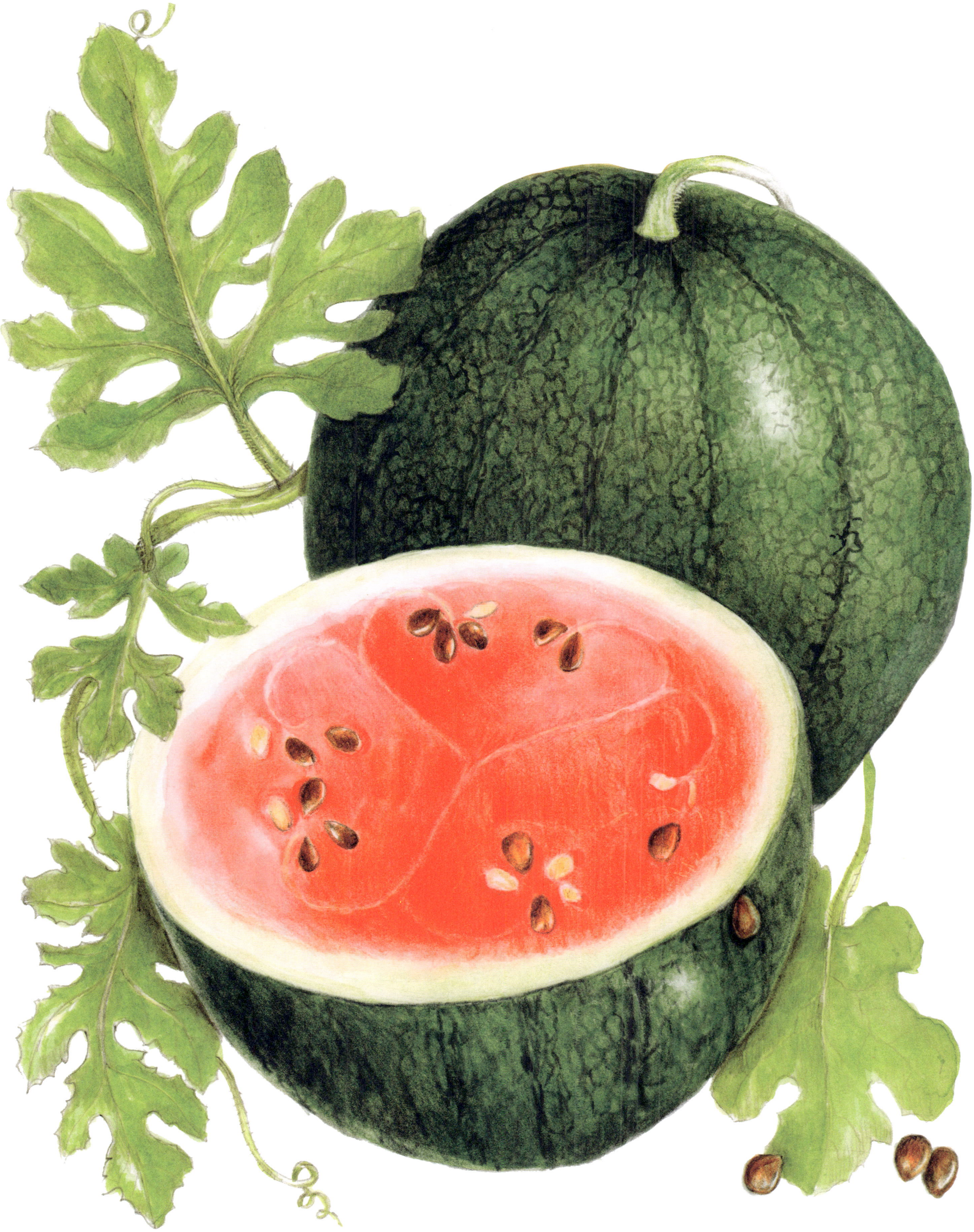

Watermelon, Sugar Baby
(*Citrullus lanatus*)

# 10
# Grass Family

## Poaceae
## 11,000+ species

The grass family is large. With over 11,000 species, it is currently considered the fifth-largest plant family, behind Asteraceae, Orchidaceae, Fabaceae, and Rubiaceae, in that order.

Economically, this family is probably the most important. From these plants we get all our grain crops, a significant part of the human diet worldwide. The history of agriculture started with this family when cultivation of wheat began.

I started to write about the grass family by reflecting on what the family means to me. Then I studied the painting to see what memories it sparked. Studying and drawing this family was really interesting because these monocot plants have some botanical features I love to draw, such as long strappy leaves and hollow stems. They also provided a good opportunity to study the hard small fruits that we call grains, as well as the tiny flowers, which require magnification to even see.

Writing about the grass family is confusing in today's world, when so much controversy surrounds this ancient family, and so many people are choosing a gluten-free diet in response. Honestly, I have more questions than answers.

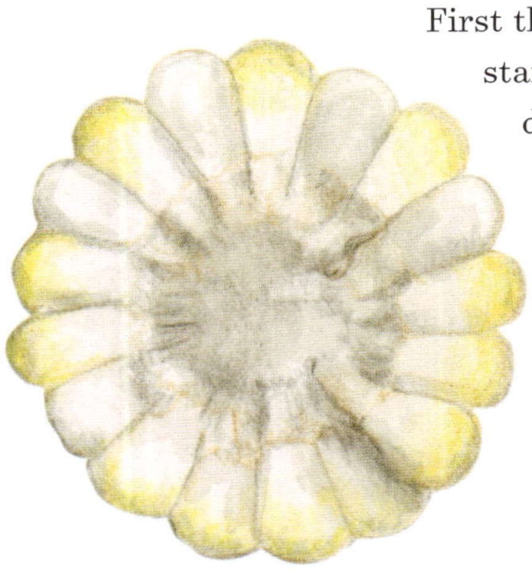

First thing I did was make a slice of twenty-one-grain/seed toast to help get me in the mood to start writing. Then I cooked whole wheat berries from a new form of perennial wheat being developed as a potentially more sustainable product called Kernza. I wanted to experience the essence of what wheat tastes like when simply cooked. It was slightly crunchy and very nutty tasting, as was a bowl of locally grown Yumepirika rice. Normally the rice I eat comes from very far away, and I wanted to see if one of the few farmers to grow rice in my region is on to something. The rice tasted very fresh and tender and was rich in flavor.

When I was drawing sugar cane in Hawai'i, I was able to get fresh-squeezed sugar-cane juice at the farmer's market to try, and I also sucked on the fibrous sugar-cane stalks.

## Plants Illustrated

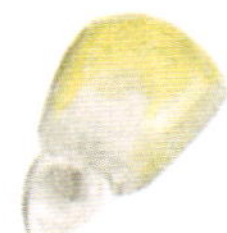

*Avena sativa*, Oats
*Bambusa vulgaris*, Golden Bamboo, Striata
*Coix lacryma-jobi*, Job's Tears
*Hierochloe odorata*, Sweetgrass
*Megathyrsus maximus*, Guinea Grass
*Oryza sativa*, Rice, Yumepirika
*Saccharum officinarum*, Sugar Cane
*Secale cereale*, Rye
*Triticum aestivum*, Wheat, Glenn
*Zea mays*, Corn, Double Red
*Zea mays*, Corn, Otto File Flint
*Zea mays* convar. *saccharata* var. *rugosa*, Sweet Corn

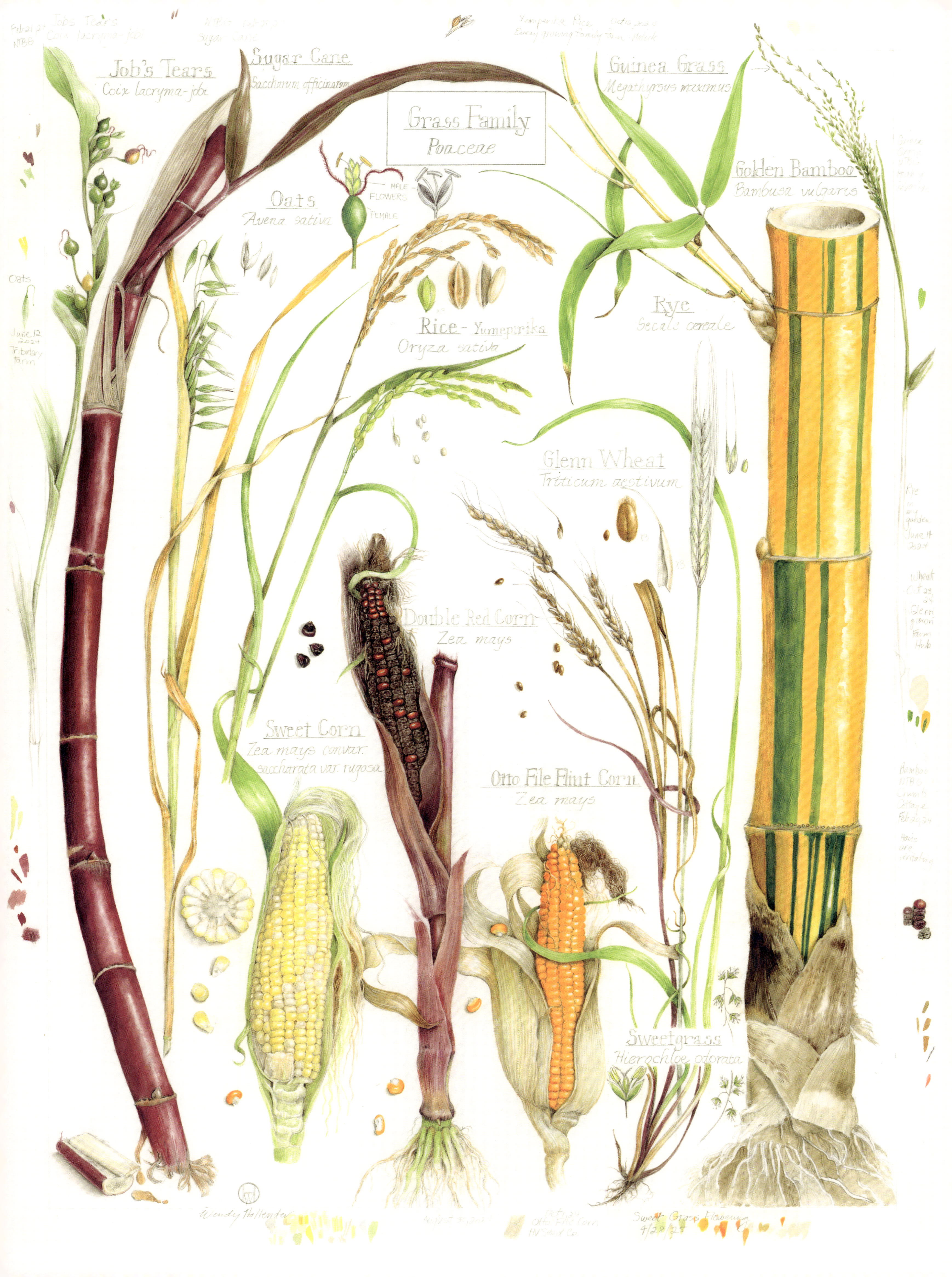
Job's Tears
Coix lacryma-jobi
Sugar Cane
Saccharum officinarum
Grass Family
Poaceae
Oats
Avena sativa
MALE FLOWERS
FEMALE
Rice - Yumepirika
Oryza sativa
Guinea Grass
Megathyrsus maximus
Golden Bamboo
Bambusa vulgaris
Rye
Secale cereale
Glenn Wheat
Triticum aestivum
Double Red Corn
Zea mays
Sweet Corn
Zea mays convar. saccharata var. rugosa
Otto File Flint Corn
Zea mays
Sweetgrass
Hierochloe odorata

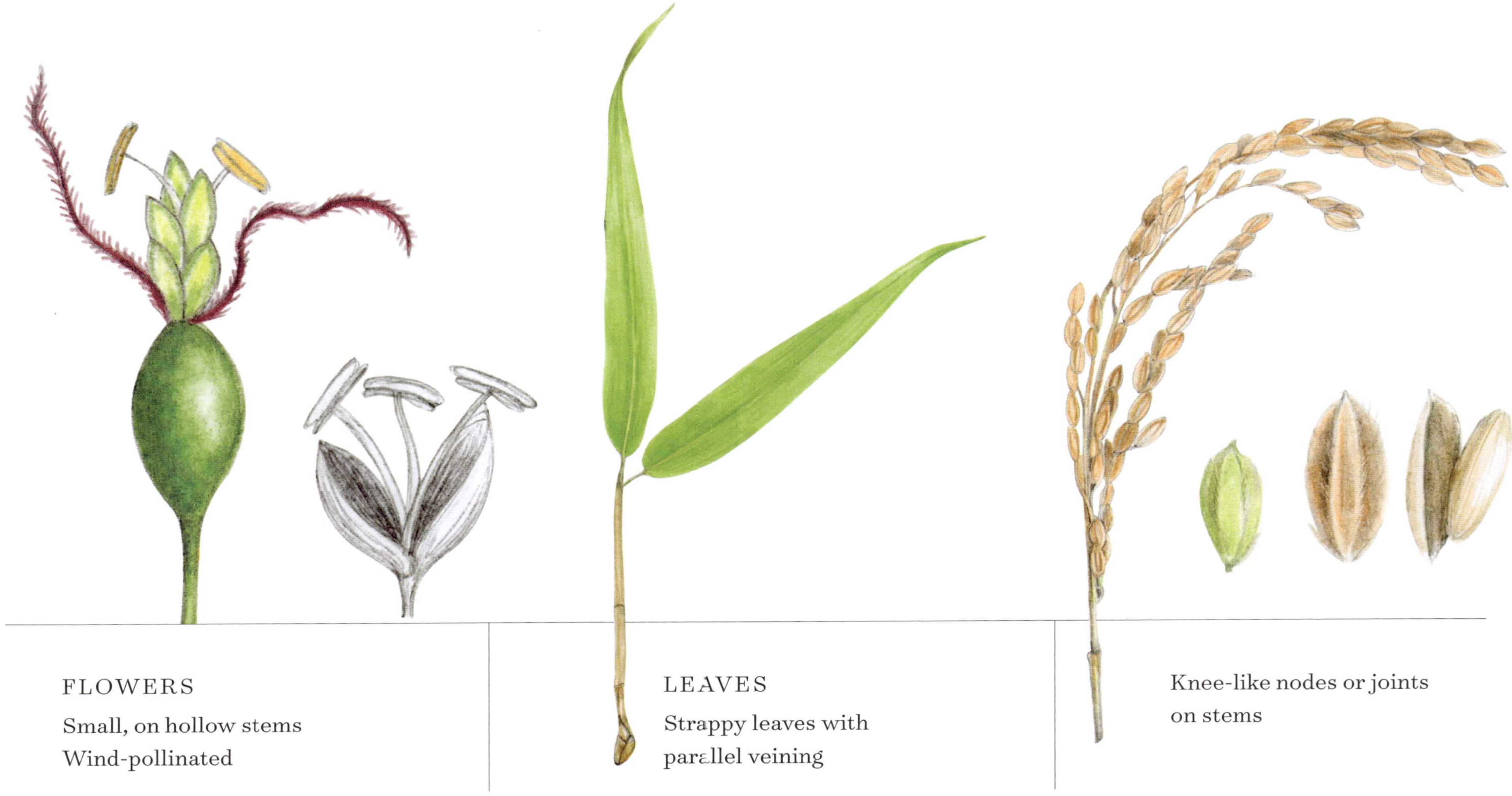


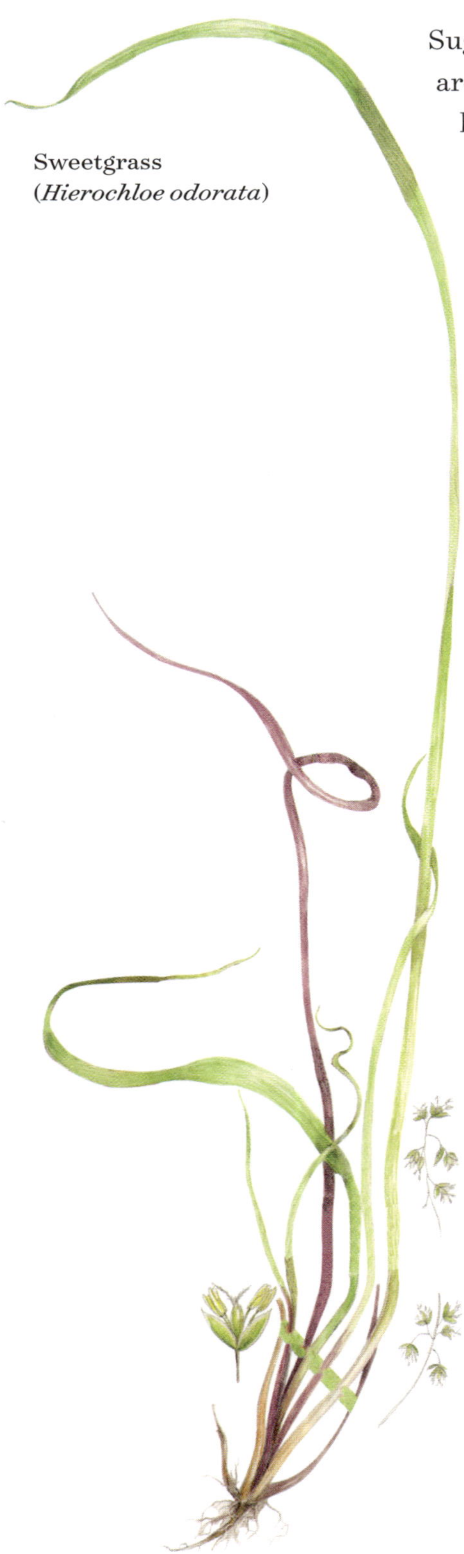

Sweetgrass
(*Hierochloe odorata*)

Sugar cane and corn stems are not hollow like those of many other grasses. There is much controversy around sugar cane, both for the impact growing it has on the environment (not to mention exploitative labor practices) and also for the way sugar is way too prominent in some people's diets to be healthy.

Bamboos grow in the tropics, like sugar cane, and some species are important for construction. They are amazingly prolific plants: some bamboos can grow up to two feet in a day—you can almost sit and watch them grow!

The controversial Guinea grass, which is pervasive in fallow fields in Hawaiʻi, where it is an invasive species, was a factor in the intensification of a devastating wildfire on the island of Maui in August 2023. For this reason I included it. Grasses have a way of moving in when land is left fallow, sometimes to devastating consequences.

The ornamental fruits of Job's Tears, another tropical grass, are used in the making of bracelets, earrings, and traditional leis.

Three kinds of corn made it into the painting, in part because I couldn't choose just one but also because drawing the seed heads (cobs), slowly toning each individual kernel, was such an enjoyably meditative process. I highlighted an heirloom variety called Otto File, used for cornmeal, as well as a local sweet corn that everyone devours around here in the summer and a spectacular-looking double red corn that was developed in recent years. Also known as eight-row flint corn because it has eight rows of seeds on a cob, Otto File originated in New England, where it was grown by Native Americans. During the nineteenth century it was introduced to Italy (*otto file* means "eight rows" in Italian), where it was popularly used for polenta. It was then lost in our country but eventually reintroduced here. I first came across it growing in Transylvania, where people were trying to protect it from encroaching hybrid corn. I picked my ripe cobs at the Hudson Valley Seed Company, where it is grown along the Rondout Creek in the fertile land there, which is ideal for farming. It is a magical setting at the foothills of the Shawangunks, a mountain range famous for rock climbing the world over.

Oats, which are planted at Tributary Farm as rotation crops, and rye, which I grow to help keep down some of the invasive weeds, came next in my painting. Rye has a beautiful sea-foam color that makes my field look like a beautiful Caribbean Sea swaying in the breeze.

And finally I drew sweetgrass, a North American native grass used for centuries by Indigenous peoples for weaving, ornamentation, and the delicious smell it gives off when braided and burned in rituals and ceremonies. I visited the garden and studio of local artist and basketmaker Katie Grove to get sweetgrass, which she grows. I got some dried, braided samples as well, and they smell delicious, an aroma reminiscent of vanilla and cinnamon.

I needed to take a break and get a hot beverage. My late morning beverage of choice in the winter is brewed toasted barley, another plant in this family. Coincidence? Are the plants helping me on my journey?

# A Visit with Don Lewis of Wild Hive Farm, Clinton Corners, NY

Since there is so much controversy and conversation about wheat and grains today, I decided to put together questions and interview a local icon in the movement to grow heirloom and sustainable grains.

Don Lewis of Wild Hive Farm started out as a wild landrace beekeeper in 1982 and then started to make baked goods to go along with his honey. He discovered that using ancient grains and milling his products using traditional stone-grinding methods created a product that was fresh, delicious, and nutritious. He markets his fresh-milled grains to bakers and chefs all over the Hudson Valley region and New York City.

Growing older varieties of wheat, harvesting and milling them quickly, and using the products within three to six months safeguards consumers from ingesting products that are rancid or unhealthy for our digestive systems. Don explained that in our current industrial approach to grains, the seeds start to break down immediately once harvested and milled, so that keeping flour for a year or two, as is typical, is a large part of the gluten-sensitivity problem.

Don described the birth of agriculture as the moment humans, after collecting wild grasses for consumption, noticed that dropped seeds grew into new plants. Thus the idea of collecting seeds and planting crops began. I mentioned my own experience discovering Otto File flint corn in Transylvania and then finding it had been returned to our country (from which it originated) in recent years by Don himself, who had collected its seeds in Italy. Don's mill creates exceptional cornmeal, corn flour, and polenta from this corn.

We also discussed the use of pesticides, how this contributes to the problem of a contaminated product, and that growing varieties without pesticides is better for people and the environment. He explained that commercially grown sugar cane is not only treated with pesticides but also processed to create white sugar that removes most if not all of the nutritional content. He mentioned a traditional unrefined sugar product from Mexico, called panela, that is much more nutritious and can be found in dehydrated bricks in specialty and Mexican food shops.

Don is so aware and tuned in to grains that he can immediately taste if a seed is fresh or rancid, and when it is delicious and will make a superior product. He describes the taste of rancid seeds as bitter and sour.

Finally, we discussed the need for grain varieties that can withstand drought or excessive rainfall. He has been developing use of several heirloom varieties of wheat planted together called Maslins, for example, some of which are drought-tolerant and others that do better with an abundance of rain. With such plantings, each season will yield a successful crop from the varieties planted, even when conditions are challenging for some crops.

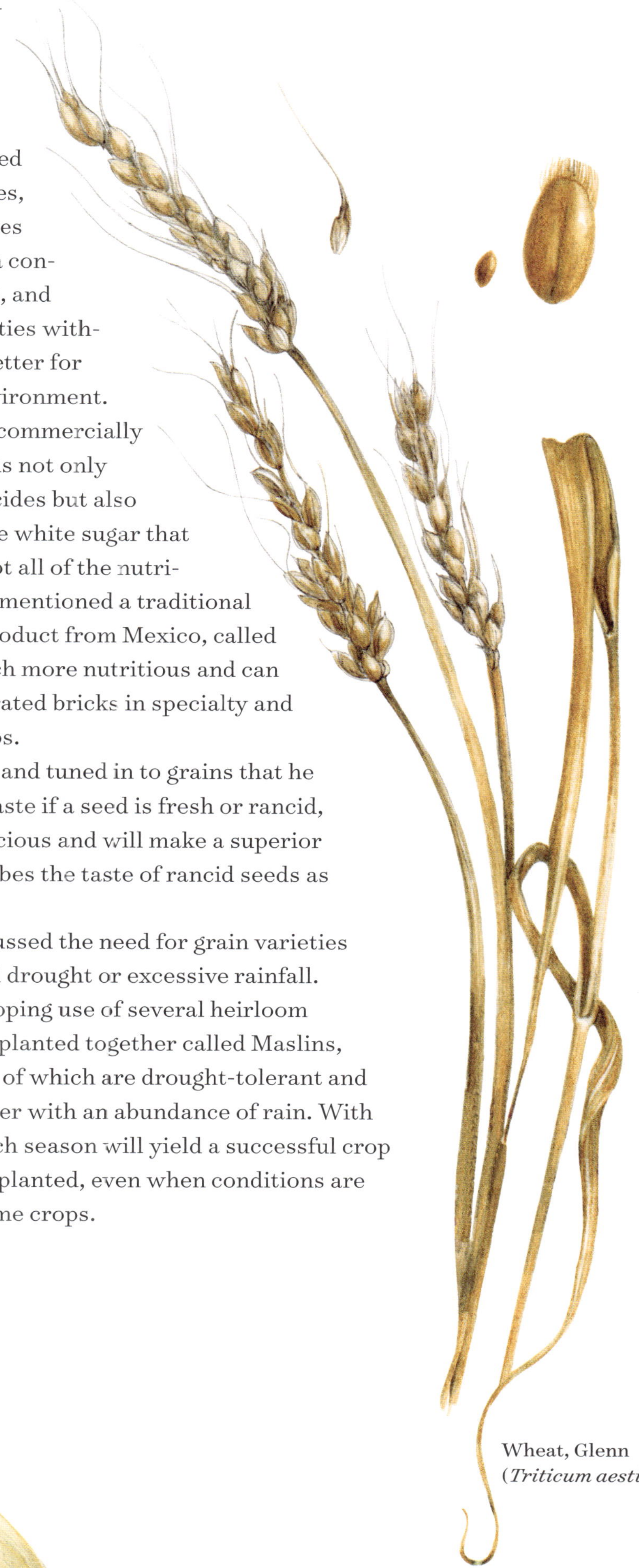

Wheat, Glenn (*Triticum aestivum*)

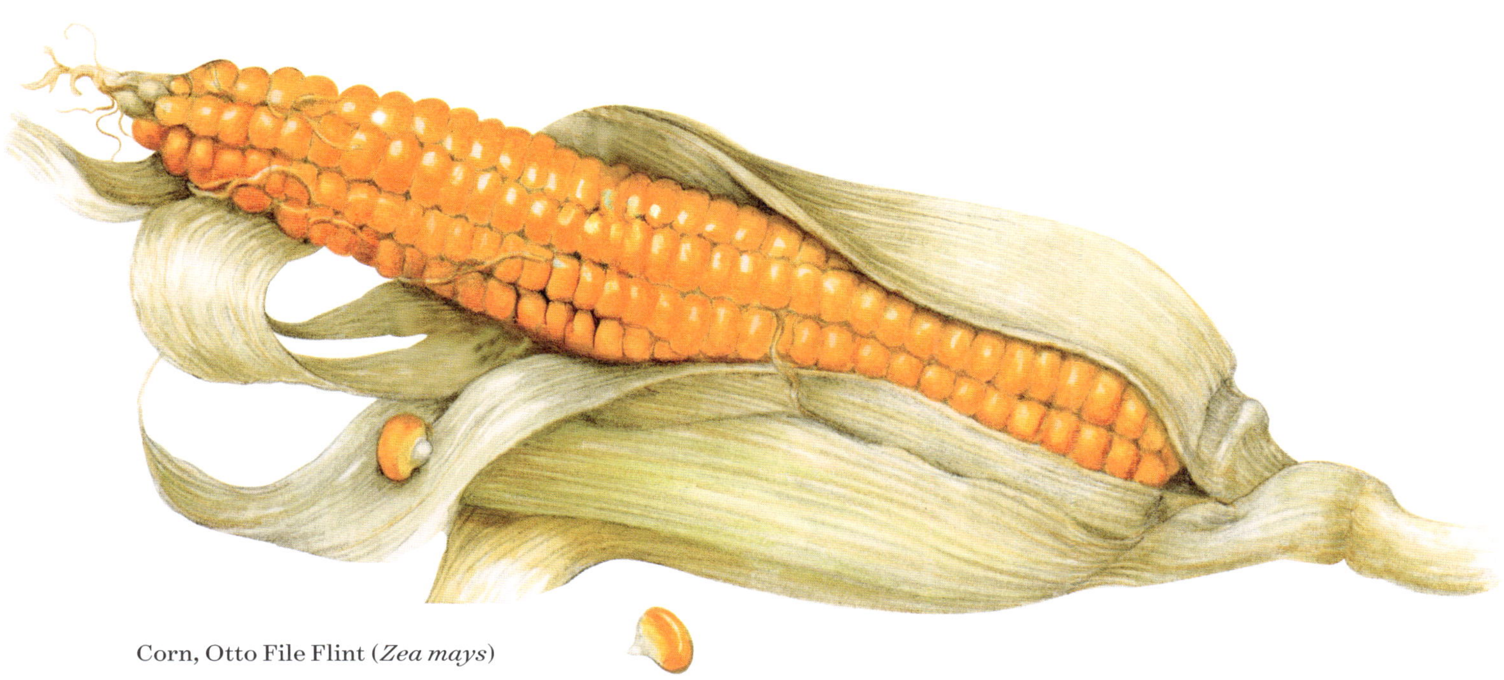

Corn, Otto File Flint (*Zea mays*)

# 11

# Lily Family

## Tulips (*Tulipa*)

### Liliaceae

### 600+ species

Tulips have always reminded me of my mom, Selma, who was proud of the tulips she grew in her garden. Each year she planted more and more tulips. On Sundays during tulip season, families would slowly drive by our garden admiring her beautiful flowers. We were all so pleased with the attention the garden got.

In high school I first tried my hand at botanical painting. I drew my mom's tulips many times, including the exotic parrot tulips that she grew. She was way ahead of her time. No one else grew these tulips in our suburban neighborhood. These early paintings of mine weren't very detailed, but they did start me on my path of enjoying nature up close by trying to draw and paint it.

In my early years of studying botanical drawing, the tulip was the first flower I drew. It showed me how to create a three-dimensional-looking flower by turning a cup shape into a tulip. It became the cover of my first book, *Botanical Drawing: A Beginner's Guide*. I considered calling the book "How to Draw a Tulip" but opted for a more encompassing title.

In 2022 I finally made it to the Netherlands to see the tulips. The history of tulips, especially during the Tulip Mania that swept the Netherlands in the 1630s, is fascinating. The idea that someone would pay as much for one tulip bulb as they would for a house at the height of the craze is hard to believe but apparently true. The demand for those early tulips was fueled by an unusual variegated pattern, caused by a virus, that gave rise to multicolored, dramatic flowers. Tulip Mania is referred to as the first economic bubble, brought about by people paying exorbitant prices based on speculation of the potential future value of tulip bulbs. Eventually, the tulip market crashed, and many people lost their fortunes.

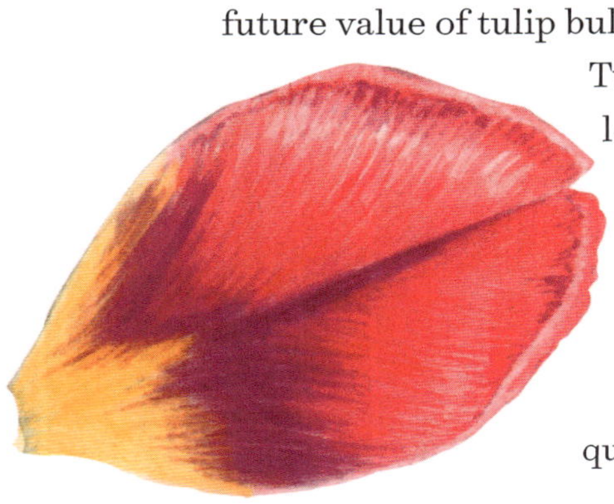

Tulip growers have since replicated numerous varieties of dramatic and exotic-looking tulips, this time without a virus. I was able to buy hundreds of bulbs developed in the Netherlands when I visited the world-famous Keukenhof gardens. The bulbs were shipped to me in the fall in time to plant in my garden. The following spring, when these tulips emerged in my garden, I created this family painting. There were so many to choose from. I must have planted twenty varieties. Each day I would walk outside and look at the developing tulips and ask questions such as: "Who wants me to paint them today?" It was hard to choose.

## Plants Illustrated

*Tulipa clusiana*, Species Tulip, Peppermint Stick
*Tulipa*, Tulip, Apricot Parrot
*Tulipa*, Tulip, Black Parrot
*Tulipa*, Tulip, Flaming Parrot
*Tulipa*, Tulip, Giant Orange Sunrise
*Tulipa*, Tulip, Pretty Princess
*Tulipa*, Tulip, Rem's Favourite

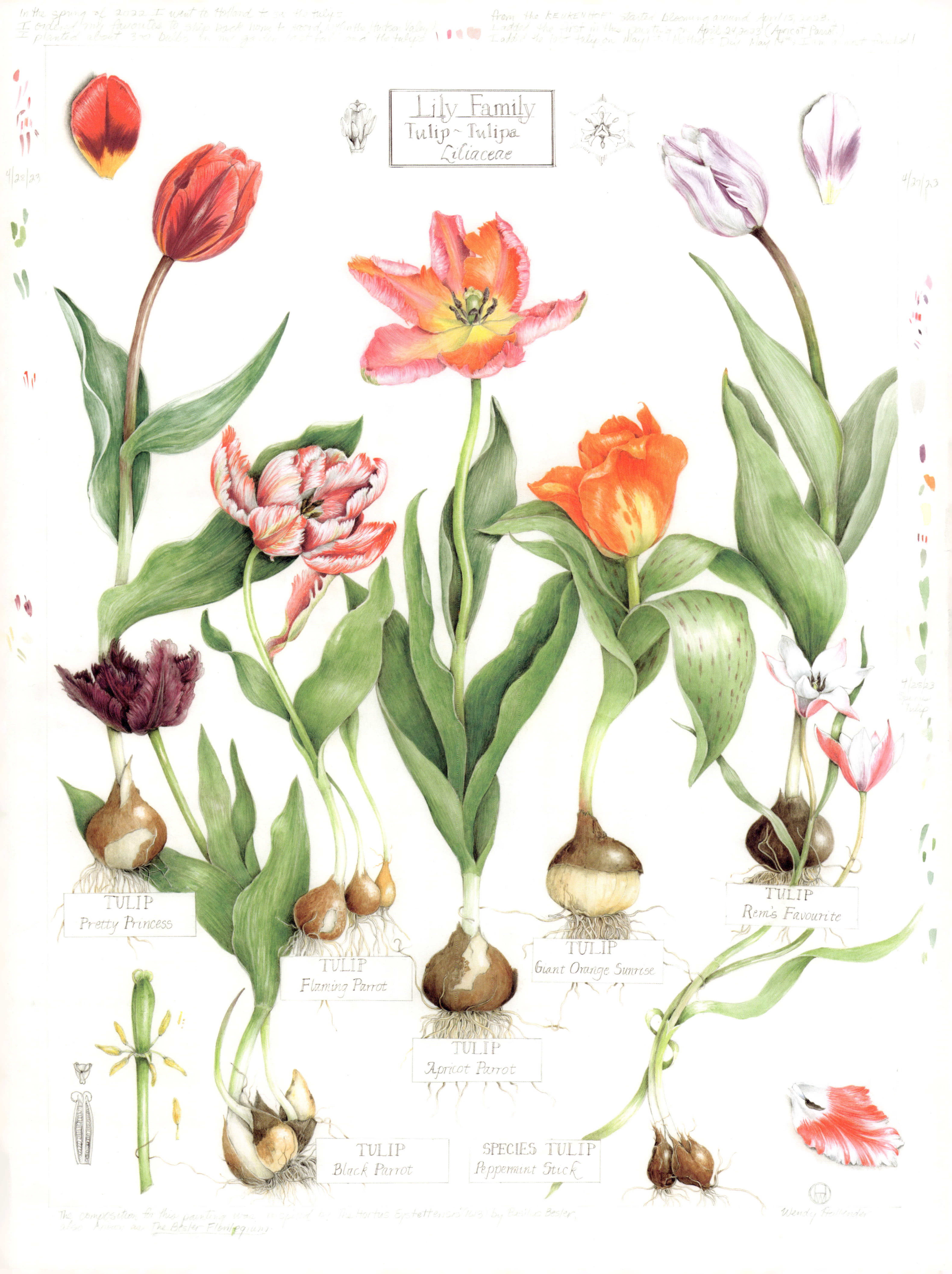
In the spring of 2022 I went to Holland to see the tulips.
from the KEUKENHOF started blooming around April 15, 2023.
Lily Family
Tulip ~ Tulipa
Liliaceae
TULIP
Pretty Princess
TULIP
Flaming Parrot
TULIP
Apricot Parrot
TULIP
Giant Orange Sunrise
TULIP
Rem's Favourite
TULIP
Black Parrot
SPECIES TULIP
Peppermint Stick

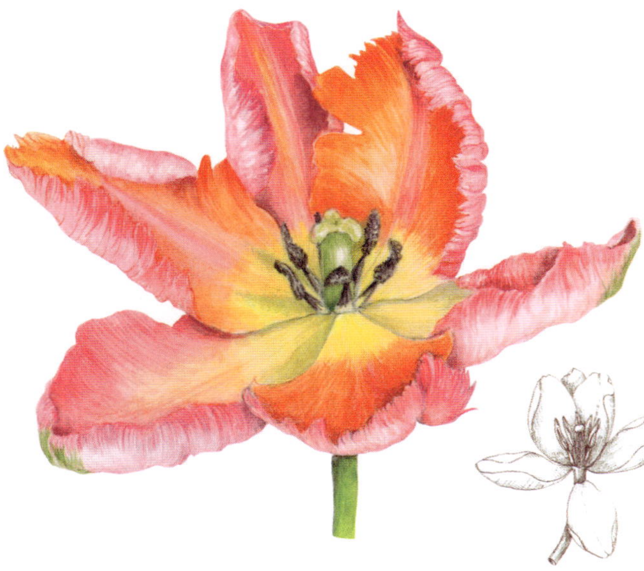

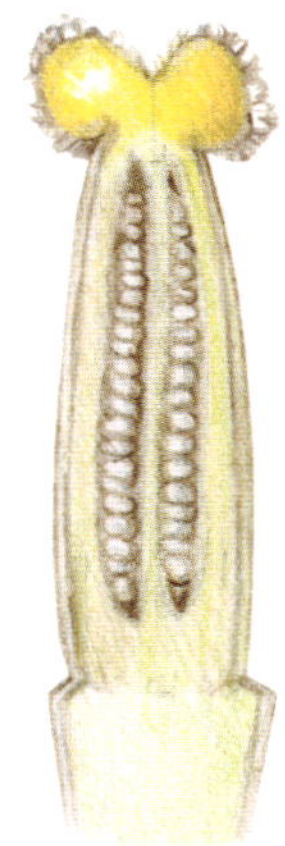

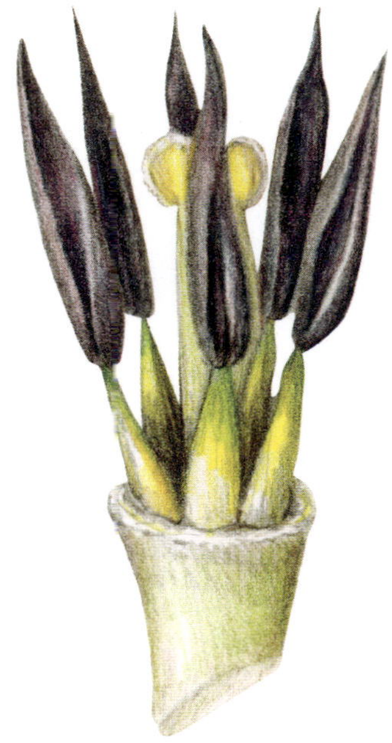

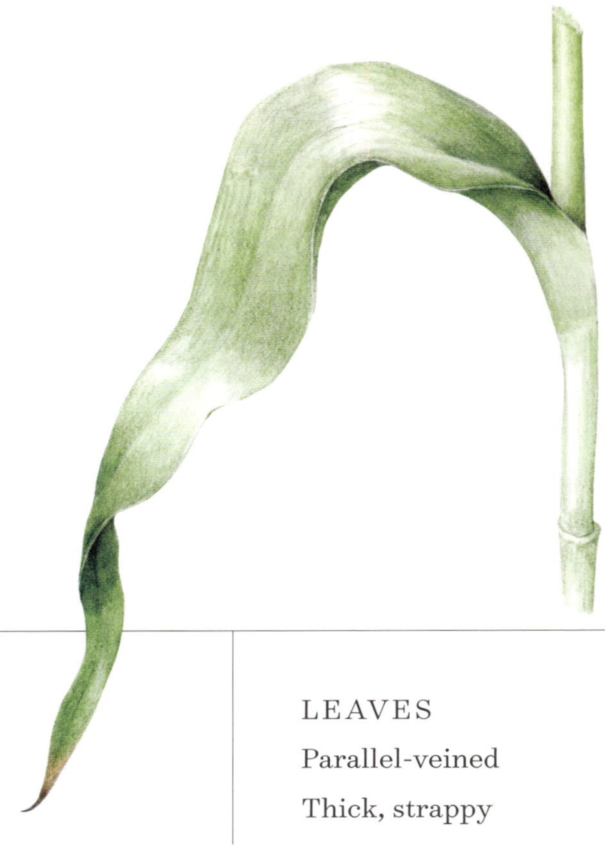

FLOWERS

Petals and sepals (tepals) look alike

3 petals and 3 sepals

6 stamens

Pistil with 3 carpels (sections) terminated by a sessile 3-lobed stigma

LEAVES

Parallel-veined

Thick, strappy

I planned this composition ahead of time, something I don't usually do, basing it on some of the layouts in the *florilegium* (book of plants) by Basilius Besler (1561–1629). I liked the idea of a semi-symmetrical layout, giving the painting a slightly formal look. I would cut a tulip and bring it into the studio and look at it in my composition to help me decide what to paint.

I learned a lot about tulips by studying them so closely each day in my garden. Each flower grows for many days, even a week or two, and gets taller and bigger each day. At night the blooms close up. I had to spray my tulips with a foul-smelling mixture called Deer Away to keep all the woodchucks, rabbits, and rodents from nibbling on them. I had a deer fence so I knew it wasn't only the deer that would munch on these apparently tasty treats. The following season I did forget to close my deer fence for a couple of days, and guess what? All my tulips got eaten before they had a chance to grow!

The red and white Flaming Parrot tulip that I painted looked just like one that had grown forty years ago in my mom's garden. I know this because I found a painting I created back then of the same or a very similar tulip. The tulip flowers were so spectacular that instead of painting several genera in the Liliaceae family, such as *Lilium* (lilies) and *Fritillaria* (fritillaries), I focused only on the tulips. I did it for my mom!

It is unusual to find seedpods on tulips, but I always hope to see one. One year in my garden, sure enough, a beautiful large tulip seedpod developed. I quickly cut it and drew it. I wondered if anyone grows tulips from seed. I have read that it takes three to five years to grow a tulip from seed as the bulbs need to grow large enough first. However, growing tulips from bulbs is a straightforward process and highly gratifying, as you plant the bulbs in the fall before frost and watch your tulips develop in the spring.

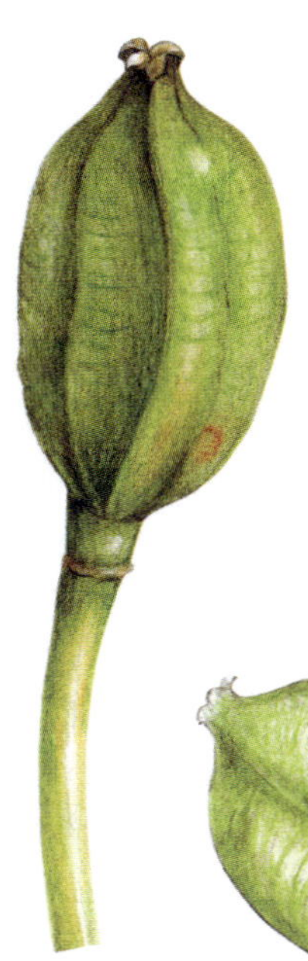

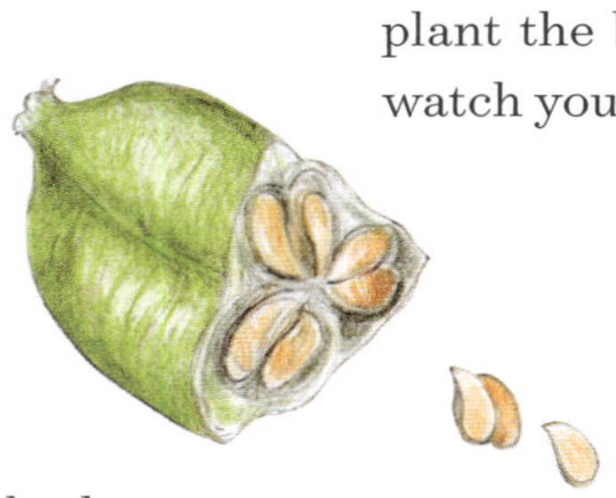

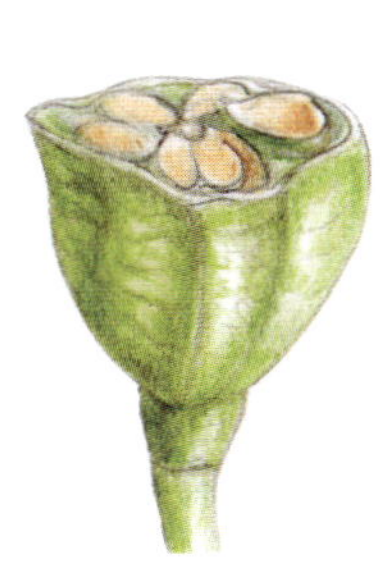

Tulip seedpods are a rare but fascinating find.

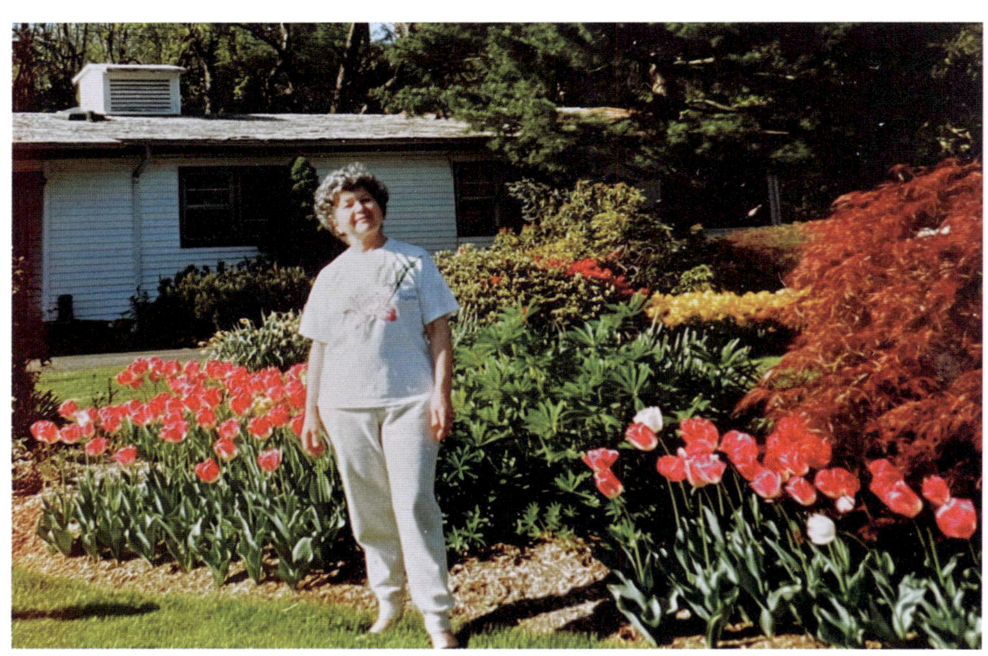

A photo of my mother standing by her tulips.

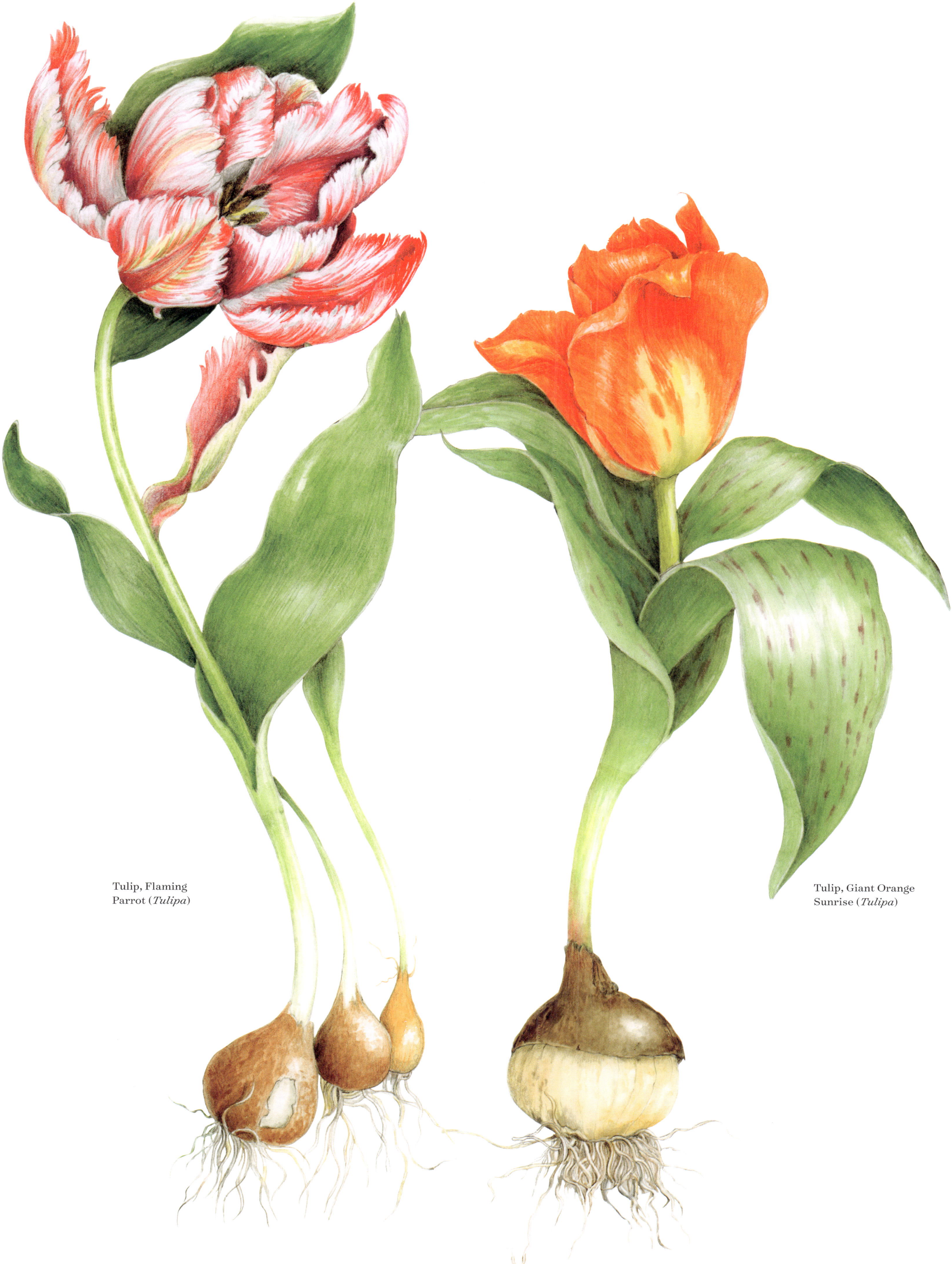

Tulip, Flaming Parrot (*Tulipa*)

Tulip, Giant Orange Sunrise (*Tulipa*)

# 12
# Magnolia Family

## MAGNOLIACEAE
## 220+ species

Deciding to include the magnolia family in my series of family paintings brought up memories for me of my early years as a botanical artist. I have focused thus far mostly on plant families that we use broadly for food and sustenance. The magnolia family is not something we grow to eat, even though the petals are edible, I believe. We enjoy magnolias for their beauty, especially as they flower early in spring when most trees do not even have leaves yet. They are beautiful, and belong to one of the oldest flowering-plant families on the planet. Their structure is so interesting too, from my point of view as an artist. The blossoms are fragrant and voluptuous, and the seedpods, when they emerge, are not to be believed. It is hard to imagine that they grow from these flowers. The clue is to look closely at the center of the flower to see the resemblance to the seedpod.

What tipped me immediately into deciding to paint this family was the realization that the tulip tree, *Liriodendron tulipifera*, represents the only other genus besides *Magnolia* in the family! I have been in love with these majestic, native American trees ever since I first discovered their blossoms on the ground at the New York Botanical Garden over twenty-five years ago. I was new to botanical illustration then, studying the practice in the NYBG's continuing education program. The beauty and unusual quality of tulip tree flowers took my breath away. They have unusual seedpods too. One of my instructors, who became a huge influence on me, was Dick Rauh; he created large paintings of these seedpods, highlighting their structure and beauty for all to see. Tulip tree flowers are literally dripping

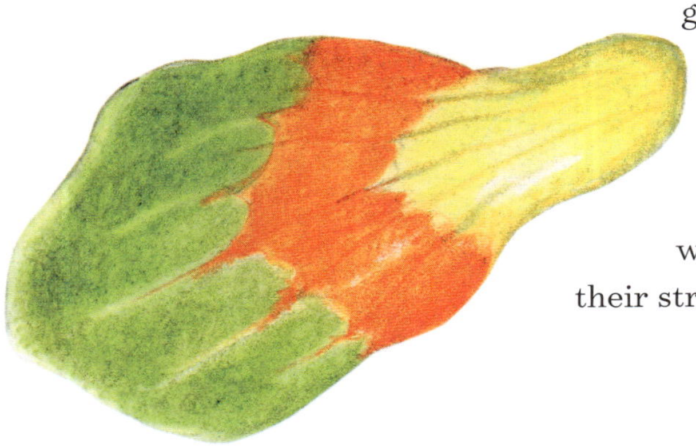

## Plants Illustrated

*Liriodendron tulipifera*, Tulip Tree
*Magnolia grandiflora*, Southern Magnolia
*Magnolia × loebneri*, Loebner's Magnolia
*Magnolia sieboldii*, Siebold's Magnolia
*Magnolia × soulangeana*, Genie Magnolia
*Magnolia × soulangeana*, Saucer Magnolia
*Magnolia stellata*, Star Magnolia
*Magnolia virginiana*, Sweetbay Magnolia
*Magnolia*, Magnolia, Pastel Sunset
*Magnolia*, Magnolia, Yellow Sun

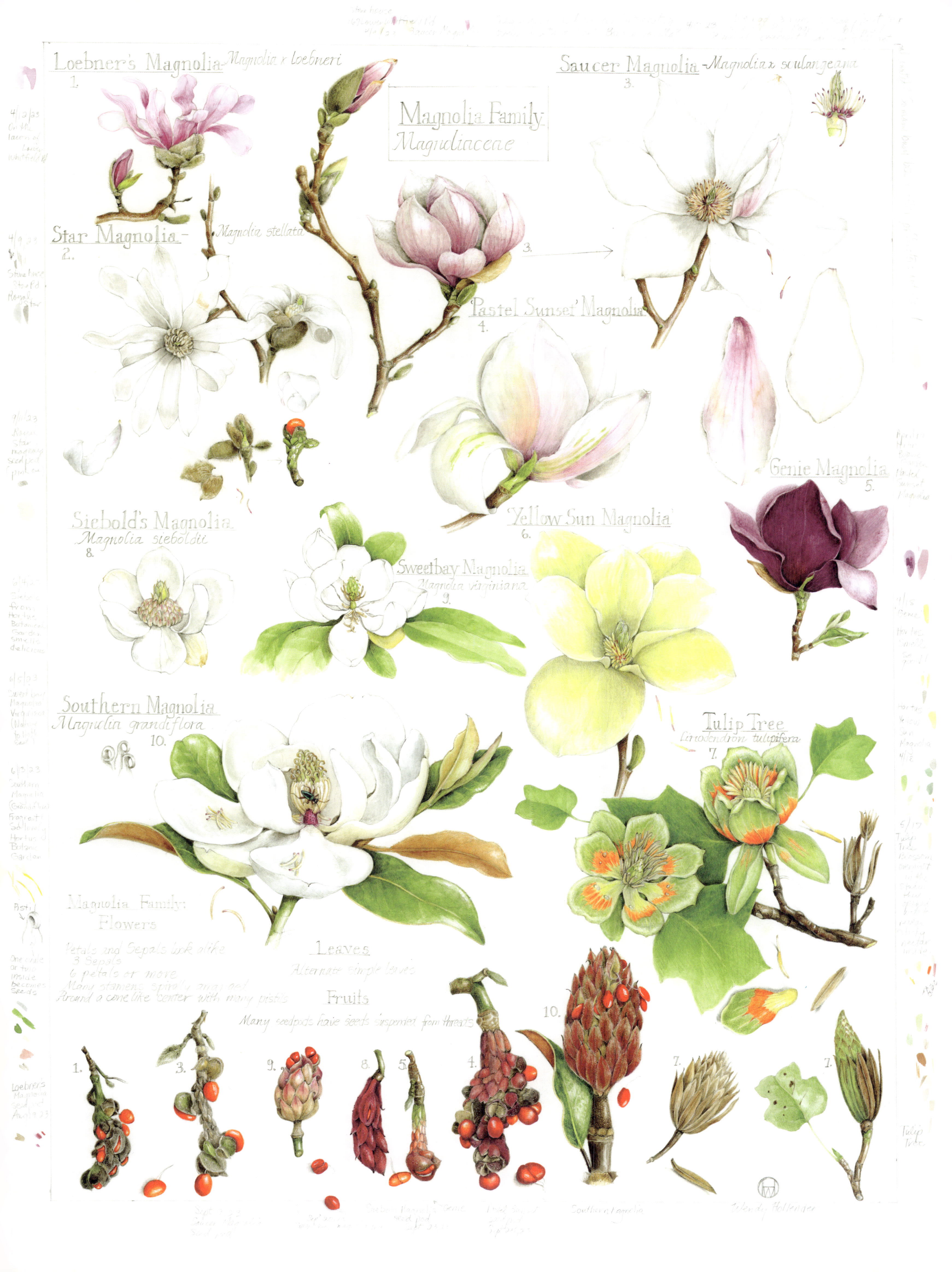

Loebner's Magnolia - Magnolia x loebneri
1.
Magnolia Family
Magnoliaceae
Saucer Magnolia - Magnolia x soulangeana
3.
Star Magnolia - Magnolia stellata
2.
3.
'Pastel Sunset' Magnolia
4.
Genie Magnolia
5.
Siebold's Magnolia
Magnolia sieboldii
8.
Sweetbay Magnolia
Magnolia virginiana
9.
'Yellow Sun Magnolia'
6.
Southern Magnolia
Magnolia grandiflora
10.
Tulip Tree
Liriodendron tulipifera
7.
Magnolia Family:
Flowers
Petals and Sepals look alike
3 Sepals
6 petals or more
Many stamens spirally arranged
Around a cone like center with many pistils
Leaves
Alternate simple leaves
Fruits
Many seedpods have seeds suspended from threads
10.
1.
3.
9.
8.
5.
4.
7.
7.

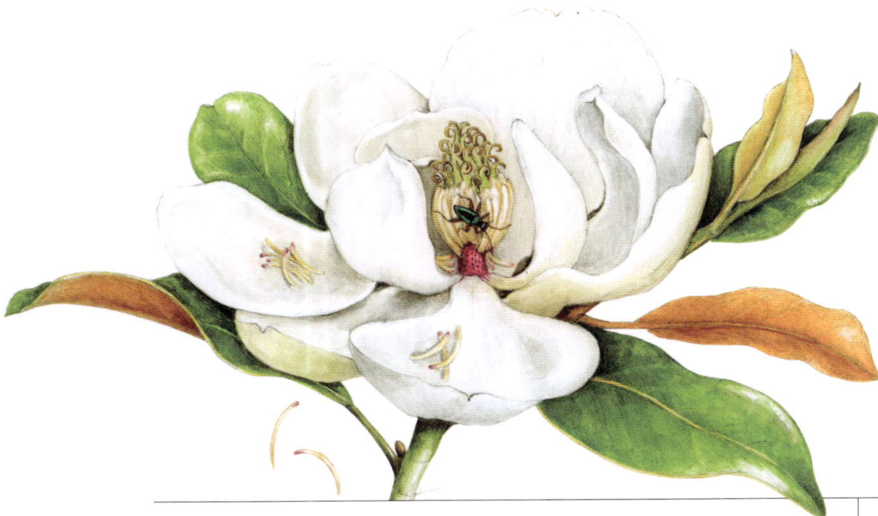

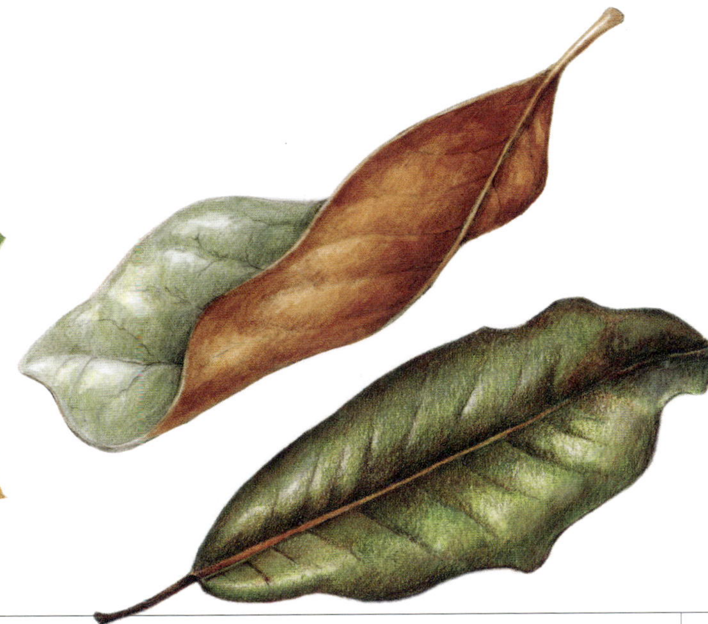

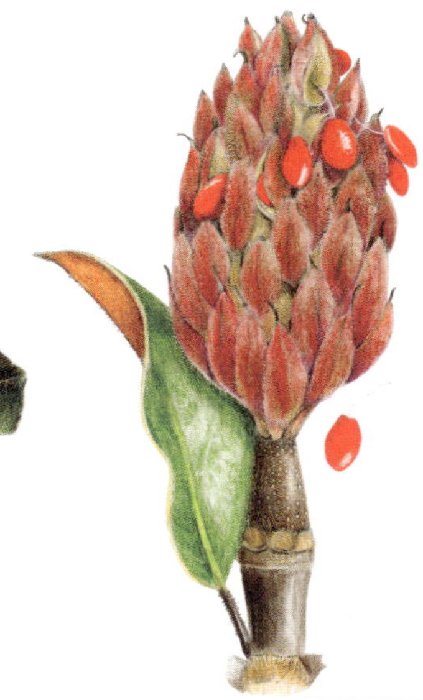

FLOWERS

Petals and sepals (tepals) look alike
6 or more sepals
6 or more petals
Many stamens spirally arranged around a conelike center with many pistils

LEAVES

Alternately arranged on the stem

FRUITS

Conelike, with seeds suspended from threads

with nectar and delicious to taste. The squirrels at the NYBG loved them so much that they clipped all the blossoms so they would drop to the ground. Normally, the blooms are high up on these trees, which are so tall and straight their trunks have been used to make canoes by Native Americans. I first found the flowers on the ground, and now I look up in the springtime to find them. Fortunately, in the woods surrounding my home, several trees have low branches where I can collect my subjects.

I put tulip tree flowers in my Magnoliaceae painting, having saved them a spot next to a yellow magnolia, so the colors would go nicely together on the page.

I invite you to run out now and see if you can find these blossoms, perhaps to draw but especially to sample their delicious, honey-flavored nectar. You can just lick the petals! Let me know what you find!

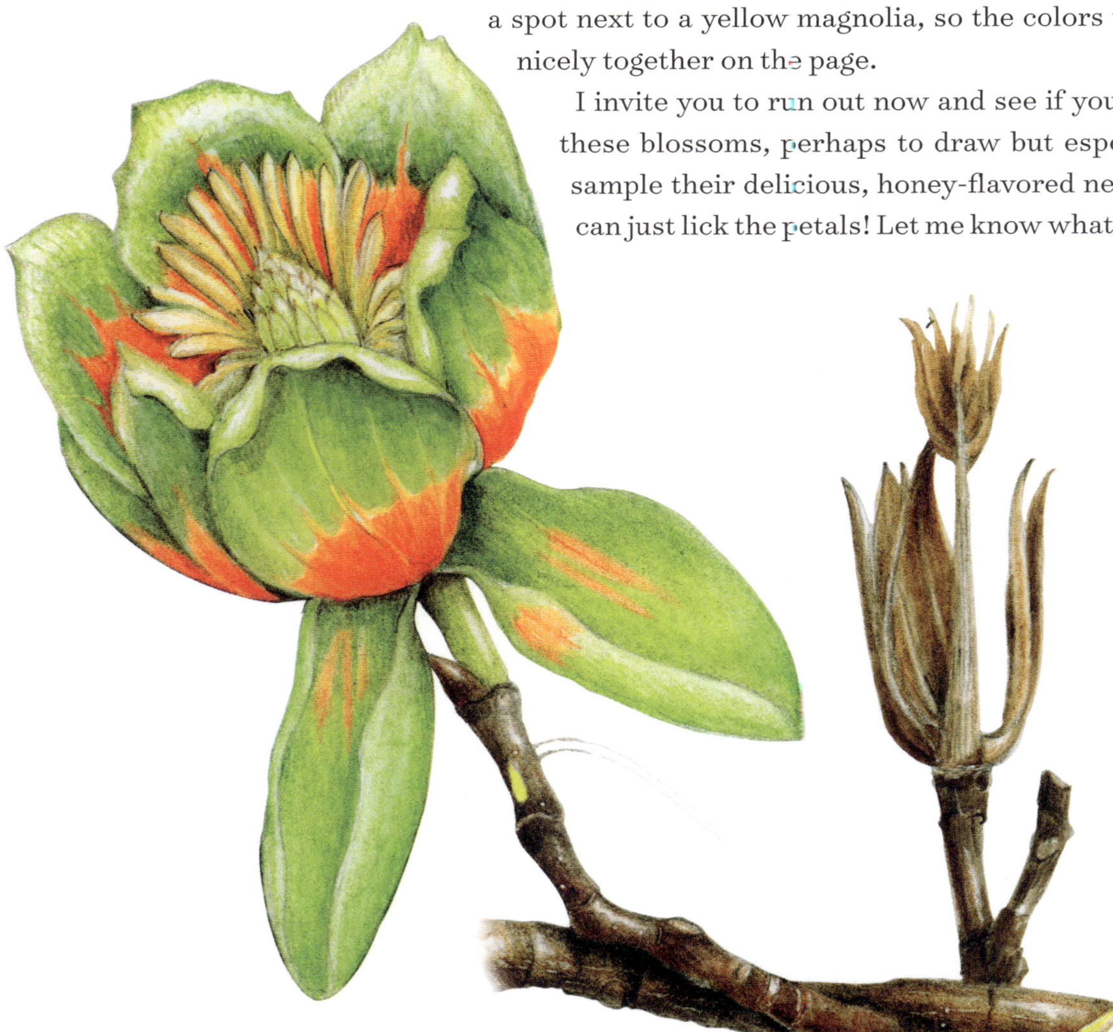

Tulip Tree (*Liriodendron tulipifera*)

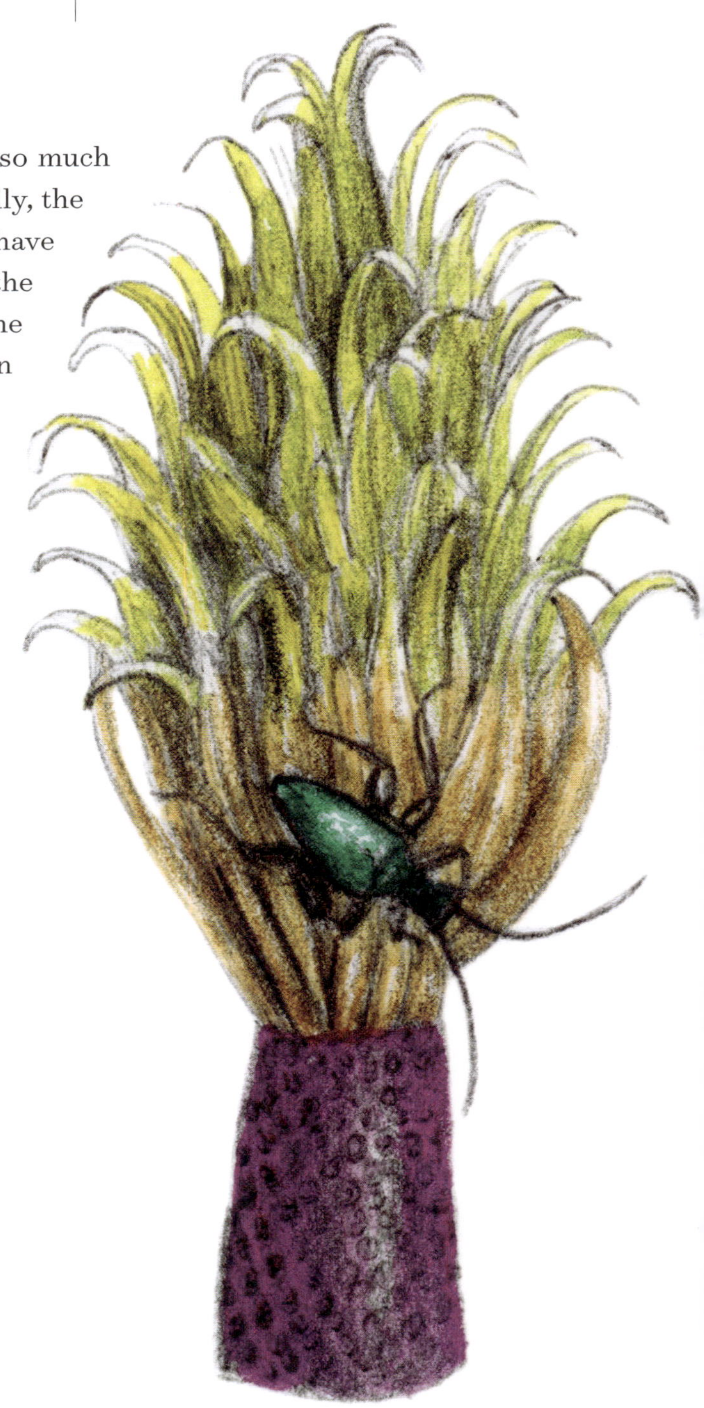

Southern Magnolia (*Magnolia grandiflora*)

# Magnificent Magnolias

BY ALLYSON LEVY

Without a doubt, the magnolia family (Magnoliaceae) is one of my favorites—perhaps because it is considered a "living fossil," whose ancestral lineage dates back to when dinosaurs roamed the earth, making magnolias truly prehistoric trees. Having existed for over 100 million years, they are among the most ancient flowering plants still in existence today, with their natural range encompassing eastern North America, Central America, and Asia. There are about 220 species worldwide, not to mention numerous hybrids, selections, and cultivars. Ten species of *Magnolia* are native to the United States, including two hailing from Puerto Rico. The only other genus in the Magnoliaceae is *Liriodendron* (tulip trees, aka tulip poplar).

Magnolias are considered some of the most primitive of the flowering plants, even though they have had such an extensive amount of time to evolve. The unique characteristics of their large, simple flowers with a distinctive arrangement of petals haven't changed much over all these years. Because of their ancient history, *Magnolia* species rely on beetles to serve as their main pollinators, as beetles evolved before many of the more complex pollinators.

The pollination strategy for magnolias is for some of the flowers to open in the morning, when the stigmas are receptive to the pollen brought to them courtesy of beetles, which come to feed on the different floral parts. Later in the day, the innermost tepals close, trapping the beetle. The flower's anthers become active at night, covering the insect in fresh pollen so that when it gets out, it will pollinate a different flower. This precise delay in flower function helps to ensure cross-pollination.

Botanists and breeders are responsible for hybridizing different species of *Magnolia*, and because of their efforts, there is a vast variety of flower colors, sizes, and shapes, as evident in Wendy's drawing. Flowers range from the full tuliplike blooms of the Genie Magnolia to the large fragrant flowers of the Southern Magnolia. Still other magnolias have small, upside-down-facing flowers that invite a closer view, even if it's on your knees to look up at them (Siebold's Magnolia). Breeding has also produced a wide color palette in solid whites, yellows, pinks, reds, and purples as well as species and hybrids pigmented with multiple colors.

All magnolias have alternate, entire (unlobed) leaves, but that's pretty much where the similarities end. Leaves vary from hairy to smooth and can be small or huge. Bigleaf Magnolia produces the largest simple leaf in North America, occasionally spanning up to 3 feet! While some magnolias are deciduous, others produce leaves that remain on the tree throughout winter, such as the *Magnolia grandiflora* hybrid Bracken's Brown Beauty, with glossy dark green upper leaves and a stunning brown felted underside, like a miniature fuzzy blanket for the leaf's bottom.

Often overlooked is the unique and highly decorative beauty of magnolia fruits, sometimes called "cones," though botanically speaking, they are not true cones. The fruits of hybrid *Magnolia* crosses come in a wide range of strange shapes and sizes—the lewdly shaped cones of the Yellow Sun Magnolia call to mind specific anatomical parts. The magnolia seed cone, known as a follicetum, can have ripe seeds ranging in color from bright scarlet to a muted orange red. When fully ripe, the individual seeds become suspended by tiny threads before they are dispersed to the ground and eagerly taken by birds, squirrels, and other small critters.

Magnolias are special to me because they were the first trees I ever planted, courtesy of the Friends of the Urban Forest organization in San Francisco. From their tree list, I selected two Southern Magnolias to plant because I knew one day they would make grand trees.

The *Magnolia* collection is one of the most prized at the Hortus Arboretum and Botanical Gardens, and I consider it an honor that Wendy's stunning botanical drawing includes several different magnolias from our collection. Her artwork captures the classic, elegant beauty and the diversity that make magnolias so beloved.

ALLYSON LEVY is a practicing visual artist and gardener. She is cofounder and codirector of Hortus Arboretum and Botanical Gardens, located in the lower Hudson Valley of New York. Magnolias have always been one of her favorite genera, and she plans to grow as many of the 200-plus species as possible! Allyson also loves to write about plants and is coauthor, with Scott Serrano, of the book *Cold-Hardy Fruits and Nuts*, published in 2022, and is currently writing another book on trees and shrubs (which will include a few magnolias), due for release in spring 2027.

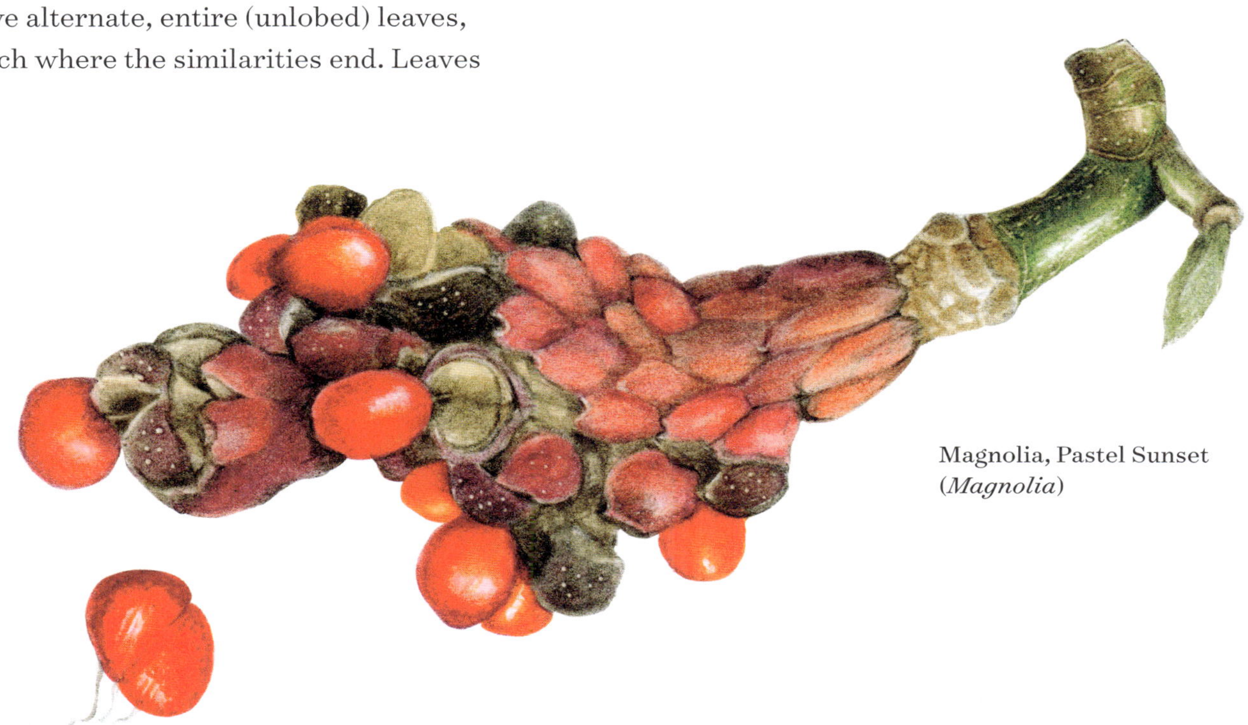

Magnolia, Pastel Sunset
(*Magnolia*)

# 13
# Mallow Family

## Malvaceae
## 4,000+ species

The mallow family was one of the first families I chose to include in this project. In my personal journey with botanical illustration, drawing the flower of a hibiscus (a mallow family member) was a transformative moment. It was early in my studies at the New York Botanical Garden, just after I had completed some of the basic classes such as flower drawing, understanding light on form, using colored pencils for understanding color theory, and plant morphology, in which I learned to dissect flowers under a microscope. I went on a vacation to the Caribbean islands of Turks and Caicos. I brought along my colored pencils and a sketchbook to try my hand at botanical drawing while traveling. In 2002, when I took this journey, the accepted conventional medium for creating botanical illustrations and paintings was primarily watercolor. Colored pencil was introduced to me primarily as a way to learn color theory, and since it was a convenient portable medium, I decided to take it along on my vacation. As a sketchbook page from the trip says, I found a hibiscus flower in a hedge along the road by an internet café. (Back then, to check email on vacation, one often had to find an internet café.) When no one was looking, I furtively cut one flower. I took the flower back to my hotel, found a lounge on the beach, and sat down to draw by the sea.

I remember vividly that I spent all day drawing that one beautiful hibiscus, trying my best to capture the rainbow-like quality of the petals along with their rippling twists and turns. The center column, typical of hibiscus flowers, was like a piece of jewelry, perhaps a dangling earring. I had to draw it in detail, including dissecting the column and studying the ovary inside. I loved the experience of spending the day in a beautiful location, on a beach, under a palm tree, with the sun shining. The hibiscus was my companion that day, as I was alone on the beach. Rather

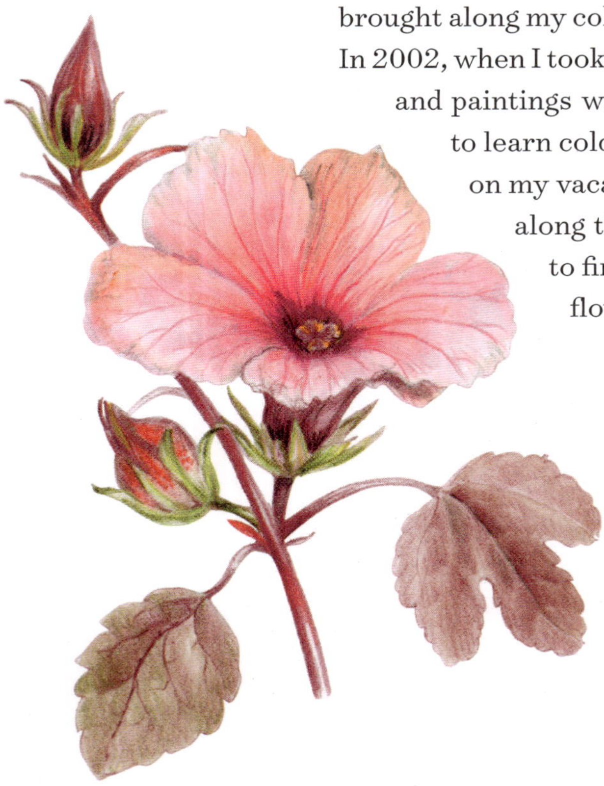

## Plants Illustrated

*Abelmoschus esculentus*, Okra
*Althaea officinalis*, Marshmallow
*Brachychiton bidwillii*, Little Kurrajong
*Ceiba pentandra*, Kapok Tree or Silk Cotton
*Gossypium barbadense*, Sea Island Cotton
*Gossypium darwinii*, Darwin's Cotton
*Gossypium tomentosum*, Native Hawaiian Cotton or Maʻo
*Hibiscus brackenridgei* subsp. *brackenridgei*, Maʻo hau hele
*Hibiscus rosa-sinensis*, Chinese Hibiscus
*Hibiscus sabdariffa*, Roselle
*Hibiscus syriacus*, Rose of Sharon
*Theobroma cacao*, Chocolate Tree, Cacao, or Cocoa Tree
*Tilia cordata*, Little Leaf Linden

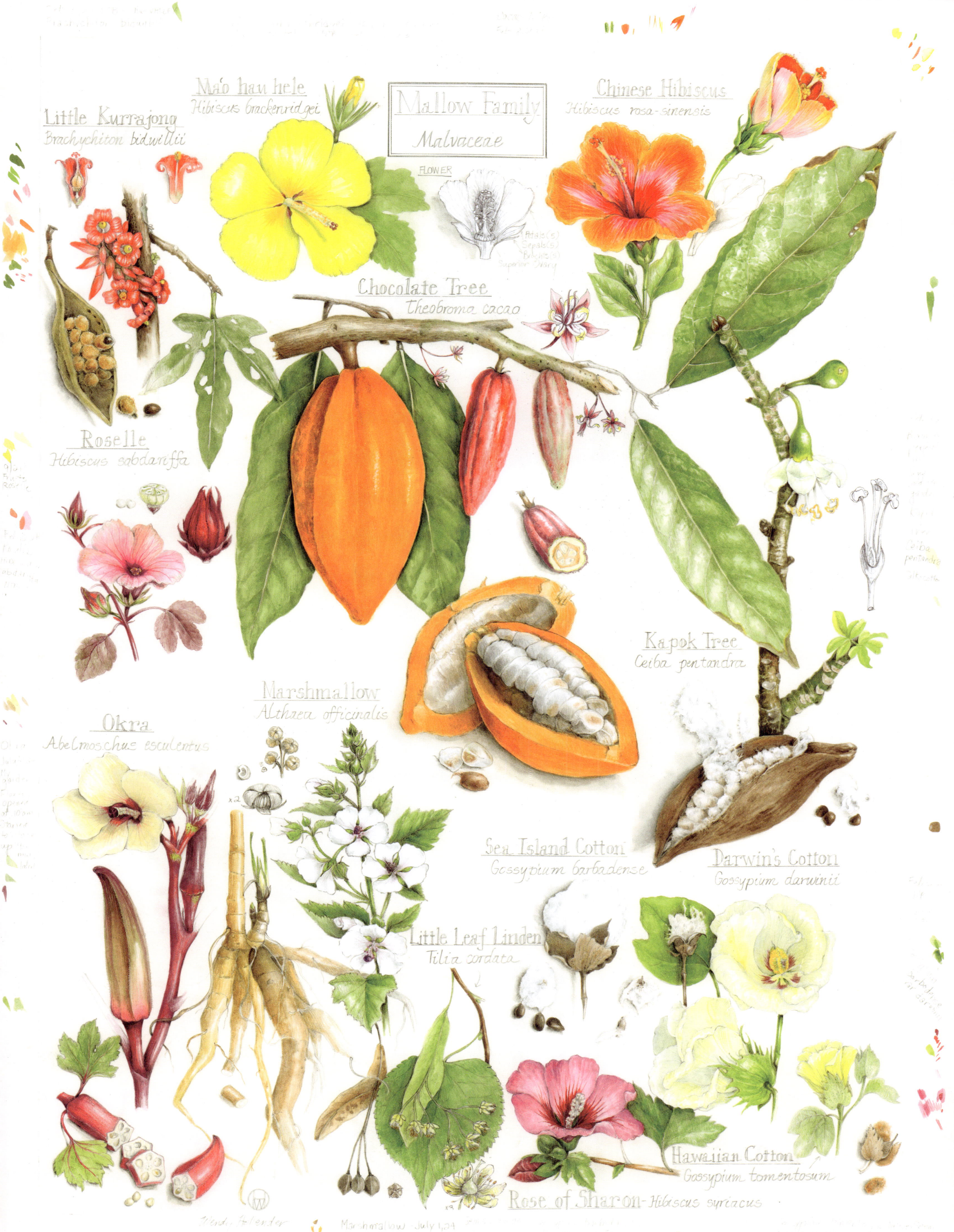

Mallow Family
Malvaceae
Little Kurrajong
Brachychiton bidwillii
Ma'o hau hele
Hibiscus brackenridgei
FLOWER
Petals(5)
Sepals(5)
Superior Ovary
Chinese Hibiscus
Hibiscus rosa-sinensis
Chocolate Tree
Theobroma cacao
Roselle
Hibiscus sabdariffa
Kapok Tree
Ceiba pentandra
Marshmallow
Althaea officinalis
Okra
Abelmoschus esculentus
Sea Island Cotton
Gossypium barbadense
Darwin's Cotton
Gossypium darwinii
Little Leaf Linden
Tilia cordata
Hawaiian Cotton
Gossypium tomentosum
Rose of Sharon - Hibiscus syriacus
Wendy Hollender
Marshmallow July 1, 24
Dina's garden

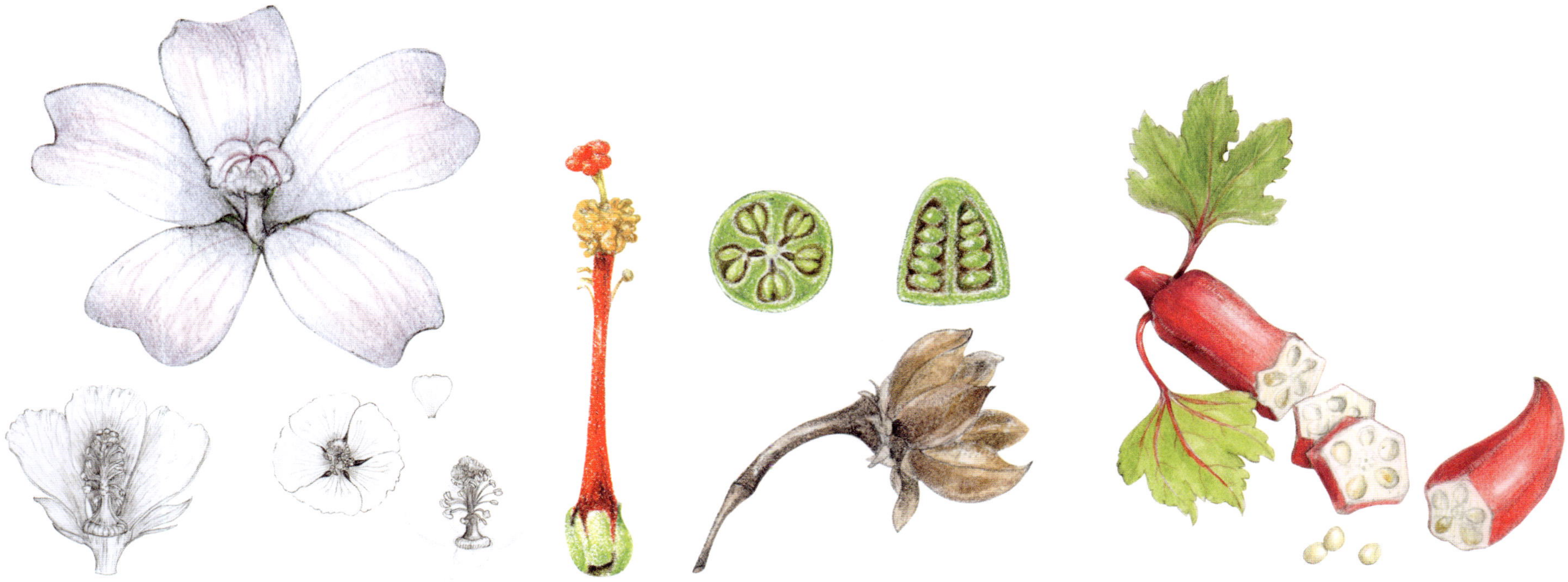

FLOWERS

5 separate petals

Stamens fused, forming a column

Mucilaginous sap

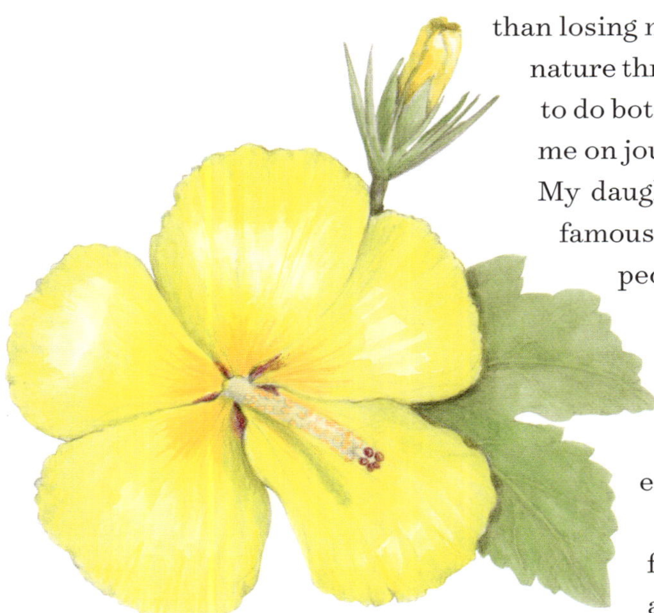

Ma‘o hau hele
(*Hibiscus brackenridgei*
subsp. *brackenridgei*)

than losing myself by reading a novel, I had descended into the magical and complex world of nature through the lens of this one stolen flower. This experience led to the idea of traveling to do botanical illustration and eventually to take students of botanical drawing along with me on journeys to so many beautiful and exotic locations around the world.

My daughter was the farming pioneer in our family, first to intern on a farm, the now-famous Stone Barns in Tarrytown, New York, right after college. She joined many young people at the time, traveling and working for WWOOF (Worldwide Opportunities on Organic Farms), which placed her on a farm in Hawai‘i. I decided to visit her there and combine the trip with an opportunity to study and draw endangered native Hawaiian hibiscus plants. The hibiscus flowers led me to the National Tropical Botanical Garden on the island of Kaua‘i, where they were propagating endangered hibiscuses from seeds collected in the wild.

There, at this magical garden, I also discovered perhaps my favorite classroom for botanical drawing in the world. I saw my first chocolate tree (*Theobroma cacao*) and experienced flowers and fruits that grow in a cauliflorous arrangement, meaning that they grow directly out of the woody trunk and branches. Harvesting ripe pods of cacao is a joy because inside are the beans (seeds) that are roasted and turned into chocolate. The outside of the beans is covered in a gelatinous (or mucilaginous) goo with a slight taste of chocolate that is delicious as a refreshing hydrating beverage.

There are many species of hibiscus, some native to Hawai‘i. I started my painting with *Hibiscus brackenridgei*, the Hawaiian state flower, which is an endangered species with fewer than 100 plants remaining in the wild on three islands. Fortunately, this plant is easy to cultivate, and it is a beam of bright sunlight to see it growing. The flower opens in the morning and, like all hibiscuses, closes by afternoon and dies at the end of the day.

Many hibiscus family plants grow in a temperate environment, and I found many closer to home. Okra grows well, producing an abundance of edible fruits for culinary uses; marshmallow is used medicinally; and rose of Sharon is an ornamental shrub that delights the landscape in late summer.

I created this composition with a center focus on the chocolate tree—because who doesn't love chocolate?—and included as many plants as I could. The multicolored Chinese Hibiscus had to be included for sentimental reasons. *Tilia cordata* (little leaf linden) is a favorite tree around the world that grows tall and majestic, producing flowers with a seductive scent that the bees and I love. The bees produce a delicious honey from the flowers, which are also used in perfumes and teas and other medicinals. My grandson Leo thinks this is a cool family because it has both chocolate and marshmallow, two of his favorite treats, which make up hot chocolate, a delicious beverage in the winter.

Explore this family wherever you live and see what you can discover too.

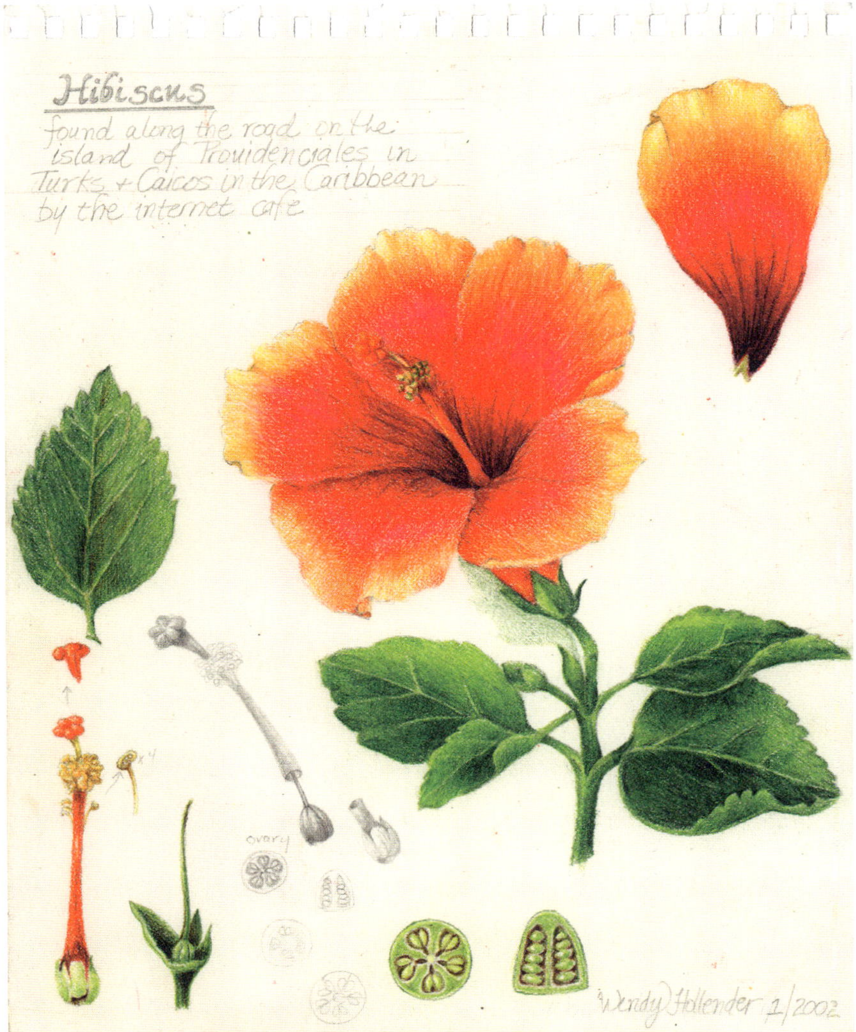


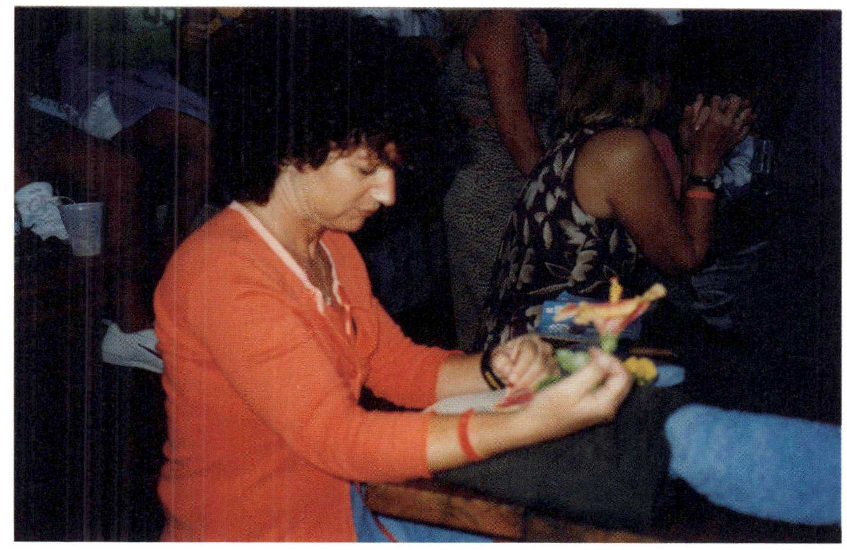

During my visit to Turks and Caicos in 2002, I spent all day drawing the hibiscus flower (*left* and *above*). This illustration and notebook sketch show my inspiration and process for planning the composition of the mallow family painting (*below*).

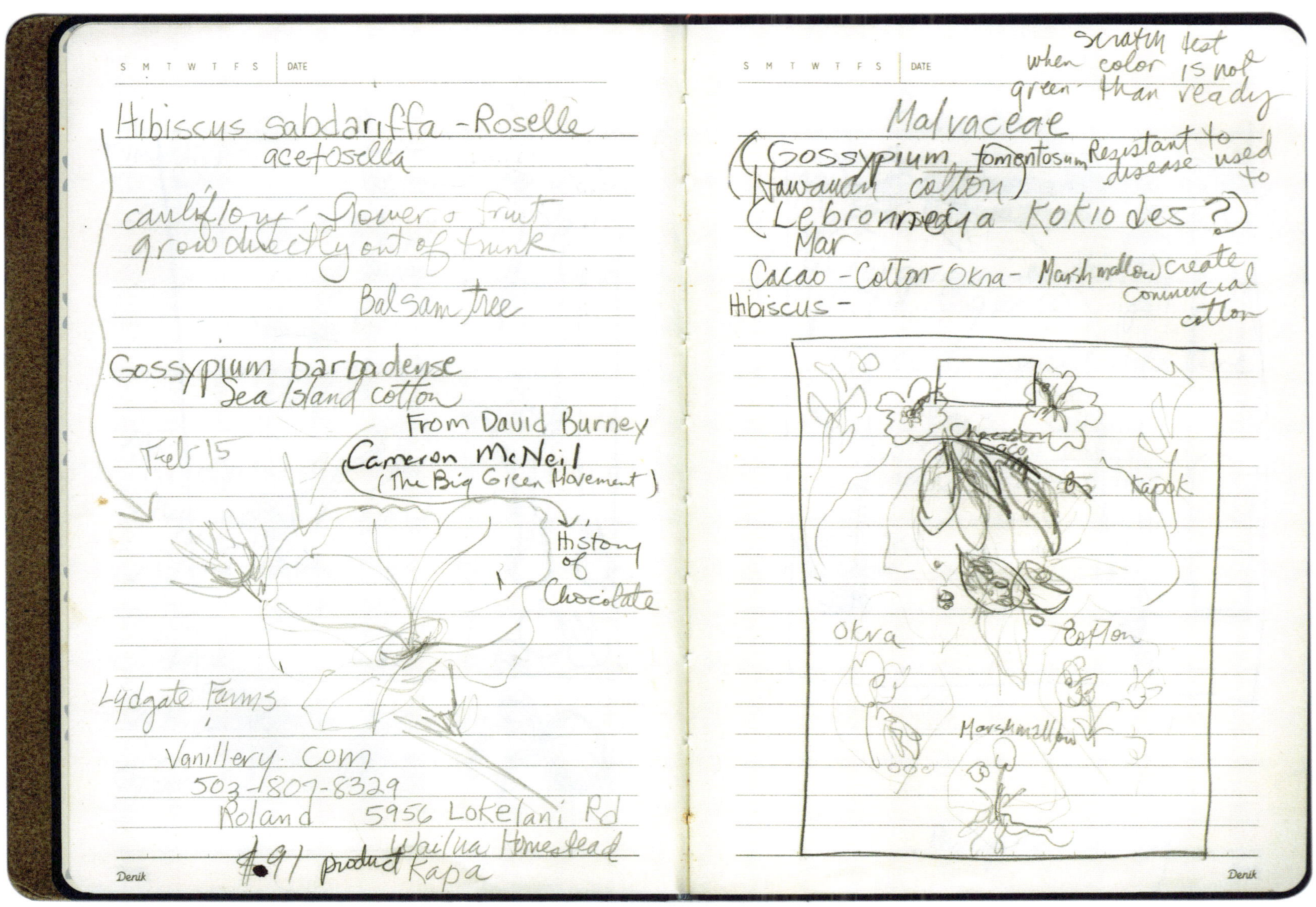

# Malvaceae: The Loving Family That Comforts!

BY DINA FALCONI

I invite you to dive into this enchanting beauty of a plant family. It's gorgeous in the garden, striking on the menu, and important in the herbal apothecary. Providing healing herbs, tree-derived medicines, a common vegetable, cottage flowers, an exalted bean, and one of the most commonly worn fibers, this family is integral to so many aspects of our lives.

If you have ever gazed into a hollyhock flower or swamp rose mallow bloom, taking in these silky, soft, soothing beauties—you've bathed in soul-filling comfort. These gorgeous cottage flowers are grown for their beauty alone, yet gentle salads and soothing medicines reside in their petals.

I fell in love with this family many years ago when learning about the humble common mallow (*Malva neglecta*). How could this healing, nutritious plant grow freely, like a weed, out of disturbed soil in the barnyard? I was impressed by its vitamin and mineral content and its ability to soothe irritations. And we can utilize all its parts: leaves, flowers, seeds, and roots. Then I learned about its relatives and fell even more in love with this family.

A star in the herbal apothecary, marshmallow (*Althaea officinalis*) reigns supreme as the cool soother whose slippery, sticky roots moisten with their copious amounts of mucilage. The root (leaf and flower can also be used, with milder effects) is an emollient—an anti-inflammatory for skin irritations; and a demulcent, used to soothe internal mucous membranes of the respiratory, digestive, and urinary tracts. It's worth noting that many of its relatives, specifically those in the *Malva* genus, share these qualities.

Digging into a bit of sticky history, we learn that this sturdy perennial of the garden or wildflower of the swamps and marshes was the origin of the famed marshmallow confection. It began as a remedy for sore throats and irritated bellies, from syrup to lozenge. To make it, the mallow herbal infusion was mixed with sugar, enough to make it into a syrup, and then whipped with tons more sugar to create a somewhat solid confection (losing its medicinal stature, from my point of view). The modern-day sweet of the same name no longer has any mallow in it and still has too much sugar, plus other less savory ingredients, sadly relegating it to the junk-food realm.

I was introduced to okra, the modest vegetable of gumbo fame, in Greenwich Village. It was a favorite of my informal food-as-medicine herb mentor Mickey Carter (both Mickey and okra share African ancestry). I was the ripe old age of eleven when this unspoken apprenticeship began. I remember relishing okra's thick, slimy gooeyness, as Mickey often cooked it for our dinner in his basement apartment on New York City's West Tenth Street, where he was the super of a handful of buildings. Little did I know that it was related to so many other plants I would fall in love with.

Linden (*Tilia* species), the medicinal tree of this family, often planted ornamentally, yields a tasty, healing tea enjoyed around the world, from Mexico to Europe. Commonly referred to as lime blossom tea, it uses the sweet-tasting fragrant flower cluster with bracts. I call it "Peace Out" tea, since it brings on a sense of calm, a peacefulness reached by settling into oneself, encouraging flow and release. That makes sense, as it is a nervine—a nervous system tonic that encourages relaxation and sleep (when needed). As a hypotensive, it brings down high blood pressure. Yes, it's potent but also gentle and very pleasant—enough so that everyone, young and old, imbibes this inviting beverage, often unaware of its healing attributes.

Most of the plants I know in this family are mild flavored, but not *Hibiscus sabdariffa*. Talk about tart! It goes by many common names: roselle, sorrel, bissap. After a vigorous West African dance class, sweat pouring off my skin, I was offered bissap in the form of a refreshing, blood-red drink. It was made with the maroon calyces (whorls of sepals, which persist after the petals have fallen off) and further enhanced with ginger and plenty of sweetener (although I prepare it now with no to low sweetener). Originally hailing from the West African tropics,

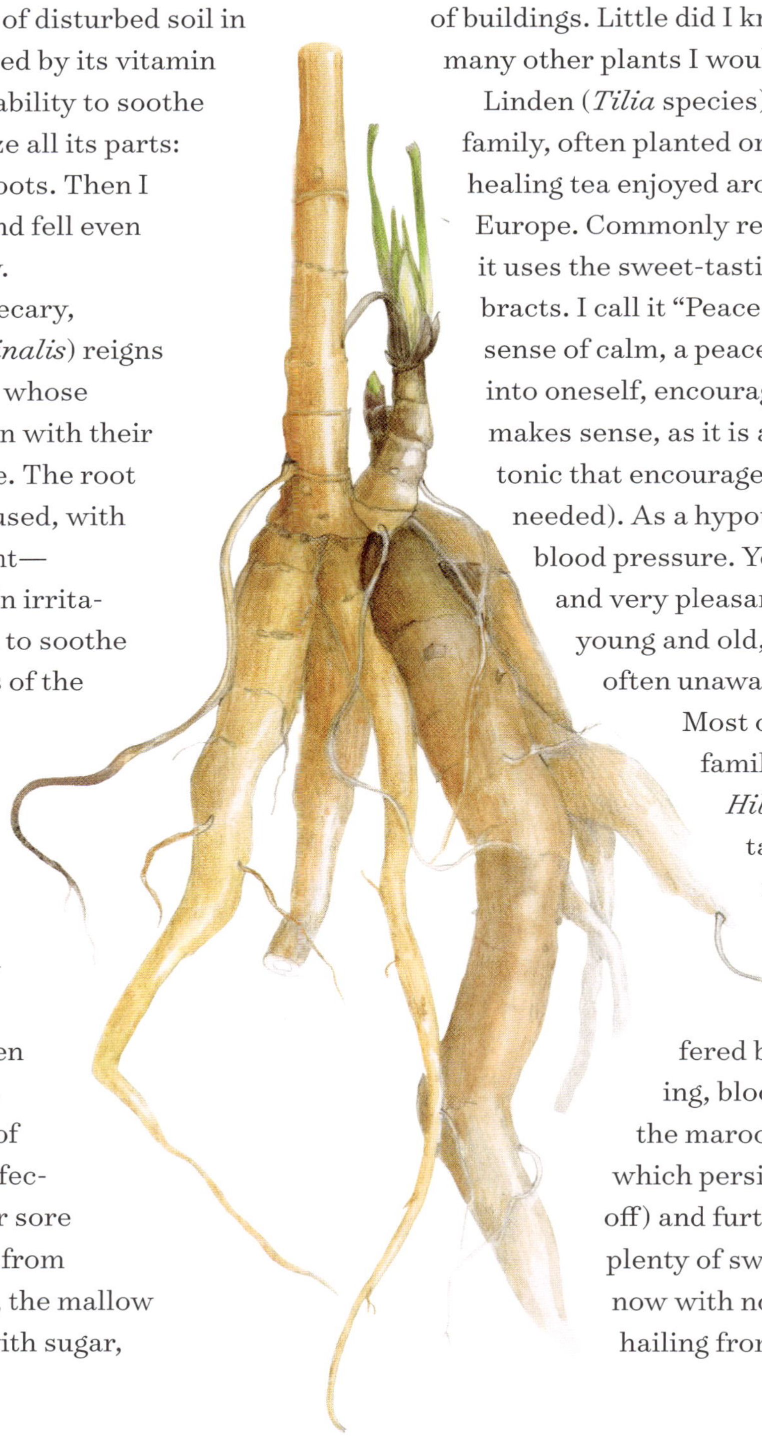

Marshmallow
(*Althaea officinalis*)

this bioflavonoid-rich cooling herb is a cardiotonic with hypotensive and anti-inflammatory actions. It's a perfect herb for cooling down!

Probably our most commonly used plant fiber, cotton, swathes us in clothing, bedding, and more. I love my cozy flannel sheets and am grateful for a luxurious, plush, highly absorbent bath towel. Yet, little is mentioned about cotton-root bark, which has been used as an abortifacient for centuries. Plants have so many sides to them!

Then there is chocolate, the family's most obsessed-over member. But keep in mind: without sweetening, most folks would not tolerate the bitter, astringent, fatty, fermented seeds of the cacao bean. *Theobroma*, the source plant's genus name, means "food of the gods," implying big stuff. (Does it make us feel like a god, or closer to god, or conjure a blissed-out-god state . . . ? Who knows?) Still, it's Valentine's Day's most popular food—love food. Its chemistry, rich in polyphenols and alkaloids (theobromine and caffeine), does seem to encourage the feeling of love, the "feel-good" chemicals in many of us. Hence, many people are having an obsessive love affair with chocolate! However, these alkaloids can become addictive and can have a negative impact, making us speedy, adrenalized, overdriven, and overstimulated, often leading to sleeplessness and burnout. So, proceed with caution!

OK, no family is perfect! Still, this is an invitation to acknowledge and fall in love with this gorgeous, nourishing plant tribe—allowing its billowy clouds of softness to calm and cool our irritations—to relish in its satiny petals, unctuous bitter seeds, and sticky roots of soothing delight.

Dina Falconi is a clinical herbalist with a strong focus on food activism and nutritional healing. An avid gardener, wildcrafter, and permaculturist, Dina has been teaching classes about the use of herbs for food, medicine, and personal care, including wild food foraging and cooking, for more than thirty years. She offers online courses at www.InTheWild.Kitchen. She is the author of *Foraging and Feasting: A Field Guide and Wild Food Cookbook* and *Earthly Bodies and Heavenly Hair: Natural and Healthy Personal Care for Everybody*. Please check out her website and YouTube and Instagram accounts: www.botanicalartspress.com, www.youtube.com/@DinaFalconi, Instagram @foragingandfeasting.

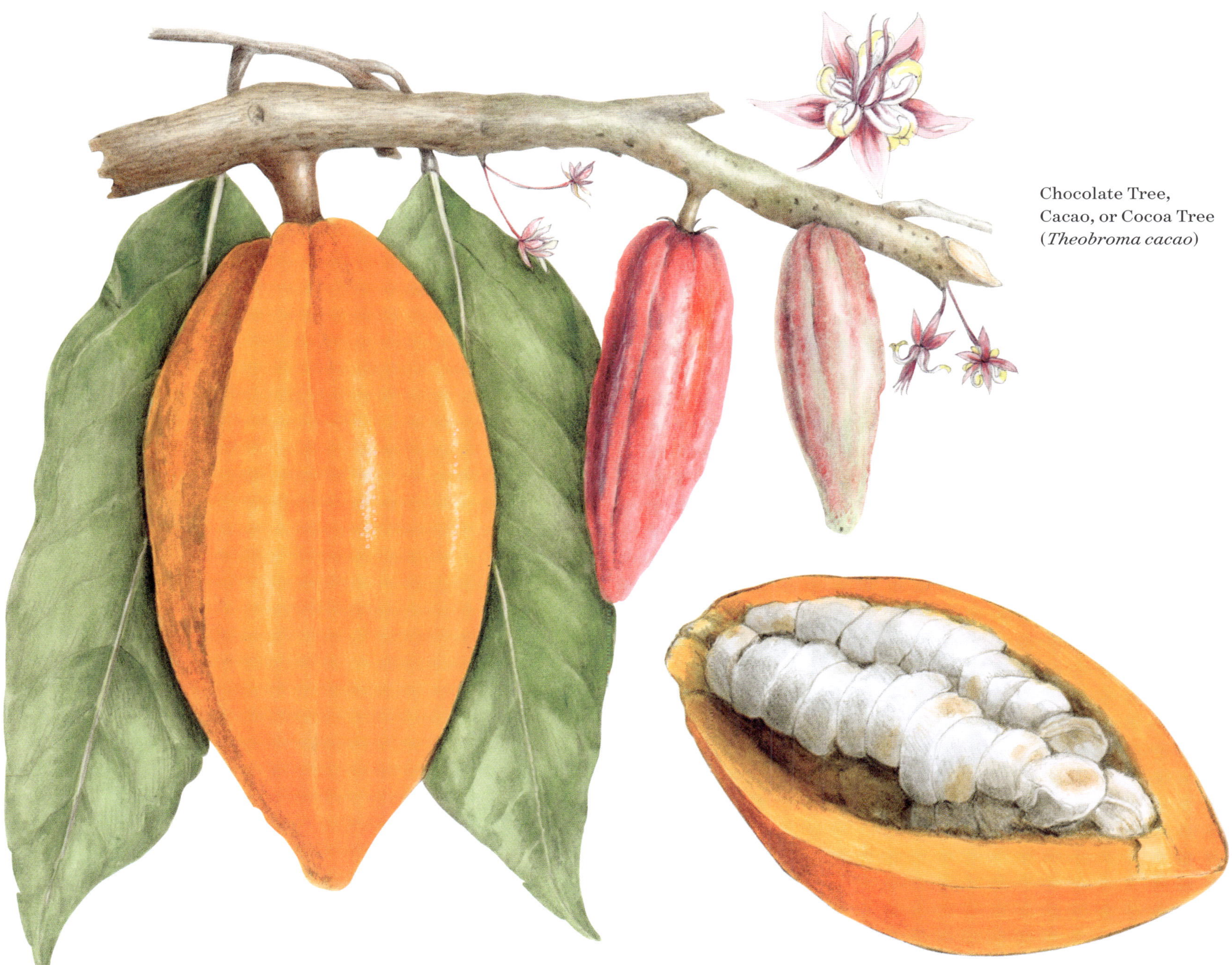

Chocolate Tree,
Cacao, or Cocoa Tree
(*Theobroma cacao*)

# 14

# Mint Family

## Lamiaceae

## 7,000+ species

The first time I drew flowers in the mint family was for an advertising print campaign for a line of organic beauty products. The body wash was made with fig and lavender. I was able to get lavender plants in the flowering stage for my drawing. This was long before I had my own garden with perennial lavender growing each year. I usually remember where I get my plants from, but that project was over twenty years ago, and I have no recollection. What I do remember clearly is the unmistakable scent of lavender as I closely studied the flowers under the microscope. This had a calming effect on me, and I was in a blissful state as I drew these flowers, at first enlarged so I could understand their structure; then I completed the full plant drawing life-size with a whole branch of flowers in place.

At first I wasn't excited to work on this plant family painting because it could be considered boring from the point of view of a botanical artist: the flowers are tiny, hard to see, and surrounded by many leaves, which are arranged in an opposite pattern on the square stems hidden beneath them. From the point of view of a forager, however, this is indeed a fun family to pursue. This is because you can easily notice a square stem and opposite-leaves arrangement. Each tiny flower in the family has a distinctive bilateral symmetry and a mouthlike petal arrangement, giving it a "mouthy" appearance. In addition, crushing the leaves usually releases an often wonderful, varied scent.

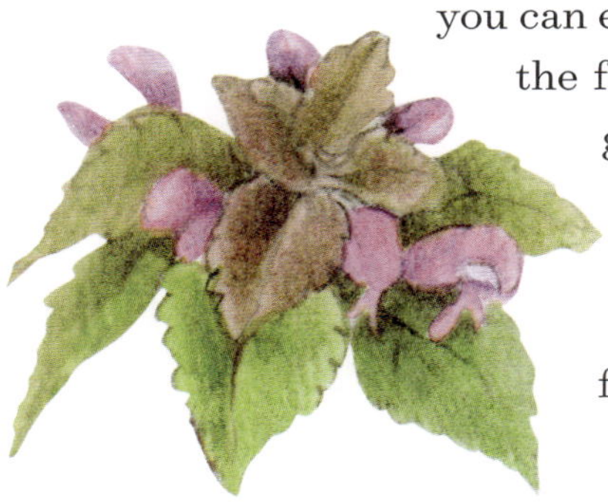

One of the earliest flowering plants in my landscape is the low-to-the-ground purple dead nettle that pops up in the grass in April. The tiny pinkish-purple flowers make it easy to see. As with most mints, it is edible. Its annual appearance

## Plants Illustrated

*Ajuga reptans*, Bugleweed
*Callicarpa dichotoma*, Beautyberry
*Glechoma hederacea*, Gill-Over-the-Ground
*Lamium maculatum*, Spotted Dead Nettle
*Lamium purpureum*, Purple Dead Nettle
*Lavandula angustifolia*, English Lavender
*Leonurus cardiaca*, Motherwort
*Mentha × piperita*, Peppermint
*Monarda didyma*, Bee Balm
*Monarda didyma*, Bee Balm, Raspberry Wine
*Monarda fistulosa*, Wild Bergamot
*Ocimum basilicum*, Basil, Bouquet
*Salvia officinalis*, Sage
*Salvia splendens*, Scarlet Sage
*Satureja montana*, Winter Savory
*Solenostemon scutellarioides*, Coleus, Wizard Scarlet

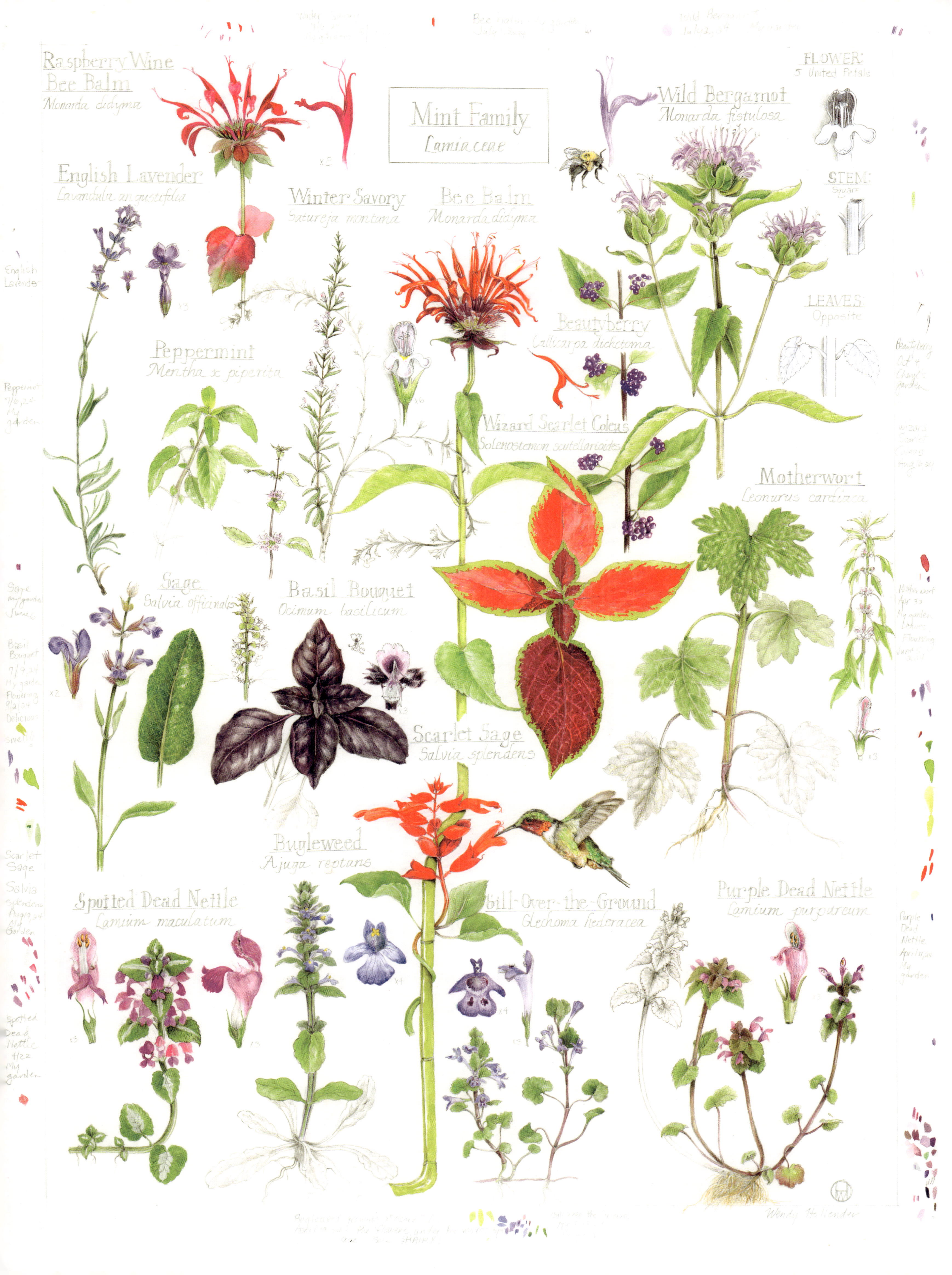
Mint Family
Lamiaceae
Raspberry Wine Bee Balm
Monarda didyma
Wild Bergamot
Monarda fistulosa
FLOWER:
5 United Petals
STEM:
Square
LEAVES:
Opposite
English Lavender
Lavandula angustifolia
Winter Savory
Satureja montana
Bee Balm
Monarda didyma
Peppermint
Mentha x piperita
Beautyberry
Callicarpa dichotoma
Wizard Scarlet Coleus
Solenostemon scutellarioides
Motherwort
Leonurus cardiaca
Sage
Salvia officinalis
Basil Bouquet
Ocimum basilicum
Scarlet Sage
Salvia splendens
Bugleweed
Ajuga reptans
Spotted Dead Nettle
Lamium maculatum
Gill-Over-the-Ground
Glechoma hederacea
Purple Dead Nettle
Lamium purpureum
Wendy Hollender

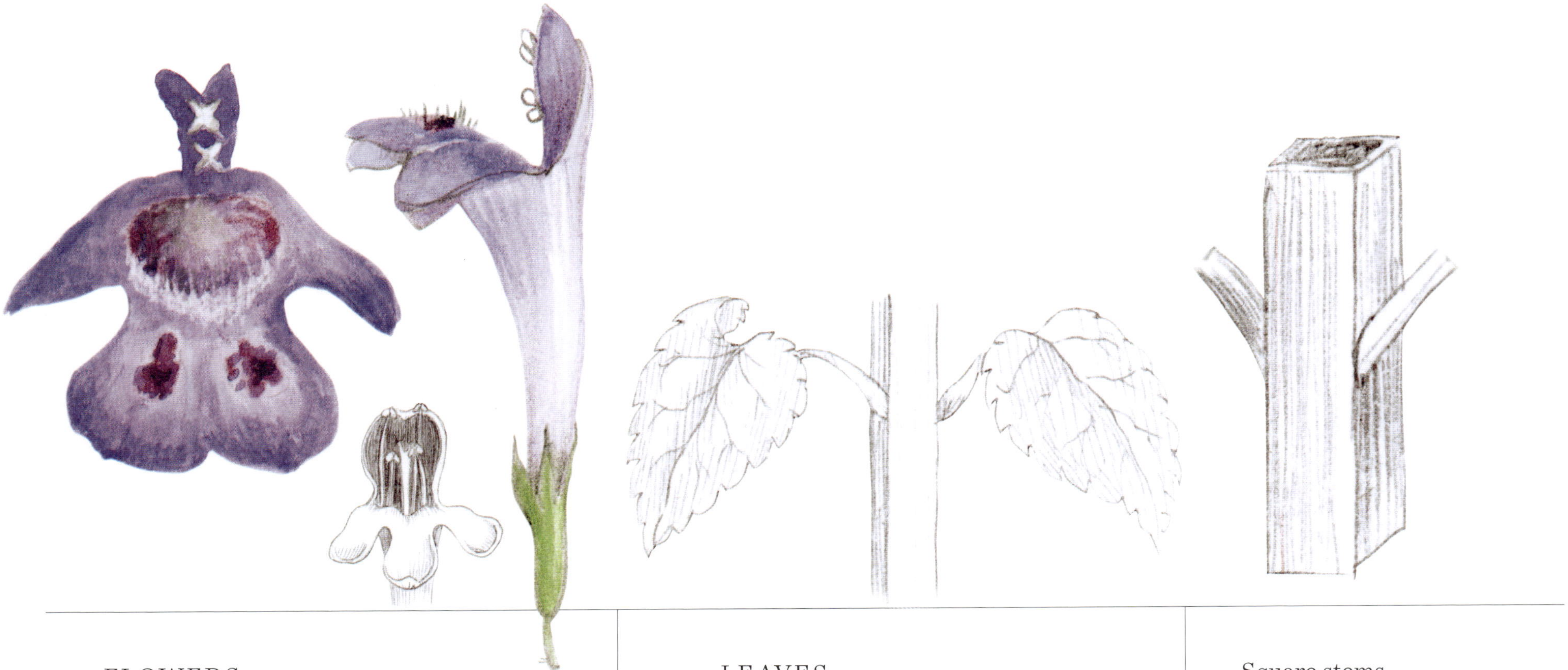

FLOWERS
5 petals
"Mouthy" form

LEAVES
Arranged in opposite pairs

Square stems
Plants usually aromatic

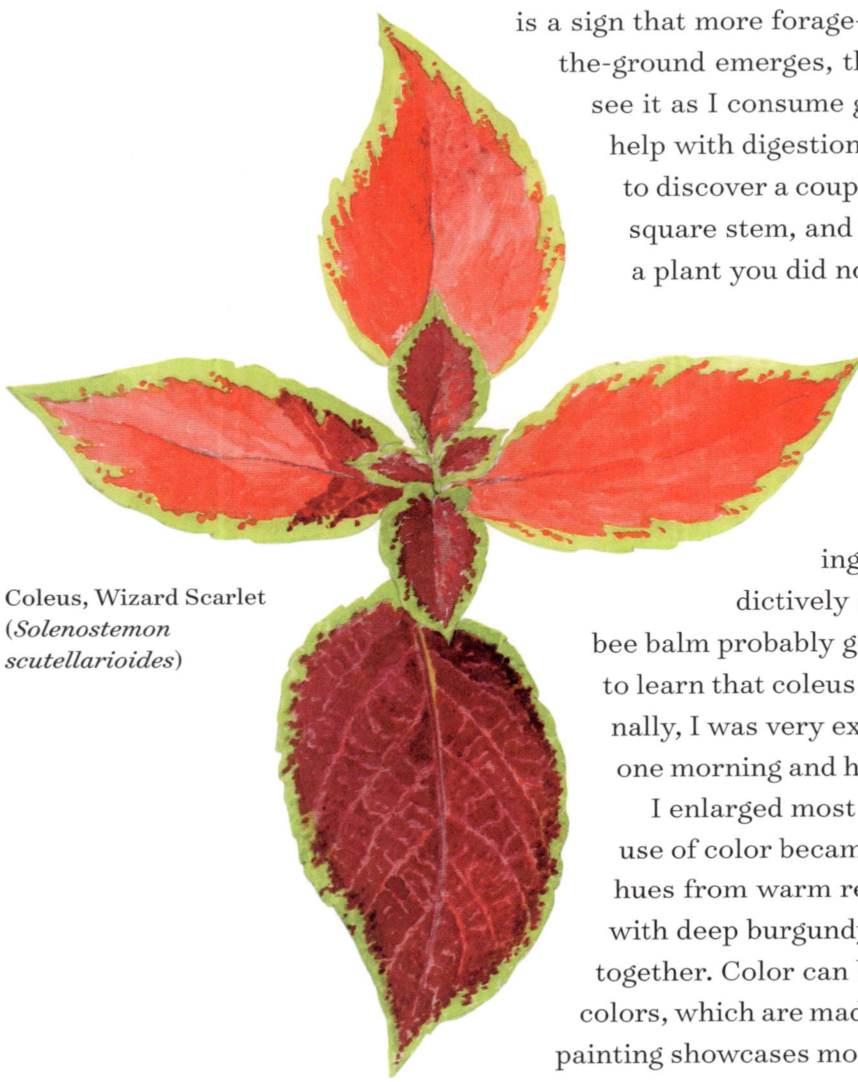

Coleus, Wizard Scarlet (*Solenostemon scutellarioides*)

is a sign that more forage-worthy weeds are on the way. Soon the low-growing gill-over-the-ground emerges, this time with a bluish-purple flower, and I am always happy to see it as I consume gill-over-the-ground regularly as a sprinkled herb in cooking to help with digestion. More plants similar to these ones come up, and I am delighted to discover a couple of new species in this family just by noticing the flower shape, square stem, and leaf arrangement. When you are able to recognize the family of a plant you did not know, it is a really empowering game. Every time I look at an unfamiliar plant and use my plant family knowledge to put the plant in its proper family, I get very excited, giving a fist pump and little dance that only I see. It reminds me of how others get when following a sports team. For me, a new plant identification puts me in the win column every time.

I made other exciting discoveries while working on this painting: I learned that basil flowers are incredibly fragrant, almost addictively so. I kept having to smell them again and again. I surmised that bee balm probably gets its name because the bees cannot stay away. I was surprised to learn that coleus and beautyberry, two stunning plants, are in the mint family. Finally, I was very excited to discover a hummingbird nuzzling into my salvia flowers one morning and had to include it in the final painting.

I enlarged most of the flowers to show clearly their bilateral symmetry, and the use of color became a subtle lesson in color theory, as I explored the differences in hues from warm red to cool red, toward pinkish violet through blue violet, ending with deep burgundy in the coleus leaves, a color created by mixing all these others together. Color can be used to create a harmonious composition by using analogous colors, which are made by combining any two primary colors in various amounts. This painting showcases mostly reds and blues.

Analogous colors are beautifully displayed in the hues of mint family flowers.

# From Garden Herb to Healing Plant

BY THOMAS J. ELPEL

The mint family, Lamiaceae (aka Labiatae), lies at the heart of fine cuisine. The sweet-savory fragrance of fresh basil in a garden salad, the zest of oregano in pizza sauce, the aroma of rosemary on roasted potatoes—these are familiar mint family flavors that spice up our daily lives. Chefs turn to lavender, marjoram, lemon balm, mint, thyme, and culinary sage to elevate ordinary dishes into extraordinary delights. These are only a few of the best-known plants of the mint family. Hundreds of other species offer new taste sensations for aromatic teas and rich cuisine.

Most mints are easily identified by their square stems, opposite leaves, and often minty aroma. Square stems provide enhanced structural stability, better distributing stress than round stems and providing greater rigidity against bending during growth spurts or windy conditions. Elongated cells with unevenly thickened walls are typically concentrated at the corners to maximize strength.

A few mint family plants lack fragrance, while some unrelated plants that have square stems and opposite leaves might be mistaken for mint family plants, but they are not aromatic. For nonaromatic plants with square stems, it is necessary to check the flowers to determine whether or not they are members of the Lamiaceae. Mint family flowers are mostly irregular, with five petals fused together to form a tube, typically with two petal lobes up and three lobes down. Every species is different, yet examine a few, and the commonalities quickly become apparent. Mint flowers primarily bloom in various shades of white, pink, purple, or blue, only rarely occurring in red or yellow.

Volatile or essential oils within the vegetation vaporize easily when the plants are disturbed, crushed, or eaten. In limited quantities, these aromatic qualities make great teas and spices, but in large doses the chemistry becomes overwhelming. This makes volatile oils an effective evolutionary deterrent against plant-eating insects and browsing herbivores. Bees and butterflies, however, are drawn to the vibrant flowers and fragrant vegetation, providing the service of cross-pollination in exchange for pollen and nectar.

Volatile oils on leaf surfaces also serve as a barrier against evaporation, conserving water in drought conditions. Slightly fuzzy leaves, such as those of lavender and sage, offer additional drought protection as the silvery hairs reflect sunlight while insulating the leaf surface. Narrow leaves, such as those of rosemary, also resist drying, thanks to reduced surface area and fewer pores in the leaf surface. With this wide suite of drought-resistant tools, mint family plants often thrive under sun-drenched skies and in rocky soils where other plants would wither.

Some species, notably mints and spearmints, are better adapted to damper conditions, rapidly spreading via horizontal stems or rhizomes beneath the soil surface to become potentially invasive, too much of a good thing in garden plots.

Medicinally, volatile oils are stimulating and warming, causing the body to open up and sweat. Many mint family plants are thus listed as "diaphoretic" in herbal books in reference to this property, which can help break a fever. A fever is the body's way of "cooking" the microorganisms that cause infections. Using a diaphoretic herb can help raise a mild fever just high enough to kill a virus, thus breaking or ending the fever. This warming property also has a vasodilator effect, opening up blood vessels to facilitate circulation. This quality is useful in cases of hypertension or for stimulating delayed menstruation. Plants used for the latter are called "emmenagogues."

Most mint family plants are aromatic and also mildly astringent; the astringent properties may constrict tissues and pores, thus having the opposite effect of volatile oils. Astringent mints are often recommended as menstrual regulators; while the volatile oils stimulate menstruation, the astringents suppress it—a balancing effect.

Some mints are bitter rather than astringent, notably pennyroyal and horehound. Bitter-tasting herbs stimulate saliva and other digestive fluids to improve digestion while also having a vasodilator effect to increase blood flow to the digestive tract. Taken together, bitter and aromatic properties accentuate each other, and herbs with these qualities have been used to purge intestinal parasites.

In addition to culinary spices and medicines, some mint family plants are favored as breath fresheners, flavoring after-dinner mints, chewing gum, toothpaste, and mouthwash. Try nibbling on any mint family plant for a fresh taste sensation.

From food to healing, the mint family gifts us with some of our finest flavors, scents, and medicines. Whether adding a savory-sweet spice to a meal, brightening a garden with vibrant flowers, or steeping a stem of mint in hot water for tea, our friends from the mint family bring a dash of pizzazz to our lives. Take a tour of your herb cabinet and see how many spices you recognize from the mint family!

THOMAS J. ELPEL is the author of *Botany in a Day: The Patterns Method of Plant Identification*, as well as *Shanleya's Quest: A Botany Adventure for Kids Ages 9 to 99*, plus a sequel book, two plant card games, and *Foraging the Mountain West: Gourmet Edible Plants, Mushrooms, and Meat* (http://www.hopspress.com). Inspired to learn about herbs from his grandmother, he homesteaded two blocks away from her place in Pony, Montana, and has remained rooted there for the past thirty-five years.

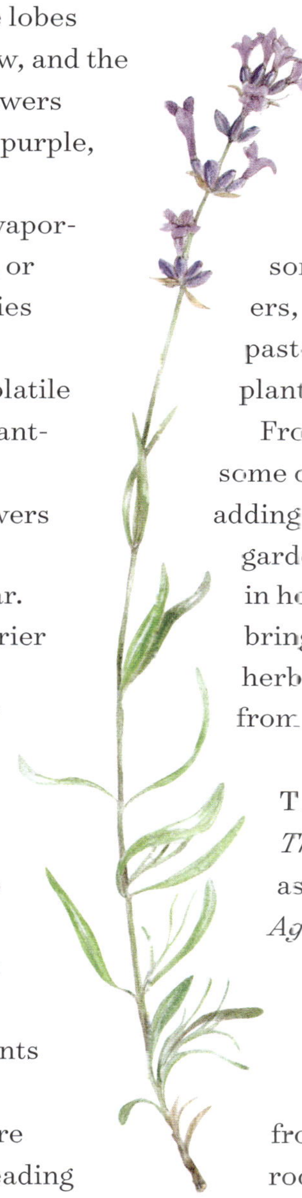

English Lavender
(*Lavandula angustifolia*)

# 15
# Mustard Family

## BRASSICACEAE
## 3,700+ species

The mustard family consists of more than 3,700 different species. The flower color and size may change, but all species have the same kind of flower, and each flower will look similar on close inspection: 4 petals in a cross pattern, with 4 tall stamens and 2 short stamens. Another name for the Brassicaceae is Cruciferae, alluding to this cross pattern. Learn this one pattern, and you can identify all plants in the mustard family.

Most of us are familiar with Brassicaceae, as many of the vegetables we eat on a daily basis are in this family. Broccoli, kale, and cabbage, for example, are all members. In the Brassicaceae, every part of the plant is edible, including flowers, seeds, roots, and leaves. The leaves of the horseradish plant, which we grow for its root, are very delicious and flavorful to eat when lightly sautéed. The same goes for the leaves of radishes. Arugula has a slightly pungent mustard-like flavor. Mustard is a condiment made from the seeds of mustard plants.

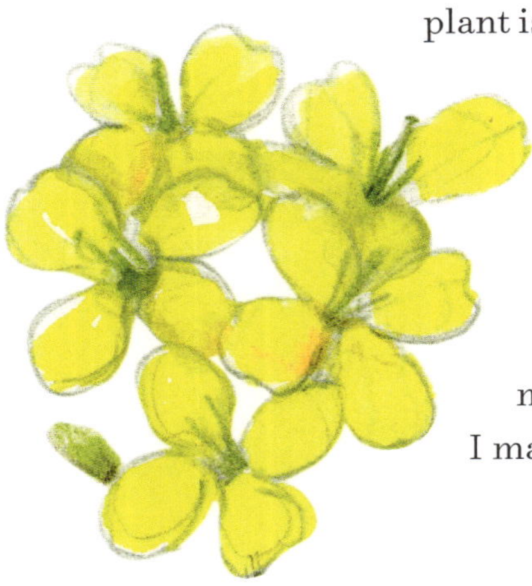

When I was planning this painting, I was surprised but delighted to discover that this is a very colorful family. I expected all green and yellow, but many Brassicaceae plants occur in beautiful hues of purples, pinks, and reds. I made sure to choose very colorful plants for my painting. There are some spec-

## Plants Illustrated

*Alliaria petiolata*, Garlic Mustard
*Armoracia rusticana*, Horseradish
*Brassica napus*, Rutabaga
*Brassica oleracea*, Broccoli
*Brassica oleracea*, Broccolini
*Brassica oleracea*, Cauliflower
*Brassica oleracea*, Romanesco Cauliflower
*Brassica oleracea*, Purple Cauliflower
*Brassica oleracea*, Kohlrabi
*Brassica oleracea*, Red Cabbage
*Brassica oleracea*, Kale, Rainbow Lacinato
*Brassica oleracea*, Brussels Sprouts
*Brassica rapa*, Field Mustard
*Eruca sativa*, Arugula
*Hesperis matronalis*, Dame's Rocket
*Raphanus sativus*, Red Radish
*Raphanus sativus*, Watermelon Radish
*Raphanus sativus*, Daikon Radish

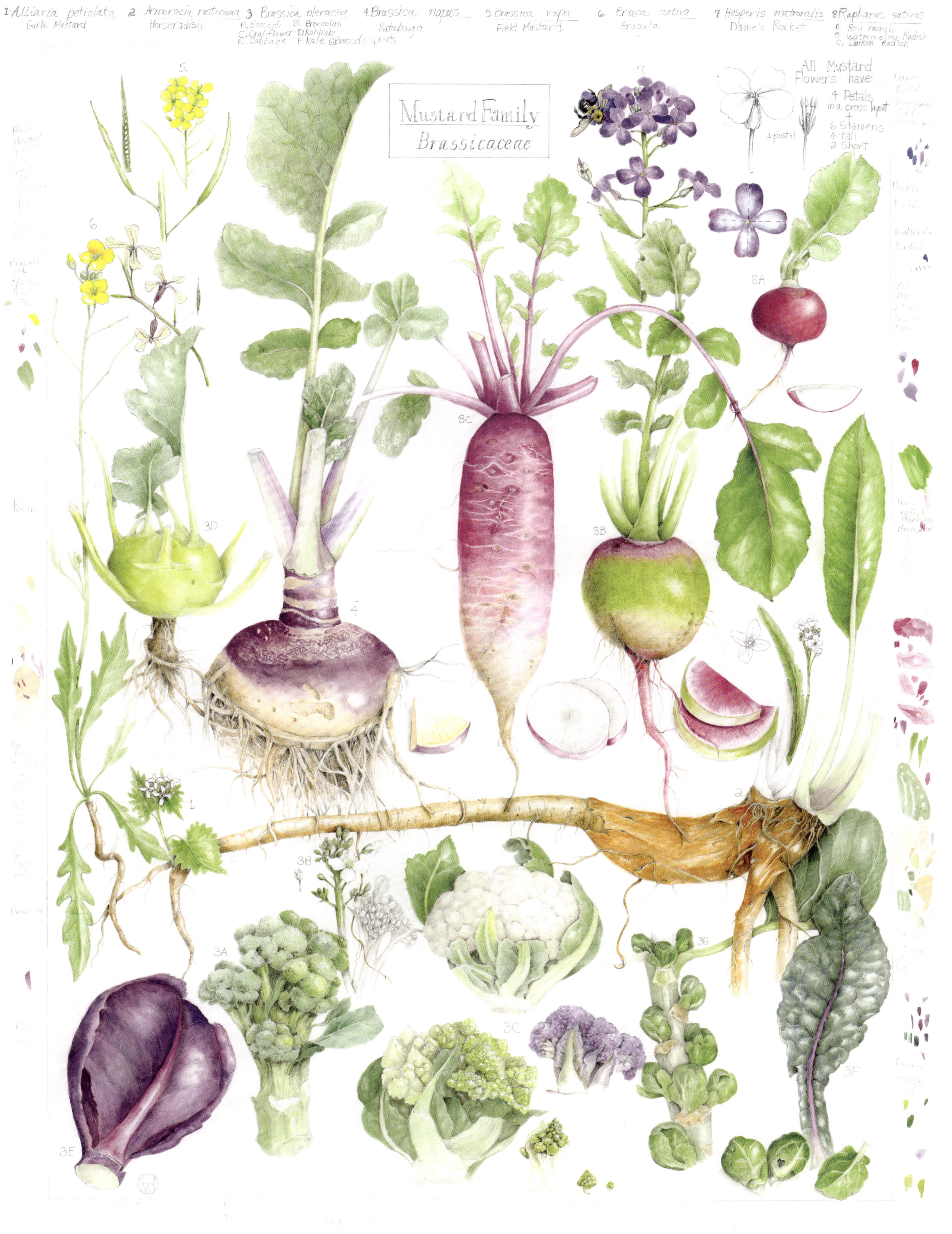
1 Alliaria petiolata
Garlic Mustard
2 Armoracia rusticana
Horseradish
3 Brassica oleracea
A. Broccoli B. Broccolini
C. Cauliflower D. Kohlrabi
E. Cabbage F. Kale G. Brussels Sprouts
4. Brassica napus
Rutabaga
5 Brassica rapa
Field Mustard
6. Eruca sativa
Arugula
7 Hesperis matronalis
Dame's Rocket
8 Raphanus sativus
A. Red radish
B. Watermelon Radish
C. Daikon Radish
Mustard Family
Brassicaceae
All Mustard Flowers have:
4 Petals in a cross layout
+
6 Stamens
4 tall
2 Short
1 pistil
5.
6.
7.
8A
8B
8C
3D
4
1
2
3B
3A
3C
3G
3E
3F

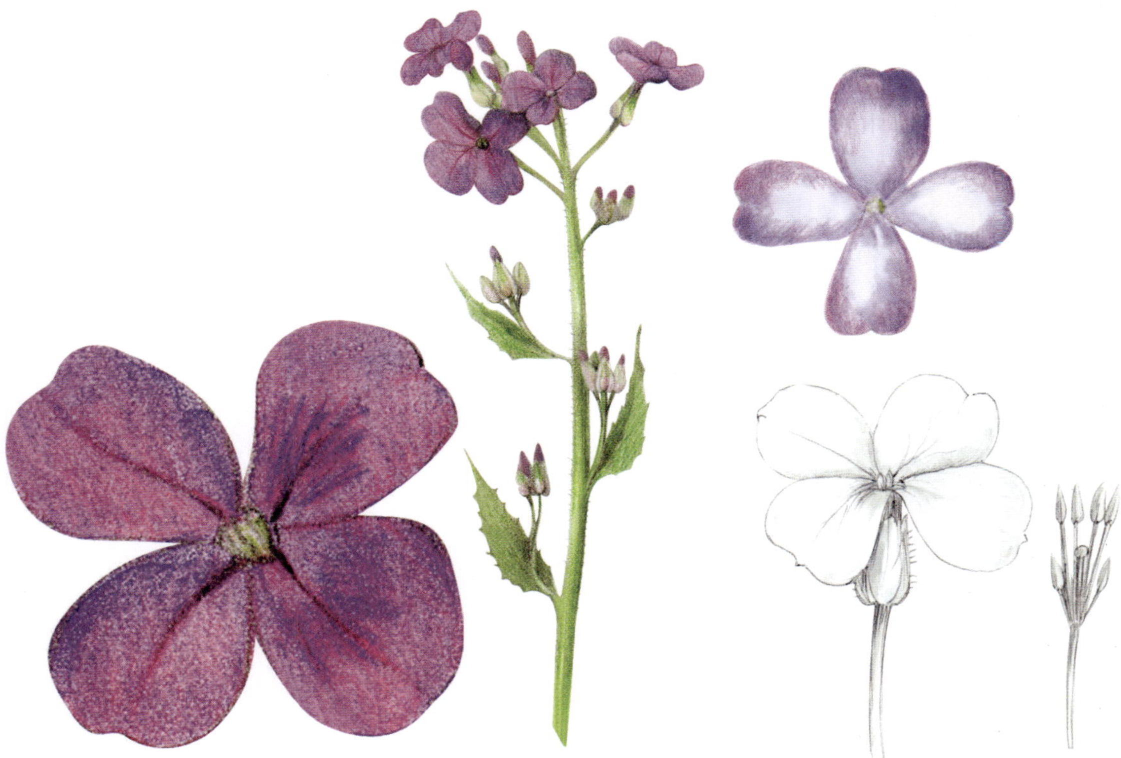

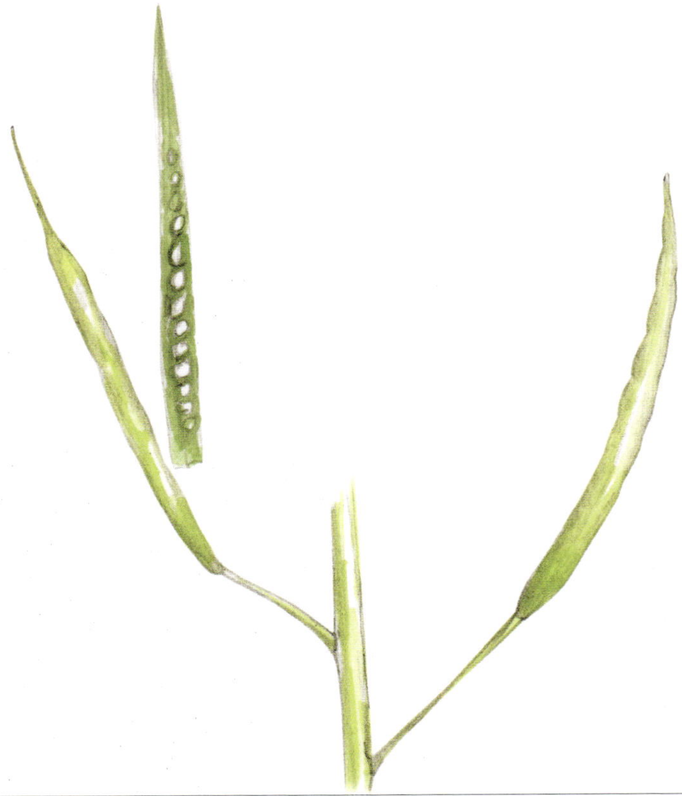

FLOWERS

4 petals in a cross pattern, 4 tall stamens, and 2 short stamens

Ovary 2-chambered

FRUITS

Seeds produced in dry podlike fruits with a partition between the halves

tacular radishes, especially the watermelon radish, which combines chartreuse and magenta colors. In cross section, this root has a pattern that looks like a slice of watermelon, thus the name. Daikon radish also comes in a similar shade of pink, and I knew these two radishes would be happy next to each other on the page. This was one of my early paintings in the series, and I could not stop adding plants, as there were so many to choose from, all growing at nearby Tributary Farm. In the early paintings, I didn't always think about labeling the text ahead of time. In the case of this painting, there was no room for the text labels, so I put them at the top of the page.

Broccoli, kohlrabi, cabbage, cauliflower, and kale originally came from the same plant species (*Brassica oleracea*). Over time, certain features of that original plant have been bred to emphasize various aspects. Visually compare the similarities of a kale leaf and a broccoli leaf. Cook the leaves to compare the taste. Note that plants in the same family share similar properties, patterns, tastes, and uses. Nowadays there are several varieties of broccoli and cauliflower, and I had to include the purple cauliflower and the Romanesco cauliflower because I love the purple color and intricate pattern, respectively.

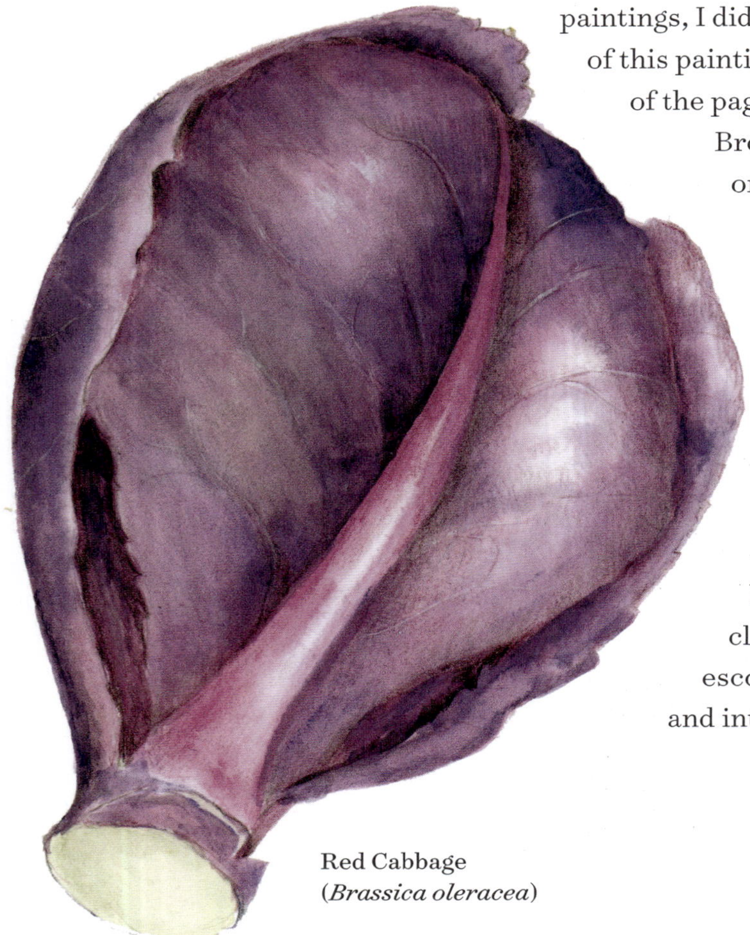

Red Cabbage (*Brassica oleracea*)

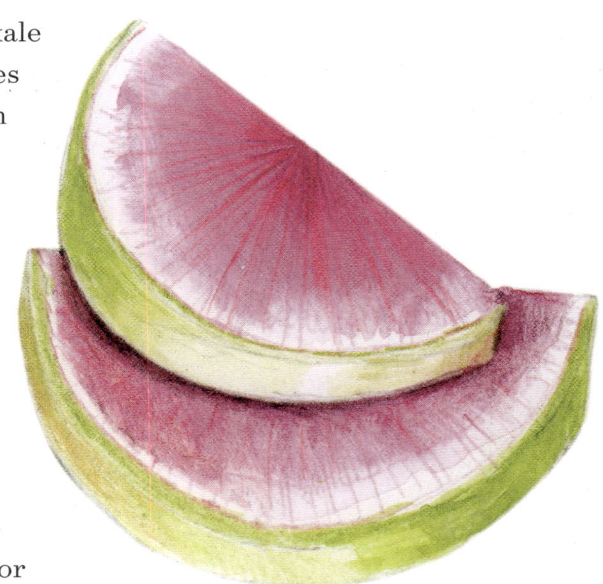

Watermelon Radish (*Raphanus sativus*)

# Bountiful Brassicas: A Vegetable Farm's Most Valuable Family

BY JESSE GOLDFARB

When it comes to the sheer number of plants eaten and known as unique vegetables today, the mustard family is certainly the most prolific. The family has also been known as Cruciferae, which in Latin translates to "cross-bearing," since all plants in the family have four-petaled flowers resembling a cross.

Since this book is about plant families and their relation to the human family, I should disclose that the artist and author is my mother! There's no doubt that her love of plants also inspired mine. While she draws them, I grow them at Tributary Farm, an organic mixed-vegetable farm in High Falls, New York. I feel lucky anytime I see her out in the field swiping a sample for her latest study.

The importance of this plant family to the human population and therefore our farm, as one small feeder of this population, is paramount. Our farm primarily sells vegetables directly to consumers, who pick them up on our farm. Thus, we have a very good understanding of the mustard family's place in the diet of our customers and the land base on our farm. Every season, the Brassicaceae family makes up about 30 percent of our vegetable fields by land area. That's more than twice the representation by land size compared with any of the other plant families we grow. We make sure we rotate our fields into different plant families each season to try to reduce disease and pest pressure while also accounting for the rate at which different families can remove different nutrients from the soil. The practice of crop rotation has been around since the early days of ancient agriculture, and it's just as important today.

The Brassicaceae contains over 3,700 unique species. The area of origin is likely Southwest to Central Asia, but plants of this family can be found endemically all around the world except for Antarctica. One of the most significant species is *Brassica oleracea*, which grew wild on the coast of Britain and the Mediterranean. It was planted in the ancient gardens of Greece and Rome, where it was initially selected for edible leaves, like those of modern-day kale. Theophrastus, the ancient Greece philosopher and naturalist, mentions three varieties already cultivated in his day (third to fourth century BCE), including a curly-leaved, a smooth-leaved, and a wild type. Over time, different varieties of this one species were selected and bred for different traits. Around the first century CE, cabbage emerged from plants selected for terminal buds. Around the same time, other *Brassica oleracea* plants were selected for wider stems, resulting in kohlrabi, the name of which comes from the German words *kohl* (cabbage) and *rabi* (turnip). In Europe, *Brassica oleracea* was selected for its inflorescence (a group or cluster of flowers arranged on a plant's main stem), resulting in early broccoli and cauliflower. In Belgium, selection for lateral buds led to Brussels sprouts, which became popular there in the sixteenth century. Broccolini, a well-loved crop on our farm, was bred as recently as 1993 by a Japanese seed company by crossing broccoli with Chinese broccoli (gai lin). Rutabaga (*Brassica napus*) was likely a result of crossing cabbage (*Brassica oleracea*) and turnips (*Brassica rapa*). The *Brassica napus* species also includes rapeseed, an annual oil crop that is a key ingredient in many foods, candles, and industrial lubricants. Canola oil is a specific type of edible rapeseed oil and is the third-largest vegetable oil consumed by volume after palm oil and soybean oil.

The culinary value of the mustard family is a result of the various flavors, nutritional value, storage life, and uses that result from such a wide variety of species and cultivars (cultivated varieties). Think of the bitter and distinct flavors of arugula (*Eruca sativa*) and broccoli rabe (*Brassica ruvo*); the spice of mustard greens (*Brassica rapa*); the storability of cabbage and its high vitamin C content (red cabbage has higher vitamin C by weight than oranges); the crunch of radish and kohlrabi; the distinctive flavor of a rutabaga mash; how roasted Brussels sprouts, broccoli, and cauliflower taste better (to me) than French fries.

These crops sure better be loved because they are not easy to grow organically. In our

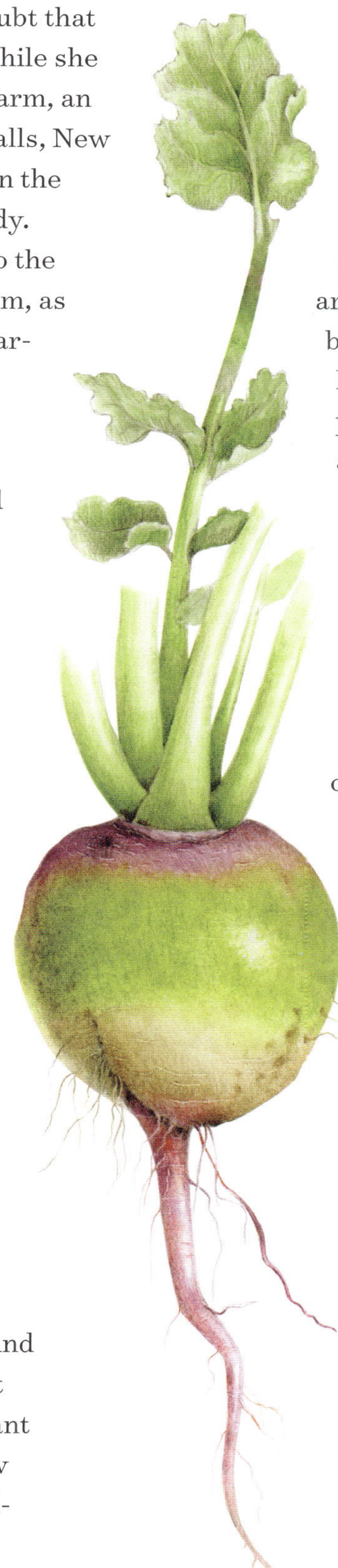

Watermelon Radish (*Raphanus sativus*)

area, these plants are up against a host of nine perennial pests that can emerge one after another throughout the season. The most damaging to our crops is the tiny, shiny black crucifer flea beetle, only about 2 millimeters long but with the ability to work as a team to defoliate entire crops if left unchecked. Since organic pesticides are not very effective against flea beetles, and we also try to avoid spraying when we can, we use massive exclusion nets to protect our Brassicaceae crops. We continually move our nets around to newly planted successions of thse crops, as the plants are more resilient once they are larger and have more leaf mass. After the flea beetle, three distinct species of moths lay their eggs on Brassicaceae leaves. When the caterpillars emerge they feed on the leaves and if left untreated might end up in your broccoli. Luckily, these can be treated with Bt (*Bacillus thuringiensis*), a soil-dwelling bacterium that produces proteins that are toxic to certain insects but not others. Bt damages the guts of the caterpillars, causing them to stop eating and die, and is the most broadly used biological pesticide in the world.

We believe in eating seasonally and locally, and although our farm is in New York State, we grow and sell produce all year around. In the winter we grow hardy greens in unheated greenhouses that we call high tunnels. About 50 percent of our winter produce sales come from fresh greens grown in our tunnels, with the other 50 percent coming from storage crops harvested late in the fall and held fresh in our coolers through the winter. The mustard family features heavily on the menu in both regards. Kale, arugula, and many varieties of *Brassica rapa* (mustard greens, bok choi, tatsoi, Tokyo bekana) do well in the cold winter conditions. From storage we sell lots of cabbage, watermelon and daikon radishes, kohlrabi, rutabagas, and turnips.

Look at the illustration of the lacinato kale leaf. The smaller leaf veins all flow into the main midrib, and the bumps along the leaves look like watershed-forming mountains, just as small tributaries flow through the landscape and come together into an ocean-going river. Our farm is located on the Rondout Creek, a tributary to the Hudson River. Each tributary plays a small role in the health of the entire watershed, as the waters all end up flowing together. Our farm is responsible for maintaining and enhancing its small role in the ecosystem, as part of a much larger whole. This is the spirit of "think globally, act locally," and I am reminded of it when I consider the global story and importance of the Brassicaceae family over millennia paired with the action of supporting a local farm to enjoy these fresh flavors all year around.

Jesse Goldfarb initially fell in love with farming when staying with a farm family in western Kenya during a college semester abroad in 2007. After that experience, he started growing food on a garden scale. In 2009, Jesse's mother moved to Accord, New York, and Jesse and his sister started a large vegetable and fruit garden there. At the end of 2009, Jesse moved to Rwanda to manage agricultural and other new product trials for a nonprofit organization called One Acre Fund, which works to support small-scale farmers throughout East Africa. Jesse stayed in Rwanda for six years and returned to New York in 2016 to work at the Hudson Valley Farm Hub in its professional farmer training program. He also worked at Pleasant Valley Farm in Argyle, New York, before starting Tributary Farm in 2018.

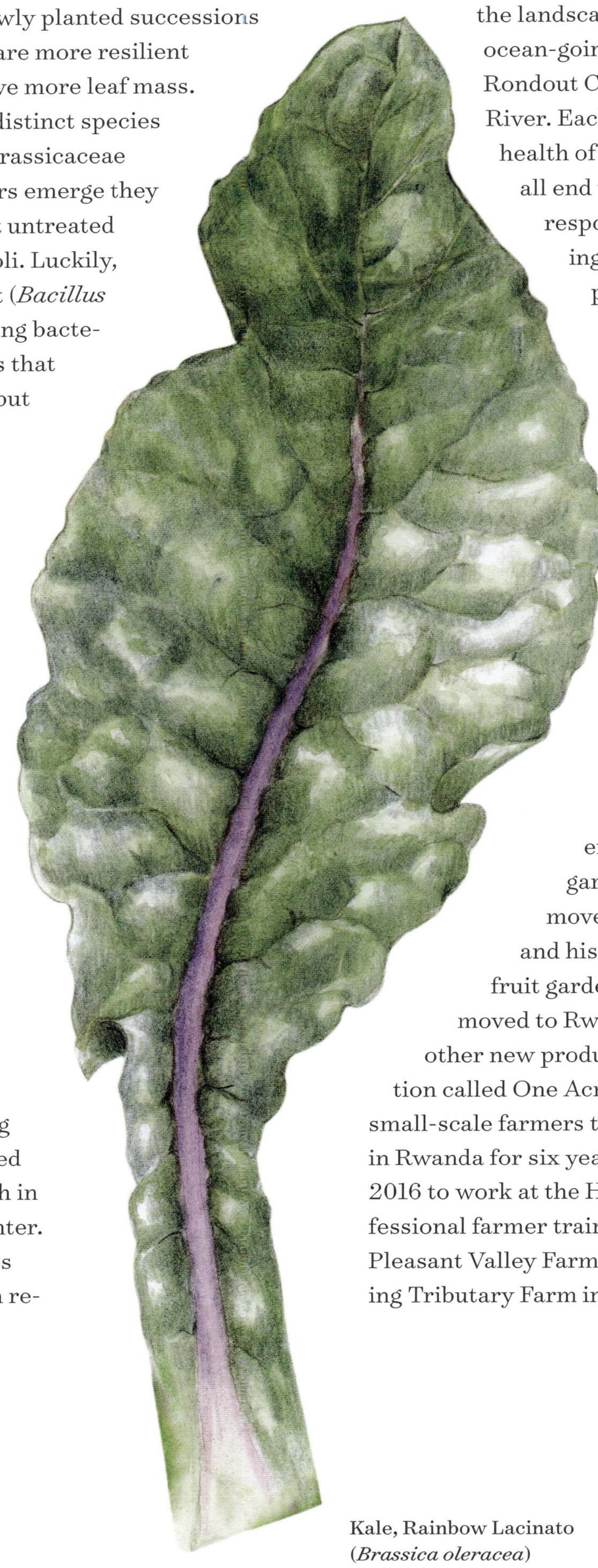

Kale, Rainbow Lacinato
(*Brassica oleracea*)

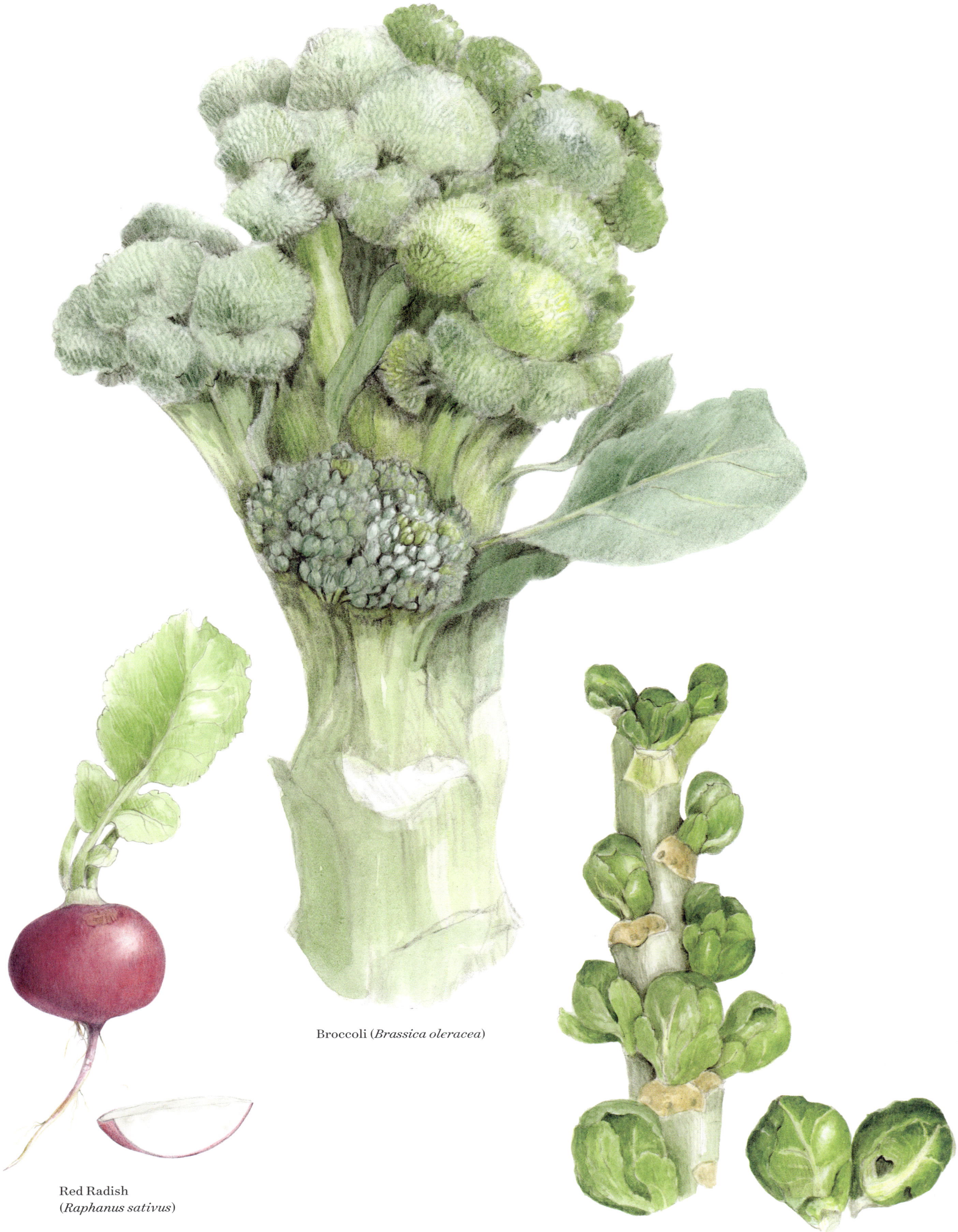

Broccoli (*Brassica oleracea*)

Red Radish
(*Raphanus sativus*)

Brussels Sprouts (*Brassica oleracea*)

# 16

# Nightshade Family

## Solanaceae
## 2,500+ species

It is hard to imagine that tomatoes and potatoes, other than their names rhyming, are related. But they are. If you look at a potato plant's flowers, you can see that they look similar to flowers of other plants in this family, such as tomatoes, eggplants, and peppers. The potato grows underground as part of the root system and is called a tuber. There are more than 2,500 nightshade species, but many of them are highly poisonous. Those we consume may be few in number, but they provide an impressive array of varied and favorite vegetables. We call them vegetables, but technically, in botanical terms, the part of the plant that contains seeds is a fruit. The potato flower produces fruits that look like tiny tomatoes, but we don't eat them, as they are poisonous—luckily, the delicious tuber is not.

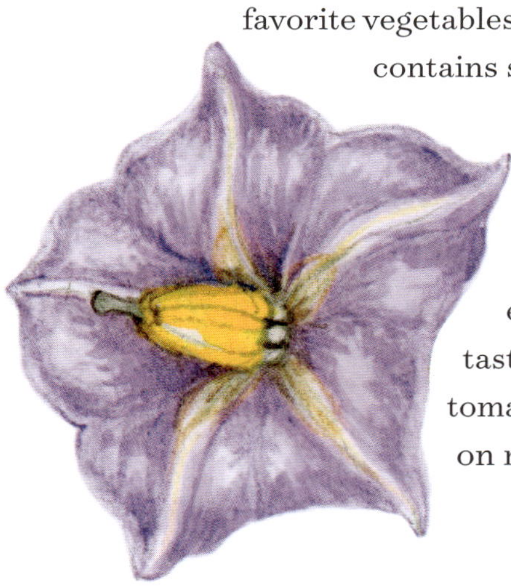

When I anticipate the joys of summer produce, I think of many of the plants in the nightshade family. I wait all year for summer to eat homegrown heirloom tomatoes, anticipating the day they start to ripen. I use them in my favorite simple recipes. I eat fresh tomatoes only in summer because the tomatoes grown commercially don't taste anything like my homegrown heirloom varieties. Once you have eaten a homegrown tomato, it is hard to eat any others. Most of the plants in this painting were grown locally on my son's farm in High Falls, New York. Tributary Farm grows organic vegetables,

## Plants Illustrated

*Brugmansia arborea*, Angel's Trumpet
*Capsicum annuum*, Pepper, Carmen
*Capsicum annuum*, Pepper, Green
*Capsicum annuum*, Pepper, Shishito
*Datura stramonium*, Thorn Apple
*Nicotiana alata*, Tobacco
*Nicotiana rustica*, Hopi Tobacco
*Petunia nana compacta*, Petunia, Alderman Blue
*Physalis philadelphica*, Tomatillo
*Solanum abutiloides*, Dwarf Tamarillo
*Solanum carolinense*, Horsenettle
*Solanum lycopersicum*, Tomato, Marnero
*Solanum lycopersicum*, Tomato, Plum
*Solanum lycopersicum*, Tomato, Speckled Roman
*Solanum lycopersicum*, Tomato, Sun Gold
*Solanum melongena*, Eggplant, Fairy Tale
*Solanum melongena*, Eggplant, Orient Express
*Solanum tuberosum*, Potato, All Blue
*Solanum tuberosum*, Potato, French Fingerling

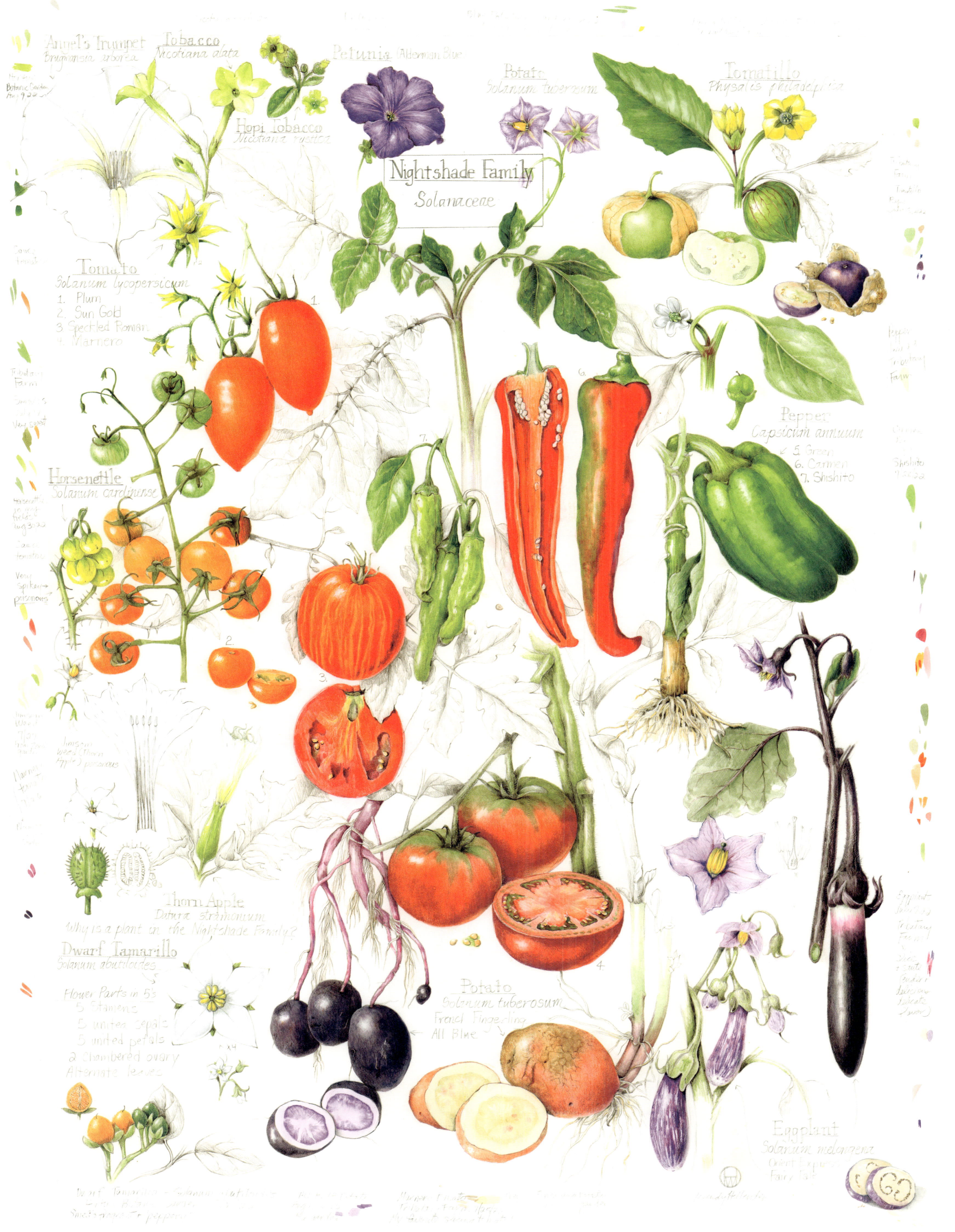
Angel's Trumpet
Brugmansia arborea
Tobacco
Nicotiana alata
Petunia (Alderman Blue)
Hopi Tobacco
Nicotiana rustica
Potato
Solanum tuberosum
Tomatillo
Physalis philadelphica
Nightshade Family
Solanaceae
Tomato
Solanum lycopersicum
1. Plum
2. Sun Gold
3. Speckled Roman
4. Marnero
Horsenettle
Solanum carolinense
Pepper
Capsicum annuum
5. Green
6. Carmen
7. Shishito
Thorn Apple
Datura stramonium
Why is a plant in the Nightshade Family?
Dwarf Tamarillo
Solanum abutiloides
Flower Parts in 5's
5 Stamens
5 united sepals
5 united petals
2 chambered ovary
Alternate leaves
Potato
Solanum tuberosum
French Fingerling
All Blue
Eggplant
Solanum melongena
Orient Express
Fairy Tale

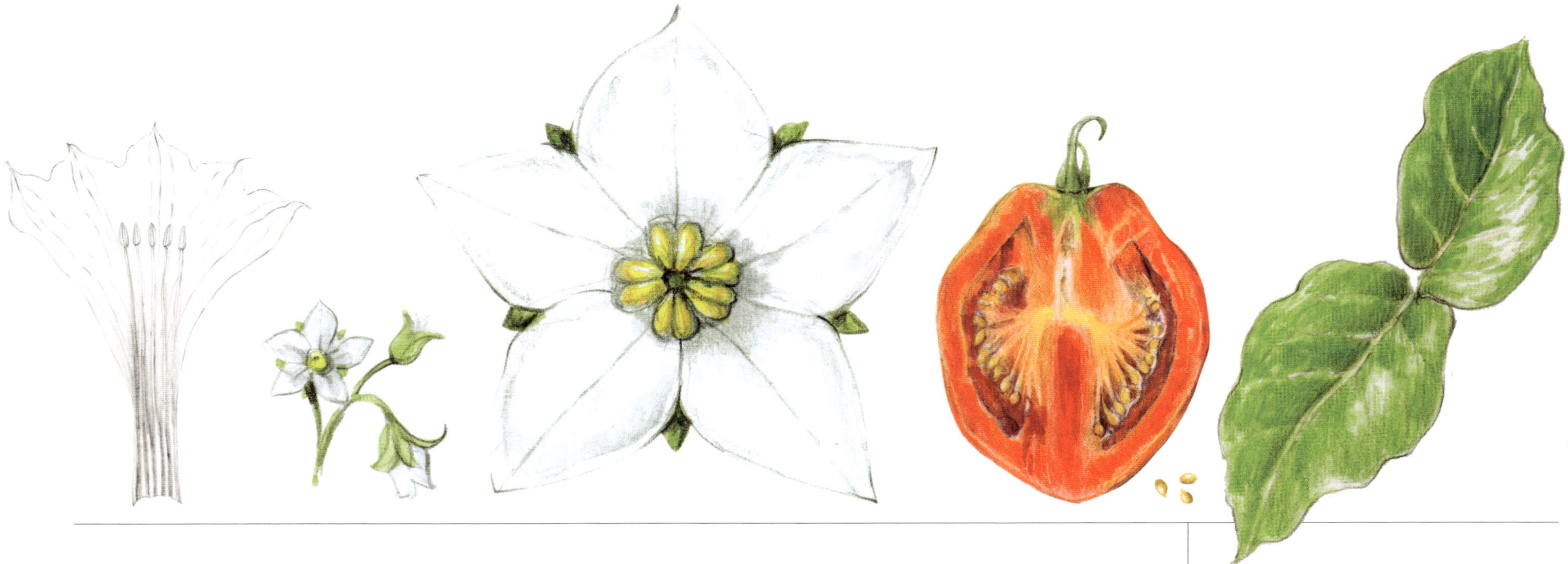

FLOWERS

Floral parts in 5s: 5 united petals, 5 united sepals, 5 stamens
Ovary 2-chambered, often with many seeds

LEAVES

Alternate or opposite on the stem

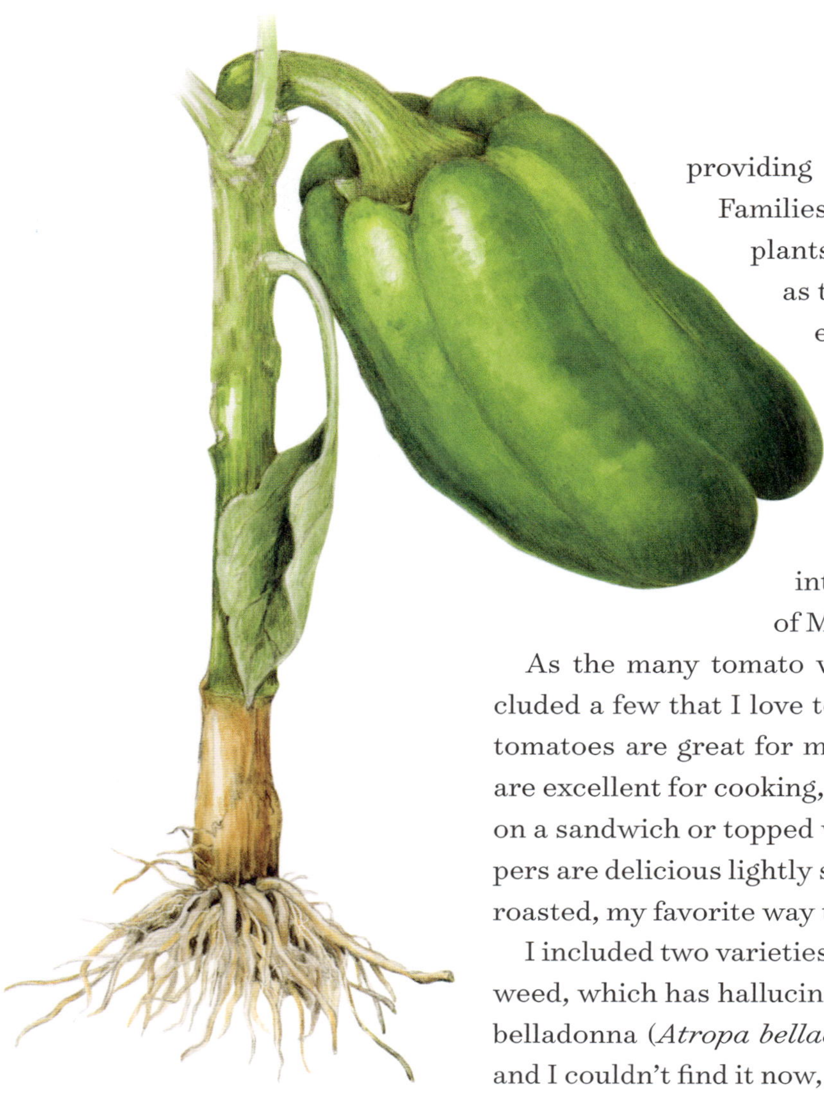

Pepper, Green
(*Capsicum annuum*)

providing food primarily through a CSA (community-supported agriculture) model. Families sign up for a weekly allotment of vegetables, herbs, and flowers. Nightshade plants are a large part of the farm's vegetable production. I started to draw the plants as they began to emerge in the spring. In June the potato plant flowers started to emerge above the leaves and became the central focal point of the painting. At the top I added a petunia, a common ornamental in the family, to illustrate the flower structure. Tomatillo, an early fruiter, is similar to a green tomato, with a flavor that is tart and lemony. Rather than eaten raw, the small fruit is best roasted and pureed into a green sauce—the salsa verde of Mexican cuisine.

As the many tomato varieties started to emerge, I included a few that I love to look at but also to eat. Sun Gold tomatoes are great for munching like candy, plum tomatoes are excellent for cooking, and Marnero tomatoes are best sliced on a sandwich or topped with slices of fresh mozzarella cheese. Peppers are delicious lightly sautéed, and eggplant can be used in so many ways once roasted, my favorite way to prepare it.

I included two varieties of tobacco as well as the poisonous thorn apple, or jimson weed, which has hallucinogenic properties. I once saw the notorious hallucinogenic belladonna (*Atropa belladonna*) growing in Greece, but since I didn't draw it then, and I couldn't find it now, I didn't include it.

I focused my painting on creating multiple focal points by using color detail and contrast on the important components such as the fruits but left most of the leaves in neutral tones to help give the viewer a place to rest on the page.

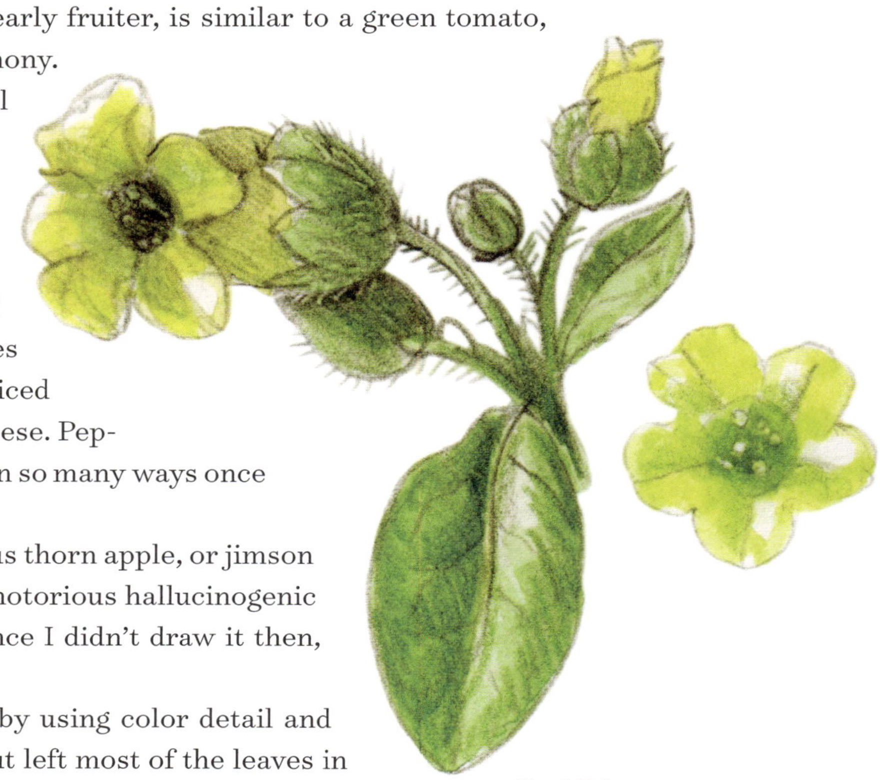

Hopi Tobacco
(*Nicotiana rustica*)

# Solanaceae: Edible and Deadly

BY NAVA ATLAS

I must really be a food nerd, because one of the most vivid memories from my adolescence isn't of a first crush but of eating sautéed zucchini at the home of a friend. It may seem wild that this is a standout memory of my thirteen-year-old self. But after a childhood of barely tolerating my mom's bland, overcooked meals, the flavorful, lightly cooked zucchini was a revelation. From then on, I was obsessed with vegetables, a love affair that continues to this day.

Fast-forward to my mid-twenties, when as a struggling illustrator and graphic designer (who liked to write and cook as well) in New York City, I started what became my first book, *Vegetariana*. Pulling together my various interests and skills, the project involved developing recipes but also, and just as enjoyably, gathering literary tidbits and food lore to accompany them. In those pre-internet days, I spent months poring over old herbals, gardening guides, and culinary history books at the main branch of the New York Public Library. So much has changed in the intervening decades, but this library is almost exactly as I left it—and somehow, that's a comfort.

The nightshade family yielded a mother lode of fascinating lore and legend to learn about and use in my book. First of all, I had no idea that the disparate vegetables of this group are botanical cousins. And just recently, I learned from Wendy's lovely illustration that petunias and tobacco belong to the nightshade group as well.

The term "deadly nightshade" surely has thrown some shade over the edible members of this enormous botanical group (numbering some 2,500 species, many of which are indeed toxic and not meant for consumption). True, there are people who don't do well digestively with some of the nightshades. But for most of us, the big four—tomatoes, eggplants, peppers (sweet and hot), and potatoes—have earned a place among the most beloved and versatile of vegetables. Before they became common fare, two of the edible nightshades, tomatoes and eggplants, had to undergo major image rehabilitation. Herbalists and botanists of the 1600s described tomatoes as "strange," even "evil." Some cultures considered them poisonous. American cooks remained nervous about tomatoes well into the nineteenth century; a recipe in *Godey's Lady's Book* recommended cooking them for at least three hours. All the while, in parts of Europe, tomatoes were enjoying a renaissance as an aphrodisiac. In France they were called *pomme d'amour*, which means "love apple."

Tomatoes were even the subject of a US Supreme Court case. According to the Tariff Act of 1883, fruits could be imported duty-free, but vegetables could not, so an importer petitioned to establish the legal status of tomatoes as a fruit. The court disagreed. Its 1893 decision concluded that although tomatoes are botanically a fruit of the vine, they're generally used as vegetables and not, like other fruits, as desserts.

Meanwhile, eggplant (known in many places more poetically as aubergine) had long-standing image troubles of its own. Its Italian name, *melanzana*, is a corruption of the ancient Latin *mela insana* ("crazy apple"). John Gerard concluded in his *Herball* (1597) that "doubtless these Apples have a mischievous qualitie, the use whereof is utterly to be forsaken." Eggplant's association with insanity lingered into nineteenth-century Egypt, where it was believed that madness was more common at the height of summer's eggplant season.

Sweet bell peppers and their spicy relatives, chili peppers, didn't have to overcome the complicated reputation of their fellow nightshades. On the contrary, many cultures have considered peppers of all kinds to be good luck charms. When dried, strung together (think Mexican *ristras* and Korean *gochu-geolim*), and hung outside doorways, they've come to symbolize prosperity as well as protection against the evil eye.

While I'm fascinated by the history and lore of the nightshade vegetables (and others, including my first love, zucchini), I like cooking with them even more. They're staples in my kitchen year-round, and even as I write this, they're on hand. So let's make a roasted nightshade medley, one that doesn't need an exact recipe.

In a large bowl, combine a diced medium-size Italian eggplant (or two Japanese eggplants, cut into half-inch slices), three or four diced medium-size golden potatoes, and a diced bell pepper or two (any color). Slice in a couple of hot chili peppers (seeds removed), which mellow quite a bit with roasting. Drizzle in a little extra-virgin olive oil and transfer to a parchment-lined roasting pan. Preheat oven to 425° F, and roast the vegetables for twenty minutes. Add a couple of diced tomatoes or a cup of grape or cherry tomatoes and give everything a stir. Roast for ten to fifteen minutes longer or until the vegetables are touched with golden brown on the outside. Season with salt, pepper, and your favorite herbs and spices. Serve hot, straight from the roasting pan, or cool to room temperature and serve over salad greens. Either way, savor your nightshades!

NAVA ATLAS is best known for her vegetarian and vegan cookbooks, including *Wild about Greens*, *Vegan Holiday Kitchen*, *Vegetariana* (revised edition), and *Vegan Soups and Stews for All Seasons*, among many others. Nava also creates visual nonfiction books, notably *The Literary Ladies' Guide to the Writing Life*, as well as limited editions that are collected by university and museum special collections libraries around the country. Visit her websites, The Vegan Atlas (theveganatlas.com) and Literary Ladies Guide (literaryladiesguide.com). She lives in the Hudson Valley region of New York State.

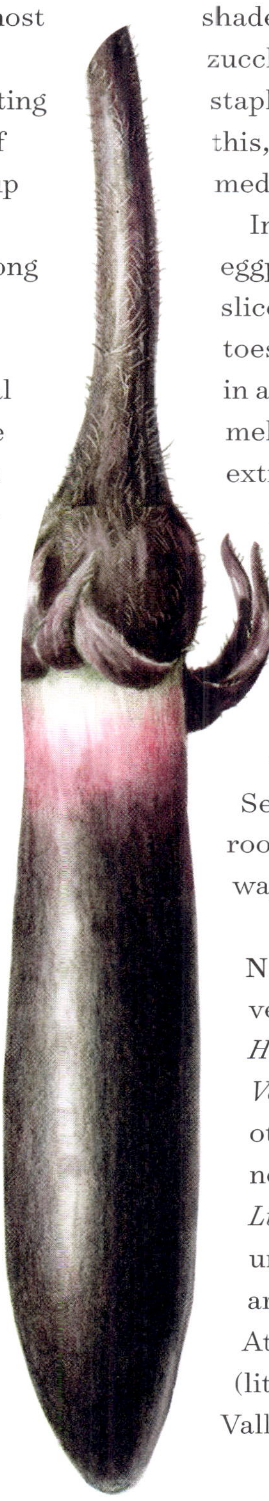

Eggplant, Orient Express (*Solanum melongena*)

17

# Orchid Family

## Orchidaceae
## 28,000+ species

I hesitated to include the orchid family in my series of paintings. My mom loved flowers and had a big garden, but she did not grow orchids, so I rarely saw them. Only one orchid is cultivated for food, the rest being primarily ornamental. Orchid flowers are often unusual and beautiful looking, but there is something intimidating about them. Perhaps it is their unique and unusual structure: the flowers have bilateral symmetry, meaning the flower can be divided in half and is the same on both sides; and the center petal is a modified lip that serves as a landing pad for pollinators. Or perhaps I was intimidated by the way many artists have depicted orchids in their natural exotic tropical environments so magnificently for centuries. In my studies of botanical artists over the years, I have become familiar with the work of several artists who painted orchids in the 1800s and early 1900s, as well as contemporary artists that have continued in this tradition. I began with late nineteenth-century artist Martin Johnson Heade, whose paintings of orchids and hummingbirds in the Amazon and the South American cloud forests are iconic. He came back to the eastern United States and painted dozens of paintings from sketches he made in the tropics, working with local nurseries to gather the orchids he put in his paintings.

By contrast, at the same time, the artist Marianne North, while traveling extensively in the tropics, painted her pictures of orchids and other tropical plants in the jungle on-site. Between 1871 and 1885, North created over 800 paintings while visiting seventeen countries on six continents. I have enjoyed her paintings immensely, and her gallery at Kew Gardens in England, with literally wall-to-wall paintings, is inspirational. She came from a wealthy family and was able to fund her own travel and exhibitions and even to fund the building that holds her gallery at Kew. For a long time, her work was not taken seriously, but because she depicted plants in their natural environments, it is appreciated now as a document of what was growing at the time.

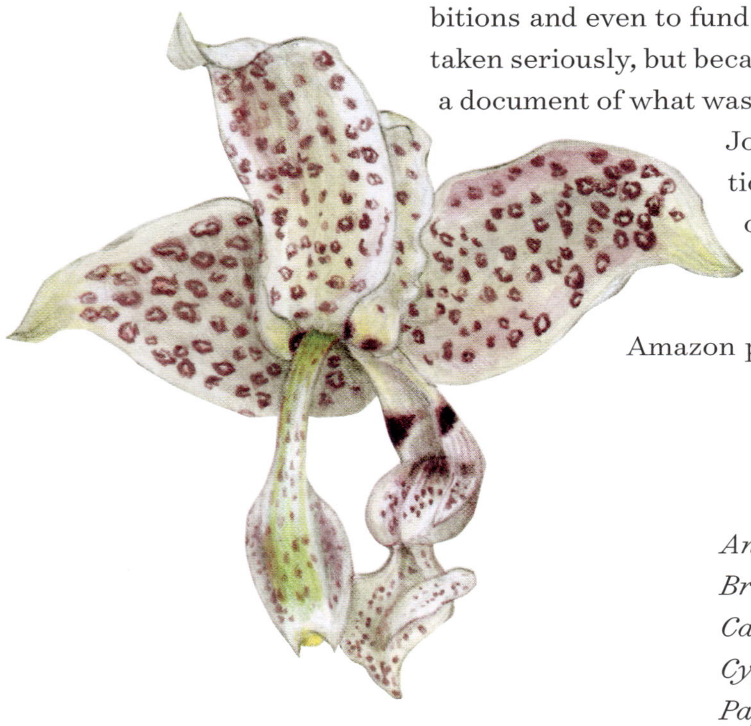

John Nugent Fitch created atmospheric backgrounds behind his orchid illustrations, which give the images a mysterious quality. He created 528 illustrations of orchids for the eleven volumes of his book *The Orchid Album* (1882). He worked closely with local orchid growers in England to get his subjects.

Margaret Mee, a botanical artist, spent her career in the 1900s in the Amazon painting tropical flora including orchids for over thirty years. She worked at

### Plants Illustrated

*Angraecum sesquipedale*, Darwin's Orchid
*Brassocatanthe*, Little Mermaid, BC Little Mermaid Orchid
*Cattleya*, Hybrid Orchid
*Cypripedium acaule*, Pink Lady Slipper Orchid
*Paphiopedilum insigne*, Slipper Orchid
*Stanhopea oculata*, Eye-Spot Stanhopea
*Vanilla planifolia*, Vanilla Orchid
*Zygopetalum*, Zygo Orchid, Artur Elle 'Tombstone'

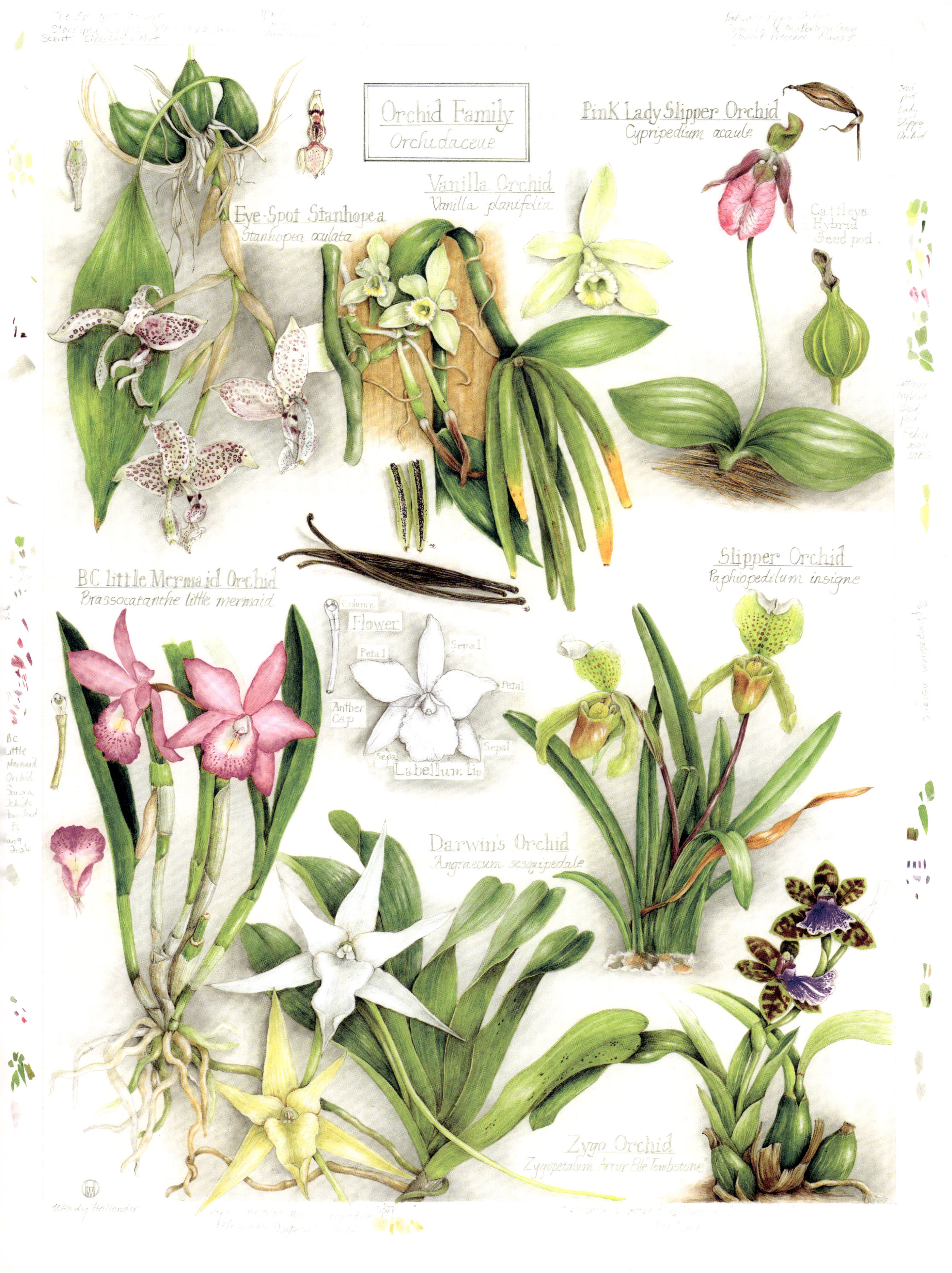

Orchid Family
Orchidaceae
Pink Lady Slipper Orchid
Cypripedium acaule
Vanilla Orchid
Vanilla planifolia
Eye-Spot Stanhopea
Stanhopea oculata
Cattleya Hybrid Seed pod.
Slipper Orchid
Paphiopedilum insigne
BC Little Mermaid Orchid
Brassocatanthe little mermaid
Column
Flower
Sepal
Petal
Petal
Anther Cap
Sepal
Sepal
Labellum Lip
Darwin's Orchid
Angraecum sesquipedale
Zygo Orchid
Zygopetalum 'Artur Elle Tombstone'

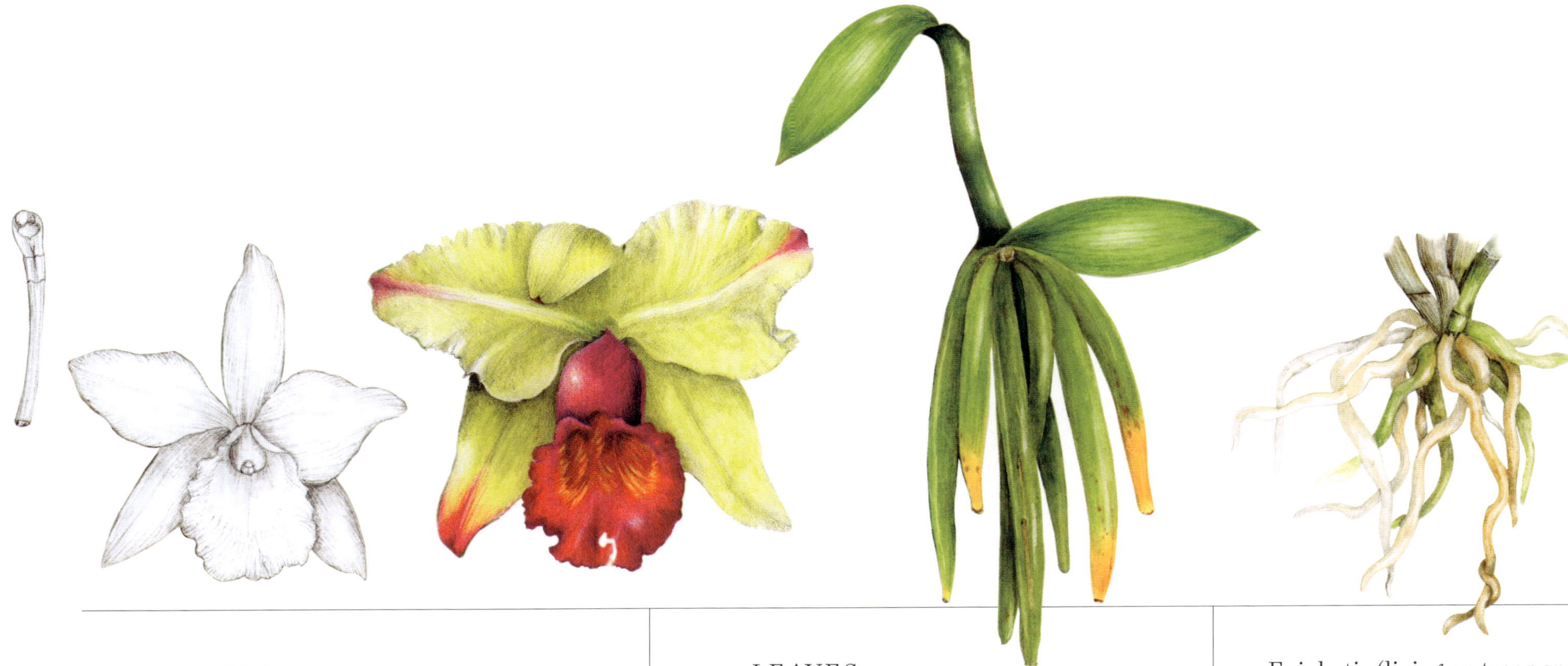

| FLOWERS | LEAVES | |
|---|---|---|
| 3 petals and 3 sepals<br>Center petal forms a lip<br>Bilaterally symmetrical<br>Often colorful and fragrant<br>Reproductive parts fused into a single column | Parallel-veined<br>Often thick and leathery | Epiphytic (living on trees or rocks) or terrestrial (growing out of the ground) |

night by flashlight to paint the night-blooming plants and became one of the first advocates for protecting the rainforest after witnessing deforestation there.

Early in my botanical art career, I became friends with another artist, Carol Woodin, whose specialty was painting orchids. After I moved to the Hudson Valley, we became next door neighbors for several years. I saw her painting orchids all the time, sometimes for botanists or collectors of exotic specimens. Carol would dash to the wilds of Ecuador and return a couple of weeks later with many field sketches, which she would then painstakingly turn into beautiful compositions with precise detail, often on vellum. I watched Carol work on one painting for months. Another contemporary botanical artist, Angela Mirro, was fortunate to have a mom who grew orchids, instilling a love she later turned into a career of painting orchids, often in situ from remote locations like Peru.

Even though the orchid family is quite large, second only to Asteraceae, many of the plants are rare and hard to find. I wasn't sure where to begin, so I started with the one orchid that is grown for its culinary use, the vanilla orchid, which grows in tropical environments. To produce vanilla beans, each flower must be individually hand-pollinated, as the one tiny bee pollinator is almost extinct. The flavor from the vanilla bean is worth it and is used extensively in baking and flavoring desserts such as ice cream. I visited a vanilla grower on Kaua'i to see how he grows and produces his vanilla. This orchid is epiphytic, meaning it needs a structure to grow on (in nature, the structure is usually a tree) and has aerial roots that draw nutrition and moisture from the air, rain, and dew. The Vanillery of Kaua'i created wooden pole structures for their vanilla orchids to grow on. I was familiar with dried vanilla beans but was surprised to see the developing seedpods resembling string beans growing.

Pink lady slipper orchids are native to New York State and grow near me in the Shawangunk Mountains. I have hiked to see them growing on several occasions, always toward the end of May. I sat on a rock to draw this orchid, happily surrounded by many of these unusual pink flowers growing on the rocky mountainside.

The New York Botanical Garden is home to many orchids, which grow in greenhouses. I was fortunate to be able to work on two orchids from the collection. The winter months are a good time to focus on some of these plants. The curator of the orchids, Marc Hachadourian, brought me a terrestrial and an epiphytic orchid to draw. Terrestrial orchids grow directly out of the soil, and epiphytic orchids are air plants, attaching to structures such as trees.

I added a bright pink orchid hybrid while visiting Homestead, Florida. The *Brassocatanthe* Little Mermaid was propagated at Soroa Orchids. They have been winning awards for over thirty years, as in addition to creating orchid varieties they also focus on floral arrangements made with their spectacular blooms.

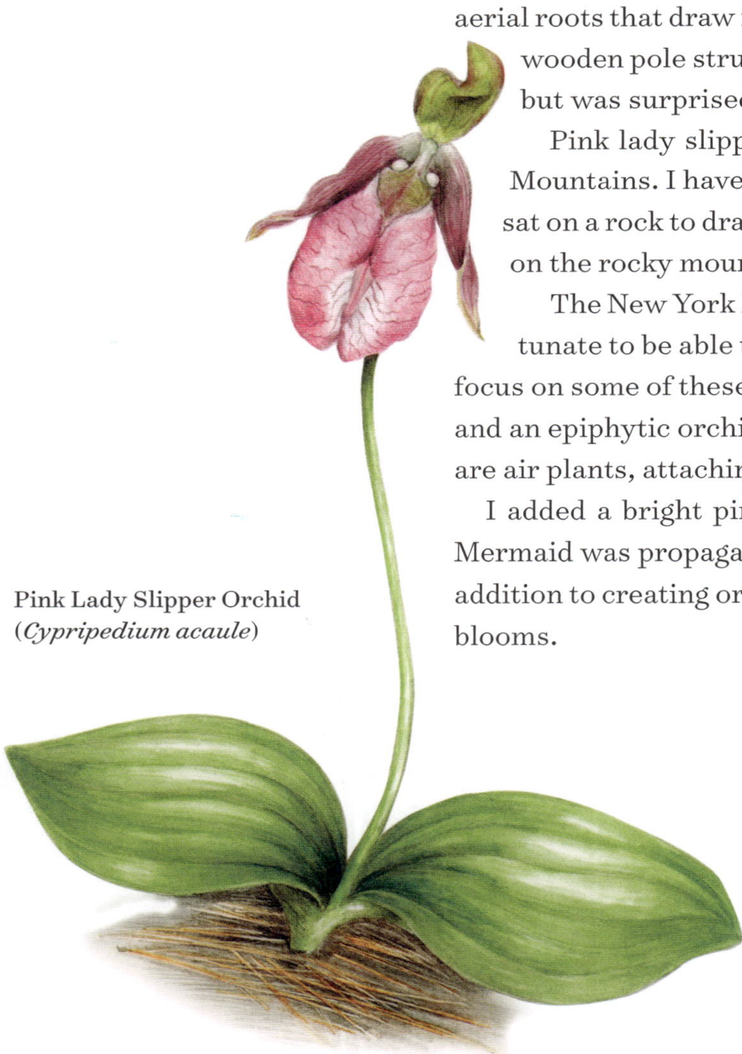

Pink Lady Slipper Orchid
(*Cypripedium acaule*)

I finished painting orchids at the National Tropical Botanical Garden on Kaua'i. Darwin's Orchid (*Angraecum sesquipedale*), native to Madagascar, is grown there and has an interesting story. In 1862, Charles Darwin proposed that there must be a moth with a long proboscis able to reach inside the orchid's extraordinarily long spur. The pollinator remained a mystery in his lifetime, but in 1934, a giant hawk moth was discovered in Madagascar that in fact had the long proboscis needed to pollinate this orchid. It was named in honor of Darwin's prediction: *Xanthopan morganii* var. *praedicta*.

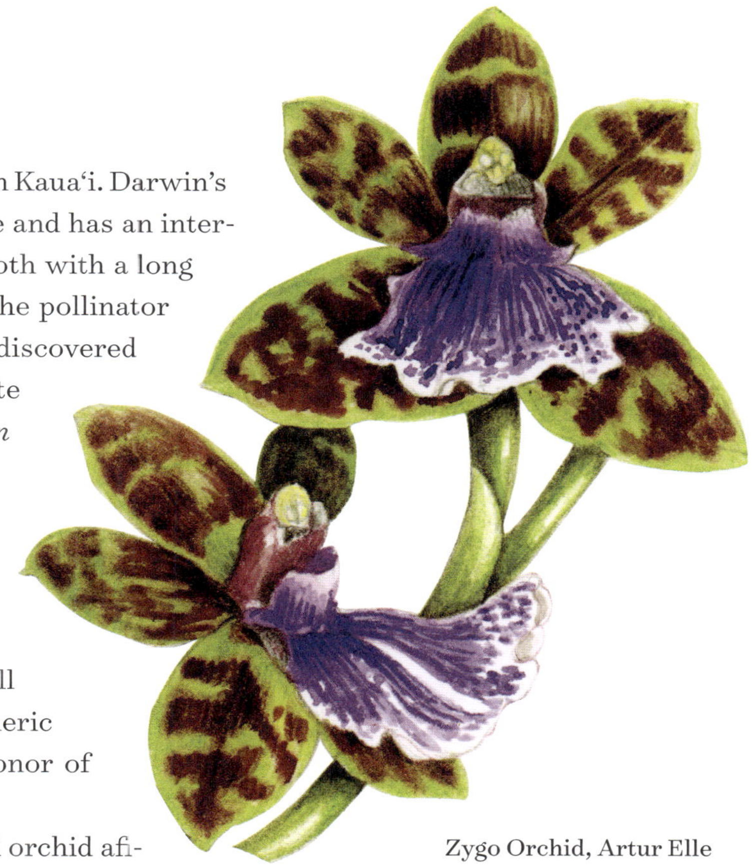

Zygo Orchid, Artur Elle 'Tombstone' (*Zygopetalum*)

I had one more spot for an orchid, and I considered several subjects. I met the "Orchid Lady" at the NTBG, Grazyna Nowak, a volunteer who has been caring for the orchids in the garden. She invited me to see some of the orchids growing on her terrace. I chose the *Zygopetalum* Artur Elle 'Tombstone' for its visually dramatic petals. Once I had all my subjects in the painting, I decided to try to weave together an atmospheric background behind the orchids to create a cohesive composition in honor of John Nugent Fitch's inspiring orchid paintings.

While visiting the Fruit and Spice Park in Homestead, Florida, I asked orchid aficionado Jay Arce: "What do you like about orchids?" Arce, a recreational specialist at the park, replied:

> During my time working with orchid conservation and education, I would ask people to describe how an orchid looks, to which I would always get the words "beautiful," "pretty," "colorful." Then I would ask them to describe how an orchid behaves, to which I would always get confused looks.
>
> "They are liars, cheaters, and thieves," I would say, to their enlightened amusement. The orchid is the scoundrel of the plant kingdom. They thrive in conditions where other plants would suffer, they deceive their pollinators, they essentially predate on mycorrhizal fungi for food, and they are neglectful toward their seeds.
>
> I've always enjoyed systems and the relationships between subjects, and in this case, all the interactions between orchids and the organisms to whom they lie, cheat, and steal. Whether it is a specific host tree, pollinator, or mycorrhizal fungi, there is always a unique relationship to unfold.

## Thoughts from the Orchid Lady

BY GRAZYNA NOWAK

The orchid family is a fascinating group of plants consisting of 28,000 species that conquered every continent and grow in nearly every habitat including the Arctic Circle and tropical beaches. I am a biologist/biochemist who has been always interested in plants. Gardening is my passion. My adventure with orchids began twenty-eight years ago after I bought a *Dendrobium* orchid at a local grocery store. This orchid is still alive and blooms every year. Many more orchids followed and rewarded me with their whimsical and colorful blooms. My full-blown affair with orchids started after I moved to Kaua'i, where I have been able to grow orchids outdoors and expand my knowledge about their cultivation needs. I use this knowledge and experience while taking care of the orchids in the National Tropical Botanical Garden in Kaua'i, where I volunteer. The collection comprises more than 200 different orchid species and hybrids, including a few endangered species and the largest orchid species on earth. This captivating collection of plants is enjoyed by visitors, who admire the orchids' diverse shapes and blooms and have an opportunity to learn that orchids are epiphytes (and not parasites), can grow in soil, on trees, and on rocks, and have mechanisms to absorb water from the air. I hope to never stop learning about this fascinating family of plants with so many existing orchid species and hybrids.

GRAZYNA NOWAK, MS, PhD, is a volunteer at the National Tropical Botanical Garden, where she cares for orchids.

# 18

# Palm Family

## ARECACEAE
## 2,600+ species

The palm family is romantic and mysterious at the same time. I think of palm trees gently swaying in the breeze on a sandy beach along a turquoise shoreline. Coconuts strewn about lie ready for splitting or sprouting a new coconut tree in the sand. On my first trip to Hawai'i, my daughter, Abby, showed us how to cut open a coconut with a pocketknife to drink the coconut water inside an unripe fruit or to eat the flesh of a ripe one. We could survive for a while on the beach with just this delicious and nutritious food.

I knew I would start this painting with one tall iconic palm tree in the center, drawn much smaller than it is so I could fit it on my page. Most of my subjects are drawn life-size on the paper, but a reduced size is necessary to create a habit drawing of a large palm. Palms technically are not true trees as they do not have bark or grow wider in layers that produce rings. Rather than sprouting from branches, each leaf grows out of the top of the tree on its own stalk. Palms are closely related to grasses and like them are monocots. Part of a palm's ability to survive lies in the fact that it can bend in the wind and withstand storms without uprooting or splitting as woody trees do. It produces large inflorescences of tiny flowers and numerous fruits, usually with only one seed.

Botanical artists have always enjoyed illustrating this family, perhaps because of the romantic notion that palms grow in exotic places. Palms produce many interesting fruits, many important to our world's survival. I have chosen to illustrate some of these here.

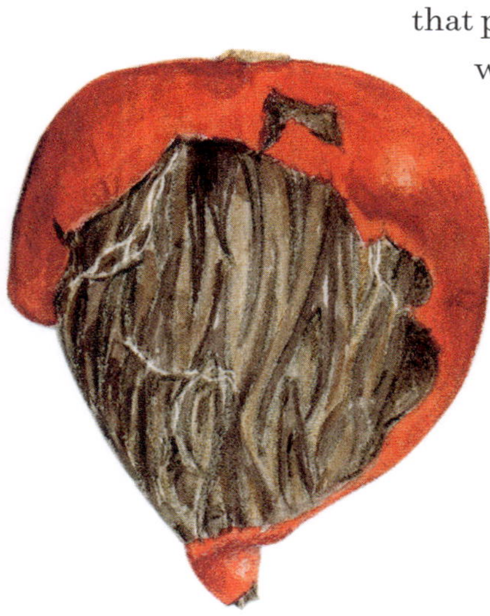

I started with two native Loulu palms (*Pritchardia*) of Hawai'i, as the National Tropical Botanical Garden has many of these in its collection. They have fan-shaped leaves and large inflorescences of very small flowers. Flowers in the palm family typically are star-shaped, with three sepals and three petals.

I chose the foxtail palm for my central, full-habit drawing, as two of these tall, majestic trees grow outside the classroom at the NTBG. The numerous flowers, which fall to the ground, and the unusual fruits are among my favorite subjects to draw and use for teaching. The fruit starts out green and ripens to a brilliant red. Underneath the colorful

### Plants Illustrated

*Areca catechu*, Betel Nut
*Cocos nucifera*, Coconut Palm
*Elaeis guineensis*, African Oil Palm
*Elaeis oleifera*, American Oil Palm
*Phoenix dactylifera*, Date Palm, Medjool
*Pritchardia forbesiana*, Loulu Palm
*Pritchardia lowreyana*, Loulu Palm
*Raphia farinifera*, Raffia Palm
*Wodyetia bifurcata*, Foxtail Palm
INSECT: *Oocytes rhinoceros*, Coconut Rhinoceros Beetle

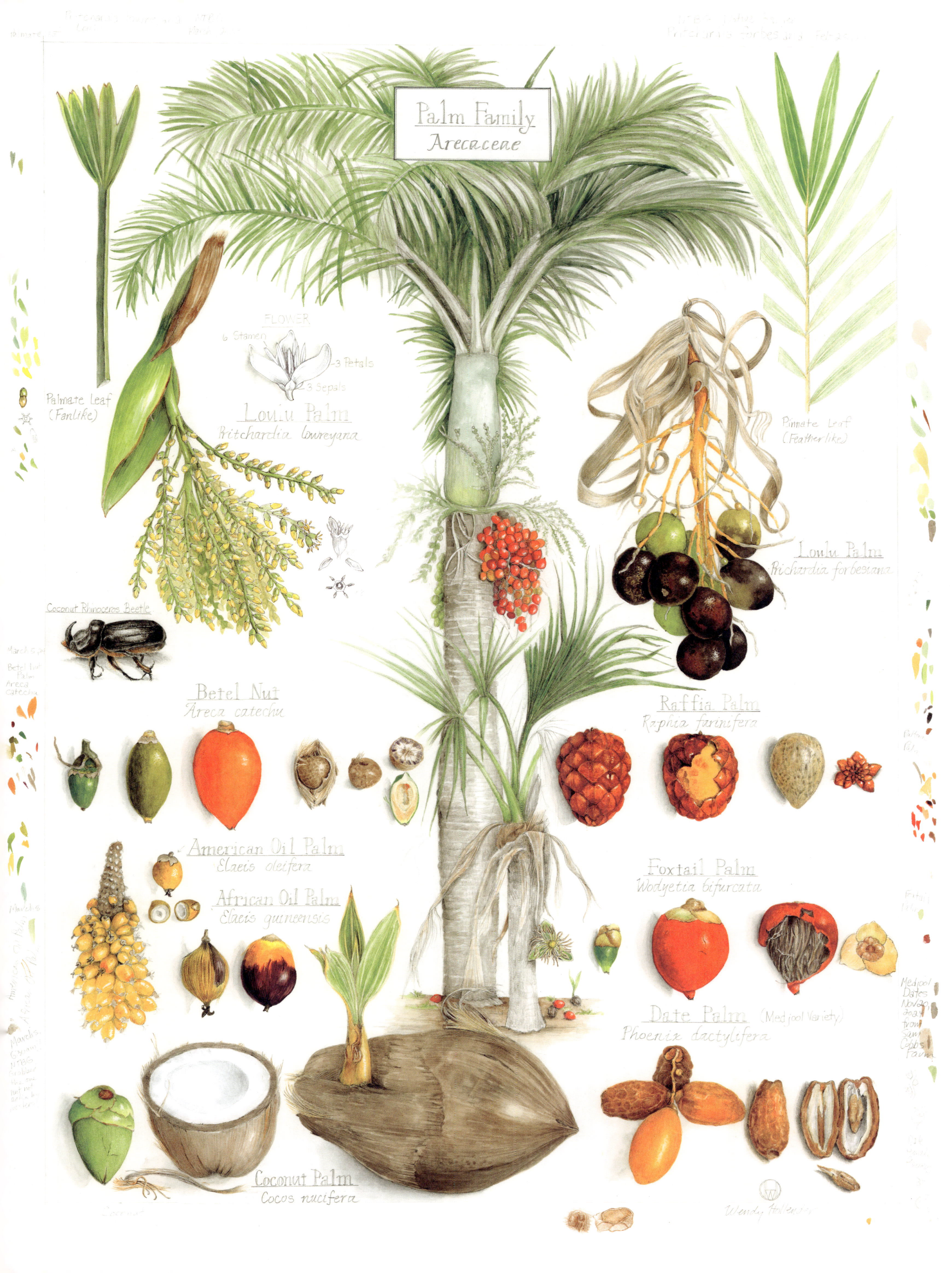

Palm Family
Arecaceae
Palmate Leaf
(Fanlike)
FLOWER
6 Stamen
3 Petals
3 Sepals
Loulu Palm
Pritchardia lowreyana
Pinnate Leaf
(Featherlike)
Loulu Palm
Pritchardia forbesiana
Coconut Rhinoceros Beetle
Betel Nut
Areca catechu
Raffia Palm
Raphia farinifera
American Oil Palm
Elaeis oleifera
African Oil Palm
Elaeis guineensis
Foxtail Palm
Wodyetia bifurcata
Date Palm (Medjool Variety)
Phoenix dactylifera
Coconut Palm
Cocos nucifera

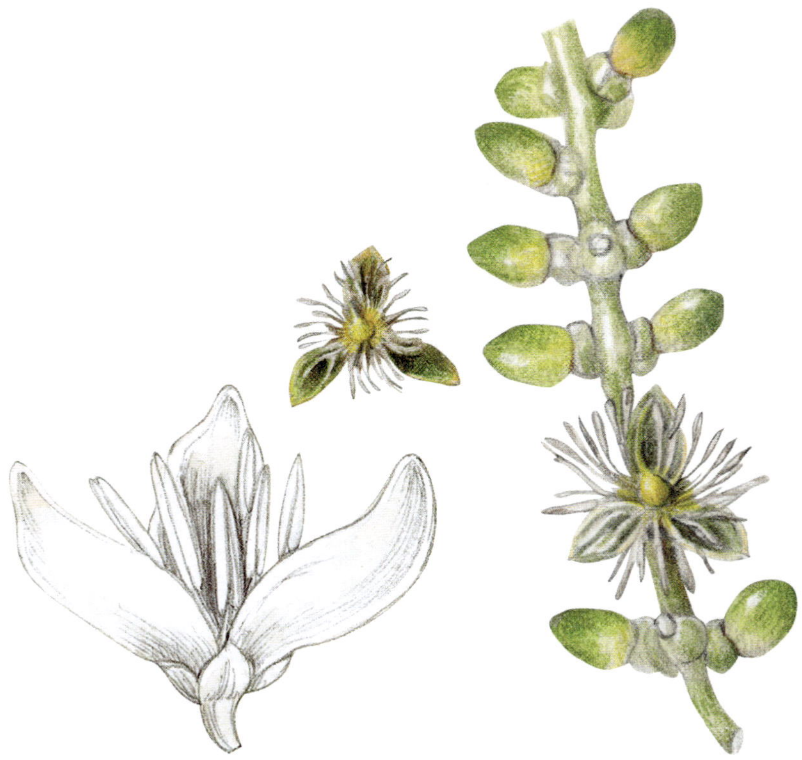

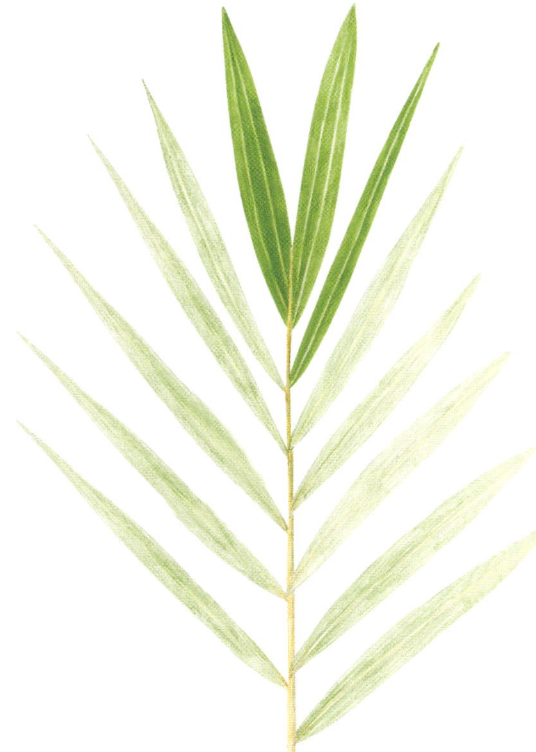

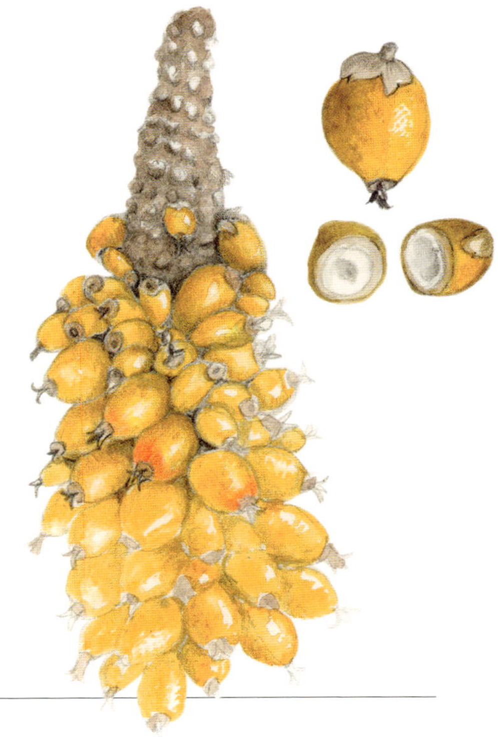

FLOWERS

Small, white, in large inflorescences
Typically 3 petals, 3 sepals
6 or more stamens

LEAVES

Pinnately or palmately compound

FRUITS

Usually a drupe, with an outer skin over a fleshy layer, and a hard inner pit holding 1 seed

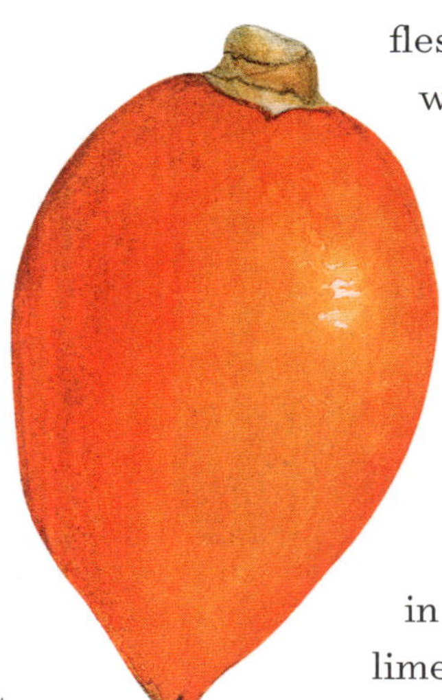

Betel Nut
(*Areca catechu*)

flesh is a seed structure that adds to the mystery of this family. Why does it have its unusual woven structure, and what is its purpose?

I collected fruits from each of the subjects I decided to include in this painting, knowing I could do only a few. I covered my painting with clear plastic and then laid the actual fruits in various stages across the page until I came up with a composition that I liked. I often work this way, laying the actual subject on the paper, usually protected by a plastic sheet. Sometimes I lay it right on the paper, leaving occasional spots staining the paper. Bugs can often be found dancing across the page too, and perhaps once or twice an unsuspecting insect thought my drawing was the real thing.

The betel nut palm (*Areca catechu*) is popular in many tropical parts of the world. People in Southeast Asia chew betel nut, wrapped in betel leaf (*Piper betle*) and treated with slaked lime (calcium hydroxide), as a mild narcotic. In the spirit of ethnobotany, I decided to give a betel nut a try. I engaged Dave Lorence, high priest/botanist at the NTBG, to help me sample this fruit, along with one of my instructors, Sam, who is always up for an adventure. We collected *Piper betle* leaves and lime powder and chewed these together with the betel nut. All the curious students stood by, watching to see what our reaction would be. We got a little energetic and giddy, but overall it was a mild stimulant, not unlike a cup of afternoon coffee. Continuous chewing of betel nut causes teeth discoloration and can lead to other serious health risks, so I doubt I will be substituting the betel nut for my afternoon coffee.

The date palm grows well in California, and I made contact with a farmer there who was nice enough to send me date fruits while they were still fresh and not yet shriveling on the tree, which is when they are harvested. As part of my homework, I ordered several varieties of dates to sample. The one I illustrated is the very popular Medjool variety. Only female trees produce fruit.

The oil palms turned into a true adventure for me. I chose two species to underscore the impact, both good and bad, these plants are currently having on our environment. Unfortunately, countless square miles of land have been deforested to plant monocultures of oil palms. In addition to the damage this is doing to delicate rainforest ecosystems, the palm oil derived from these plants is not the healthiest oil for humans to ingest and has been linked to several ailments.

It is ironic that humans are destroying the environment by planting these palms that are used to make an exorbitant amount of cheap oil that is actually destroying the health of the humans that are cultivating and using it.

But let's move on to my adventure story.

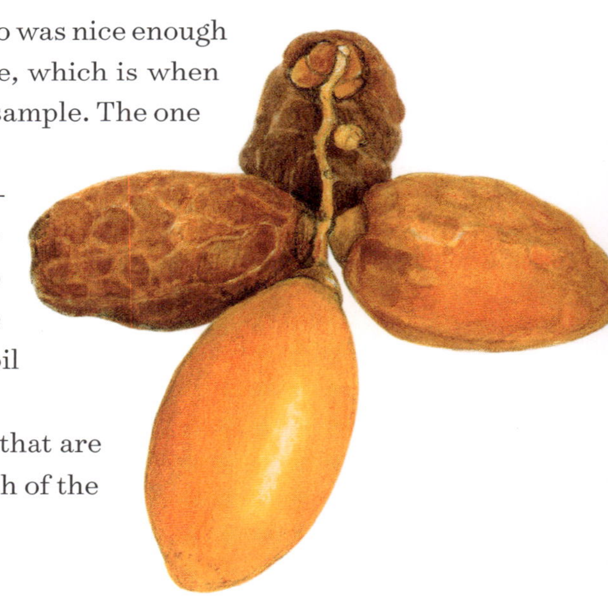

Date Palm, Medjool, fruits
(*Phoenix dactylifera*)

March 8, 2024

> 6:30 a.m. As the sun is rising over the ridge of the Lāwamʻi valley, pink clouds appear and the roosters crow. I drive down the steep road into the NTBG toward the ocean. In Hawaiʻi the surfers head to the beach for the best early morning surfing, but I am in search of the elusive oil palm fruit of the African oil palm tree. In this garden the two African oil palm trees are quite tall, over 60 feet high with a dense area of leaves covering the large mound of fruits growing nestled at the top of the tree. I see a lone rooster already scrounging around underneath the tree even though it is still dark outside. I think I am too late to get a fresh fruit. I have only seen the fruit seeds with just a bit of remaining orange flesh, and I am hypothesizing that there are fresh fruits up in the tree. But what do they look like, and what color are they? I have been going for days to find one so I can draw it and add it to my palm painting. Finally, on my last day in the garden, I decided to make one last effort and try to find a whole fruit, getting there early before the roosters have arrived. I pull up to the tree and leave my headlights on as a lone rooster scurries away. I think this is a futile effort and perhaps there aren't going to be any fresh fruits. I start to search through the leaf and seed litter around the tree. To my disbelief, I find one fruit, just slightly gnarled. I had gotten there just in time.

This illustrates the fundamental points of why I do this work and this particular project. I don't have the answers. I don't know a whole lot of science. But I get to use my curiosity and hunger to learn and understand to find answers and make connections. These roosters must love the flesh of these fruits and wait for them to fall. That was my guess, and to find one half-eaten fruit in the pale light of the sunrise was confirmation of something—at least of the color of this fruit and its flesh, a deep red burgundy on the outside and bright yellow orange on the inside.

"The early bird gets the palm fruit," Dave Lorence said when I told him I'd found the oil palm fruit.

Back in the studio, I have noticed since working on the palm family that some of my precious fruits have disappeared. Specifically, the gorgeous complex fruit of a raffia palm was gone when I arrived one morning. I started to ask students and colleagues to see if one of them might have swiped my fruits. But looking around, I noticed a few crumbs of the fruit trailing on the floor and thought that maybe it wasn't a student but a rat that was taking samples off my desk at night. I already knew the roosters love oil palm fruits, so after I finished drawing the fruit, I left it out overnight to see if it might get taken. When I came in the next morning, it was gone. My hypothesis held true! My conclusion is that the palm fruits must be quite fatty and tasty to the birds and rodents that munch on them.

Just recently, I was visiting the Fruit and Spice Park in Homestead, Florida, to draw tropical fruits there. It has a collection of oil palms, and one tree is only about seven feet tall. Its large cluster of fruits was at my eye level, and I was able to reach inside to pull one off. The ground around the tree was strewn with many half-eaten fruits, further confirmation that my drawings were accurate. Birds darted inside the trees to grab the fruit hidden there. Every time I visit a plant that I have drawn in the past, it is like coming home for a family gathering. The familiarity warms my heart.

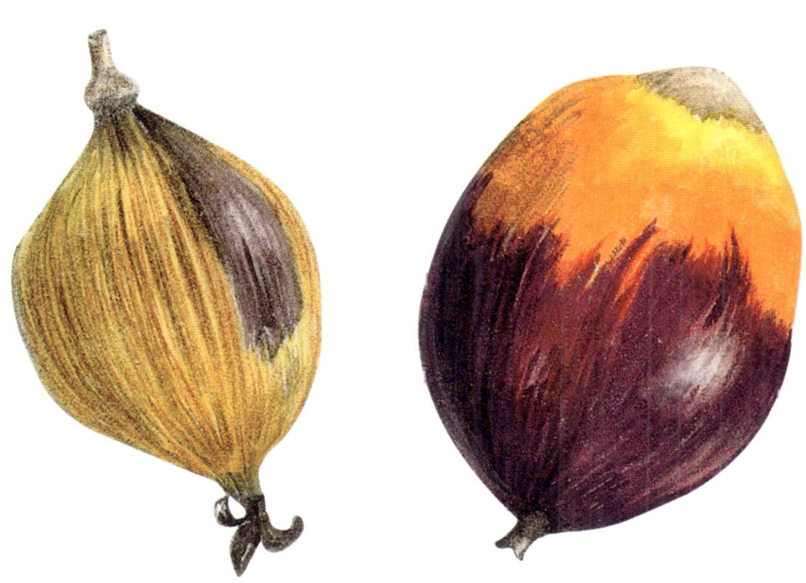

African Oil Palm fruits (*Elaeis guineensis*)

# Ethnobotany of Palms in the Neotropics

BY MICHAEL J. BALICK

Palms are among the most useful plants in the diverse and complex biological world that is the Neotropical forest.[1] For the nomadic hunter-gatherer tribes of the Amazon who depend on the forest for sustenance, palms provide a wealth of products that make their existence in this ecosystem possible. The importance of palms is continually reaffirmed by the lowland farmer who, when clearing away the forest, carefully leaves the palms to be managed and harvested on a semipermanent basis or utilized as forage and shade by domestic animals. Indigenous people and local community members use palms in thousands of different ways in their everyday life.

Harold E. Moore Jr. (1917–1980) was one of the great palm botanists of the twentieth century. During many decades of study, he logged tens of thousands of miles to all parts of the globe in search of palms and information on their biology and utilization. His reflection on the uses of palms in the Neotropics is both eloquent and precise:

> Man . . . is the animal in the South American forest that utilizes palms in the greatest number of ways. The versatility of palms in the hands of man is astonishing. Houses, baskets, mats, hammocks, cradles, quivers, pack baskets, impromptu shelters, blowpipes, bows, starch, wine, protein from insect larvae, fruit, beverages, flour, oil, ornamentals, loincloths, cassava graters, medicines, magic, perfume—all are derived from palms. . . . To whatever extent man has been involved in the tropical ecosystem, palms have certainly been a major factor in making possible this involvement, and even today, despite the advent of the corrugated tin roof and the rifle, they are of primary importance.[2]

One of the reasons palms are used so ubiquitously throughout their range is that they possess a physical construction differing from other plants that provides them with certain advantages for utilization. Within the stem is a series of small strands passing through a matrix of starchy ground tissue. These strands, up to 2–3 millimeters in diameter, comprise three parts: phloem to conduct nutrients, xylem to conduct water, and fibers that incompletely sheath the strand and offer mechanical support. The strands ring the periphery of the palm stem and account for its strength and flexible nature. Mechanically, this is an efficient way to support a heavy object such as a palm. Its durability is proven in the most powerful of tropical storms, when palm stems often survive intact while those of other plant groups are broken and scattered like toothpicks. I have witnessed the steel head of an ax chipping or breaking when used to fell a mature palm specimen, attesting to the strength of the stem. Palm stems are used in house construction, to make bows and arrows, blowguns, and blowgun darts, and for many other purposes.

This same diversity of use can be found for the palm leaf, due to its unique construction, variation of form, both pinnate and palmate, and overall size, ranging from a few inches to many feet in length. Large leaves are used as sails for canoes in the Orinoco River valley in northern South America. Seedlings of some palms are used medicinally—for example, to treat snakebite. Many kinds of palm leaves are used as thatch for houses and other structures. Evil spirits are believed to assume the form of some palm species and to terrorize local inhabitants. Heart of palm (rolled young leaves) is the basis for a substantial commercial export industry in both Central and South America. Leaves of some palm species are used to make a durable plant fiber that may be woven as rope, as was used, for example, on early sailing ships.

Palms also have specialized structures that people employ directly to their advantage. The waxy layer on some palm leaves is harvested and sold as a source of commercial wax—for example, carnauba wax from a South American palm species. Spines from leaves and stems provide protection against predators such as birds, rodents, and other mammals in the wild. People collect these materials and turn them into darts, arrow points, combs, brooms, and similar products. Fruits with oily outer coverings are used by people to produce protein and oil.

Perhaps most important, as Wendy's magnificent drawing shows, palms have great beauty and have always been showcased as ornamental plants in tropical and subtropical regions.

DR. MICHAEL J. BALICK is a renowned American ethnobotanist and economic botanist, recognized for his extensive work on the relationships between plants and people. He serves as the Vice President for Botanical Science, Director, and Senior Philecology Curator at the Center for Plants, People and Culture at the New York Botanical Garden.

1. This essay was excerpted from *Ethnobotany in the Neotropics*, edited by G. T. Prance and J. A. Kallunki, Advances in Economic Botany 1 (New York Botanical Garden, 1984), 9–23.
2. H. E. Moore, "Palms in the Tropical Forest Ecosystems of Africa and South America," in B. J. Meggers, E. S. Ayensu, and D. Duckworth (eds.), *Tropical Forest Ecosystems in Africa and South America: A Comparative Review* (Washington, DC, Smithsonian Institution Press, 1973), 63–88.

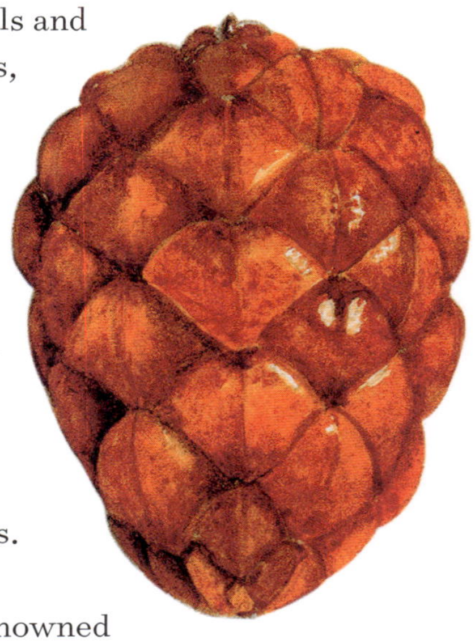

Raffia Palm fruit (*Raphia farinifera*)

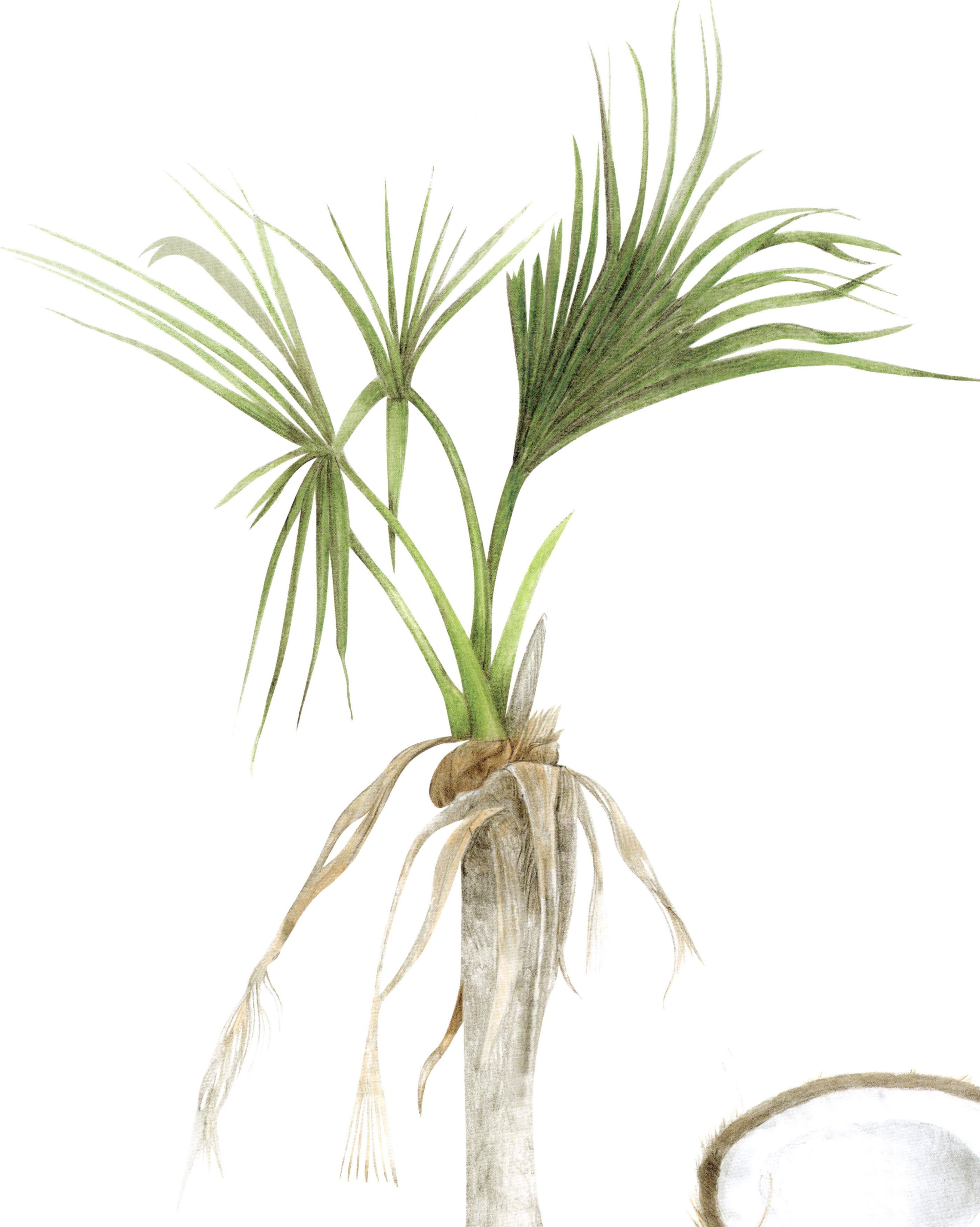

Loulu Palm stem
(*Pritchardia forbesiana*)

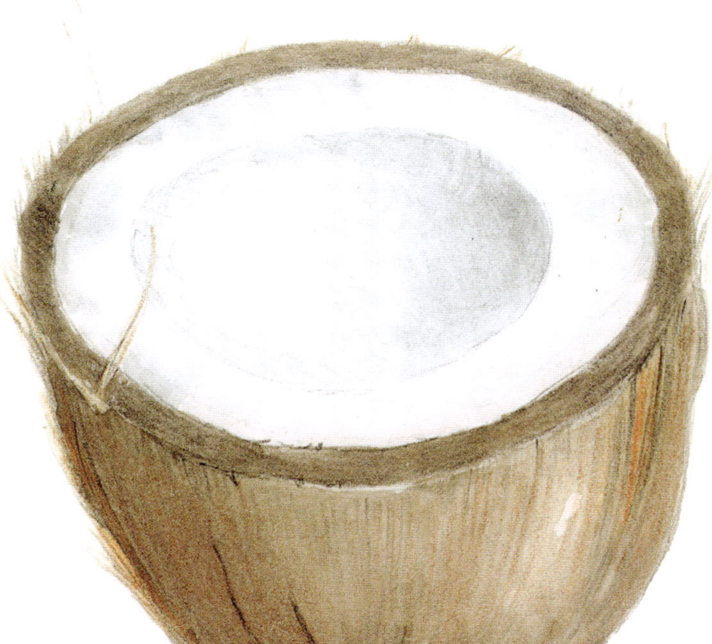

Coconut (*Cocos nucifera*)

# 19

# Pea Family

## Fabaceae

## 19,000+ species

My pea family painting is like a travel journal for 2024. It started with an early surprise: while walking in the garden one morning, I noticed that a gorgeous redbud tree, native to my region, had started to bloom. I was curious to learn what family the redbud is in, because I didn't know. The flowers are tiny and hard to see, so I did a quick plant family lookup with my phone (see how to do this in the Resources section) and found that the redbud is in the pea family (Fabaceae). I quickly cut a small grouping of flowers and ran to the studio to study them under the microscope. Pea family flowers are quite distinctive. They have five petals: one larger petal, identified as the banner, two wing petals, and one keel. Yes, "keel" is the name for the bottom centerline of a boat, and this structure does resemble that shape. Once you see a typical pea flower, you will be able to identify many plants in this family, and there are over 19,000 species!

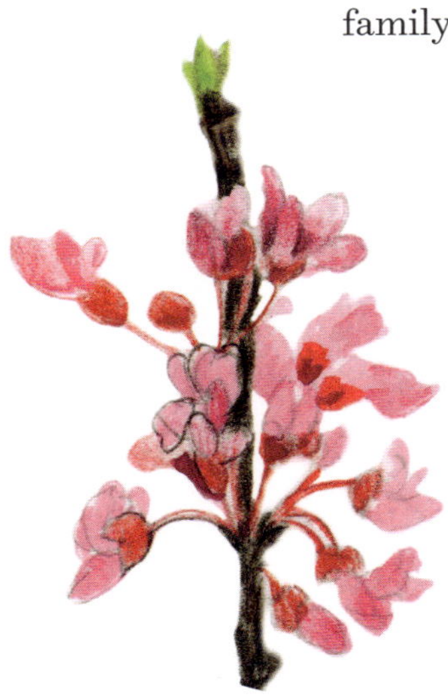

Once I identified the tiny pink flowers of my redbud tree as pea flowers, I was ready to start my pea family painting. I was waiting for the wisteria to bloom, as I knew I would be including it. Hanging down like an elegant purple necklace, wisteria's cascading flowers bloom first close to the stem, leaving the tips of the inflorescence as tiny buds, and then slowly open down the length of the cluster. In my neighborhood of small farmsteads and old farmhouses, one neighbor has a beautiful entryway covered in wisteria. I visit it every spring just to look, but this year I was planning to ask for a bloom to paint. I arrived when the wisteria was in bloom and was just looking, not cutting, when the owner came out to let me know his wife had been watching me for days when I had come by to look at the wisteria vine. They actually have a sign on the gate that reads, "Not a U-pick. Do not touch!" I wasn't the only one attracted

### Plants Illustrated

*Arachis hypogaea*, Peanut
*Cercis canadensis*, Eastern Redbud
*Chamaecrista fasciculata*, Partridge Pea
*Dolichos lablab*, Hyacinth Bean, Ruby Moon
*Lathyrus oleraceus*, Shelling Pea
*Lupinus polyphyllus*, Lupine
*Phaseolus vulgaris*, Bean, Dragon's Tongue
*Phaseolus vulgaris*, Green Bean, Provider
*Phaseolus vulgaris*, Pinto Bean
*Pisum sativum*, Sugar Snap Pea
*Robinia pseudoacacia*, Black Locust
*Thermopsis californica*, False Lupine
*Wisteria sinensis*, Chinese Wisteria

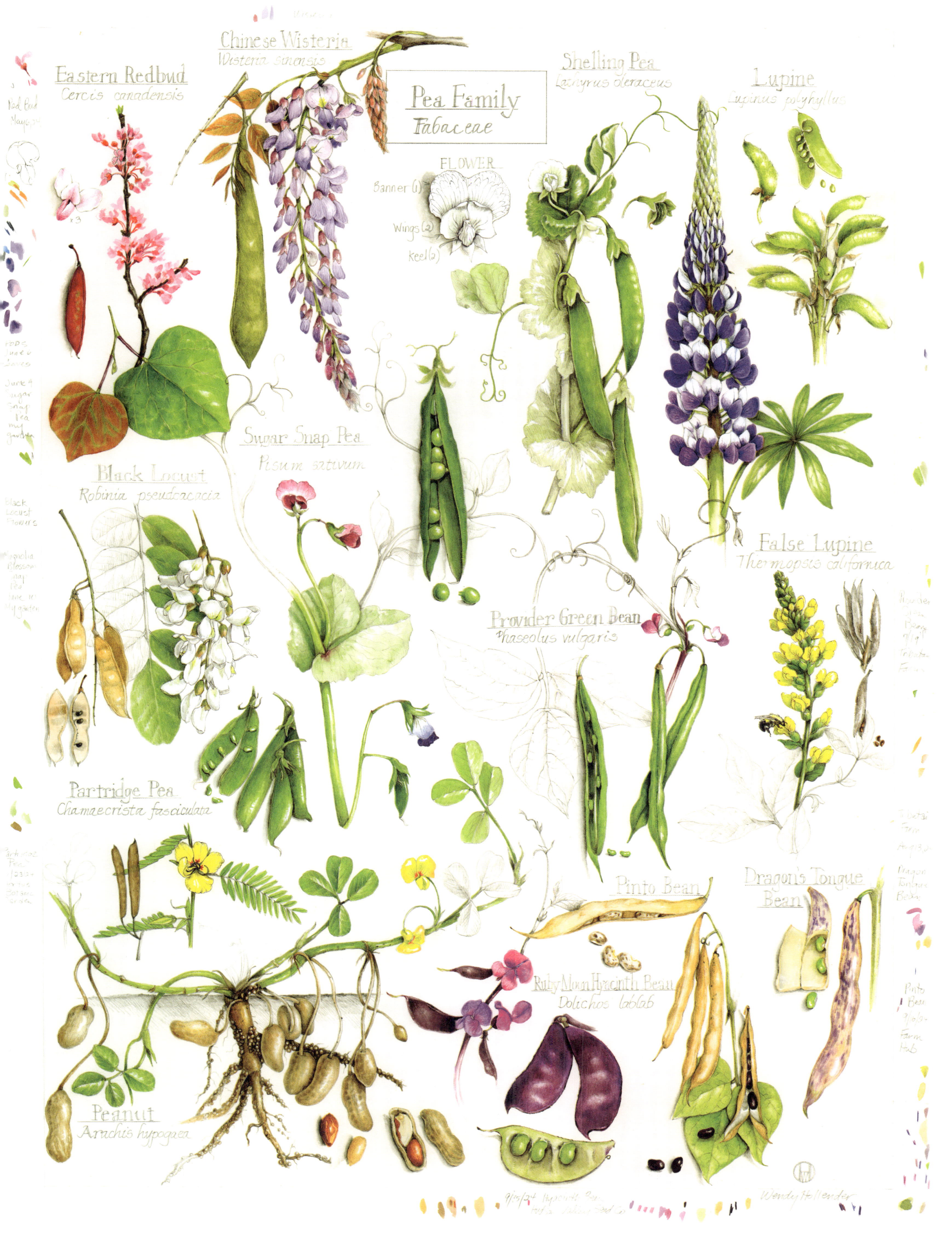

Pea Family
Fabaceae
Eastern Redbud
Cercis canadensis
Chinese Wisteria
Wisteria sinensis
Shelling Pea
Lathyrus oleraceus
Lupine
Lupinus polyphyllus
FLOWER
Banner (1)
Wings (2)
Keel (2)
Sugar Snap Pea
Pisum sativum
Black Locust
Robinia pseudoacacia
False Lupine
Thermopsis californica
Provider Green Bean
Phaseolus vulgaris
Partridge Pea
Chamaecrista fasciculata
Pinto Bean
Dragon's Tongue Bean
Ruby Moon Hyacinth Bean
Dolichos lablab
Peanut
Arachis hypogaea
Wendy Hollender

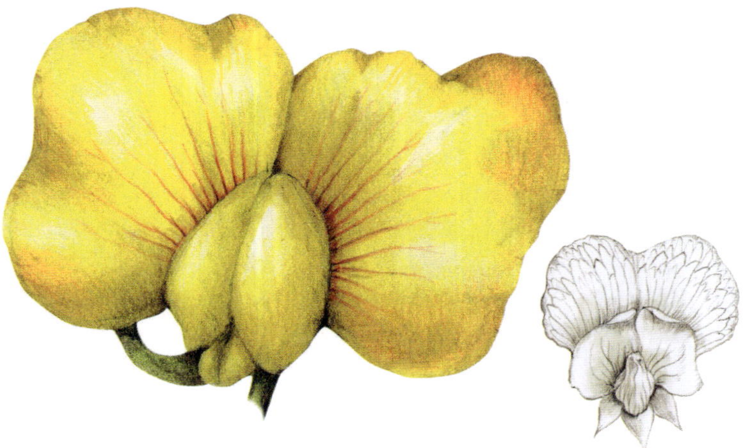

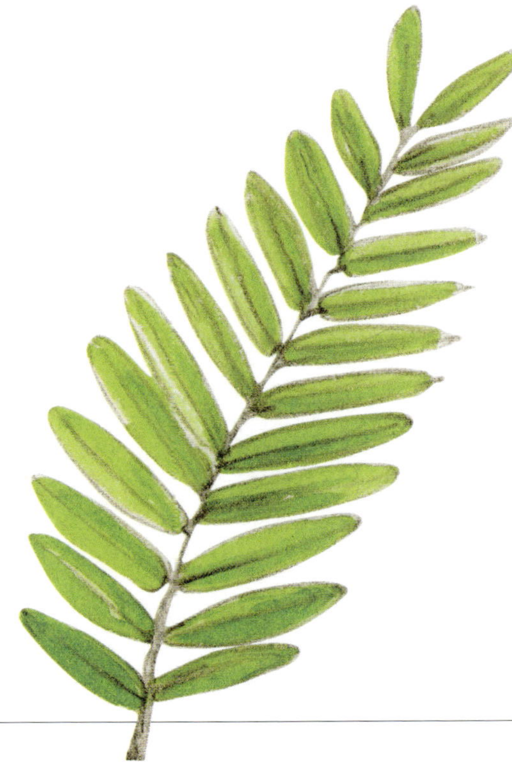

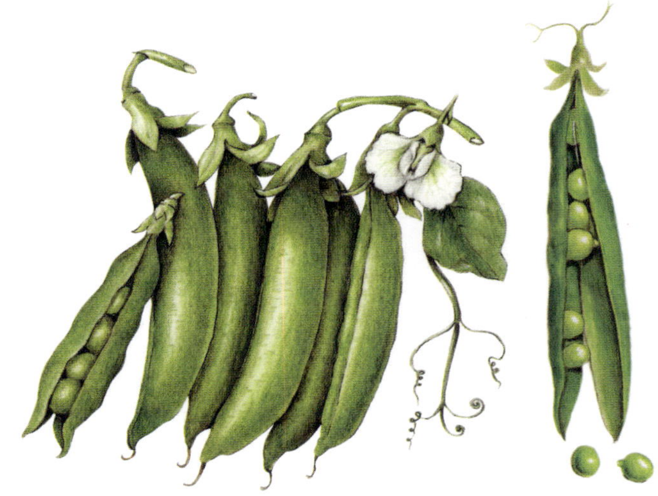

FLOWERS

5 petals: 1 banner, 2 wings, and 1 keel (see text)

LEAVES

Often pinnately compound

FRUITS

Pea-like pods

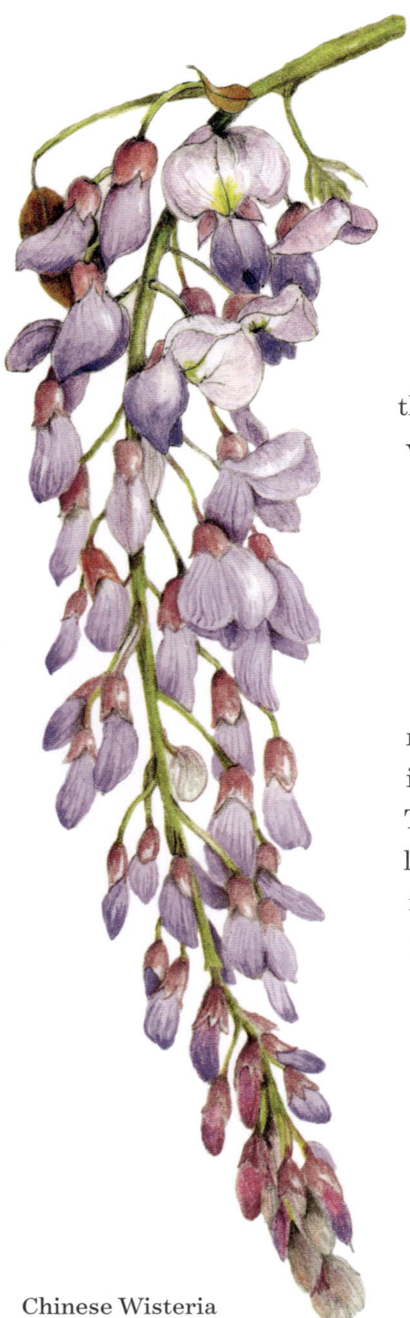

Chinese Wisteria
(*Wisteria sinensis*)

to these blossoms. I quickly introduced myself as their neighbor down the road and explained my work and what I was doing. I opened my phone and showed him some photos of my project, and he agreed to let me cut one bloom. I promised a print of the finished painting, which usually works. I also let him know that I would want to come by for one fuzzy pod in summer.

After securing the wisteria, I left to go on a monthlong adventure in Andalusia, Spain, and then the United Kingdom. In Spain I was teaching a workshop at one of my favorite venues, the Garden of Lemons, close to Granada. This beautiful, intimate retreat is the creation of Ingo Schauser and Henrik Vilain. Here they have built a beautiful hotel with small cottages and lovely gardens full of their favorite roses, which they have transported from their home country of Denmark, and also many vegetables and fruits to use in their cooking. They have written five cookbooks in addition to running this wonderful retreat center. While there, I drew the shelling peas for this painting and was able to sample them daily, popping the delicious, crunchy round peas out of the pod as well as prepared in Henrik's kitchen (a recipe follows). In the hilly landscape around the garden grow ancient olive and citrus trees as well as several pea family plants, mostly with yellow flowers. I always get excited when I identify a pea family flower.

After Spain, I headed to Northamptonshire in the English countryside. I am fortunate that my son married into a lovely British family with a wonderful home and extensive garden in this area. I had been there in the winter, but this would be my first visit in the spring. The gardens were colorful and in full bloom. The showstopper in Anna's garden was her lupine. I had never seen so many large and beautiful varieties. I grow these flowers in my garden, but they can't compare to Anna's. I finally settled on a beautiful deep purple lupine for my painting, though I admit to having trouble choosing between this purple and white flower and a bright reddish-pink one. Purple won out, which usually does with me. I am especially attracted to purple and probably would have liked purple clothes as a child, had they been more available then, but sadly I don't think they were. Currently, my five-year-old granddaughter, Lucy, considers purple her favorite color, and nowadays it is easy to find. She almost never wears anything else—OK, maybe pink, but that's it for her right now.

Back home in June, I painted black locust and the varieties of snap peas and pole beans that were growing. I had planted snap peas in my garden for this purpose, but the woodchucks or the rabbits, I am not sure which, perhaps both, completely devoured them as they emerged.

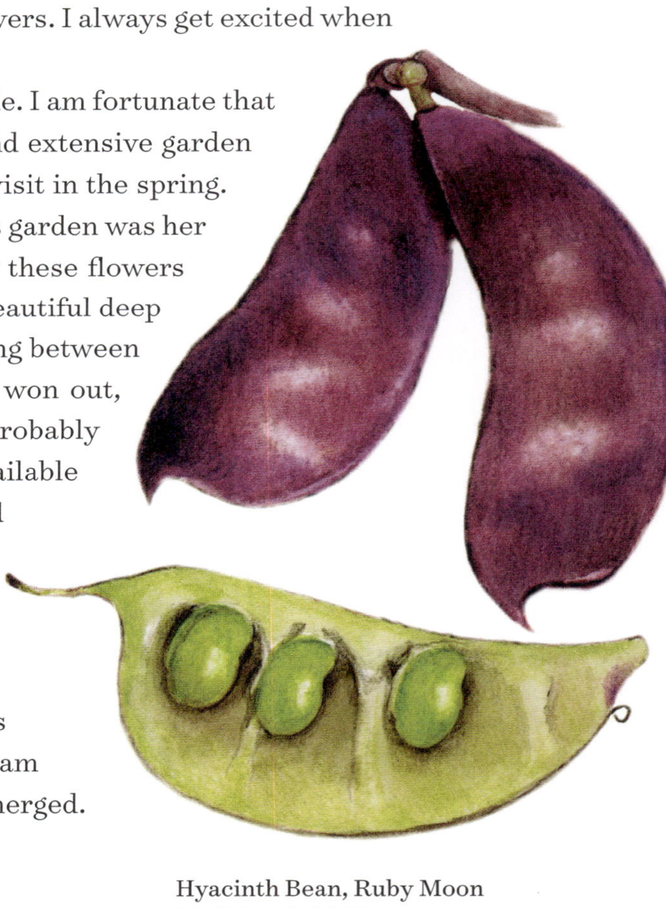

Hyacinth Bean, Ruby Moon
(*Dolichos lablab*)

Pea shoots are delicious, as many of us know, so who can blame them. Fortunately, my son's farm had rows growing for me to choose from.

I got my dried beans from the Hudson Valley Farm Hub in Hurley, New York, which runs extensive trials to choose the best varieties for growing in our region. I found a farmer in Georgia to ship me growing peanut plants, as I didn't have much luck trying to grow them here. Still, it was informative to watch them sprout in my garden but also get munched daily by the rabbits. Peanuts can be prepared, as they are in many southern states, by boiling with salt or spices. Eating a boiled peanut is more like eating a bean than a crunchy nut. This makes sense because peanuts are a legume and not a traditional nut. Boiled peanuts make a fun and healthy snack to munch on, once you get used to them. I couldn't resist adding the ornamental Ruby Moon hyacinth bean from the local Hudson Valley Seed Company; perhaps it was the purple color of the pods that attracted me.

## Pea Soup with Mint

BY HENRIK VILAIN AND INGO SCHAUSER,
THE GARDEN OF LEMONS

Pea soup is a dish we often return to at the Garden of Lemons. The soup has a wonderful taste and color and is quick to make with ingredients that are almost always in the house. Sprinkle some shredded mint or basil leaves, some grated cucumber shavings, or something completely green and crunchy on top of the soup before serving—something that signals early summer and has a freshness to it.

The soup can be upgraded by pouring a little champagne into each plate when serving, so that the soup bubbles up elegantly—then it has become a dish called crème Ninon, named after the sublimely beautiful Ninon de l'Enclos, a seventeenth-century French writer and courtesan, whose acquaintances would have enjoyed both good company and a good meal.

MAKES 4 SERVINGS

INGREDIENTS

2 onions, chopped
25 grams [2 tablespoons] butter
400 grams [14 ounces] peas, frozen or fresh
50 grams [2 ounces] fresh spinach
thyme
1 lemon, juice and zest
7.5 deciliters [3 cups] vegetable broth
a splash of fino/manzanilla sherry or dry white wine
salt and freshly ground black pepper

Sauté the chopped onions in butter until they are soft and transparent. Add peas, spinach leaves, thyme, lemon zest, and broth. Let the soup simmer for a few minutes. Before serving, blend it to a completely smooth consistency. Season with sherry/wine, lemon juice, salt, and pepper. The soup is blended close to serving to maintain the beautiful green color.

Recipe adapted from Henrik Vilain and Ingo Schauser, *Citroner* (Copenhagen: Muusmann), 2019.

HENRIK VILAIN and INGO SCHAUSER are both trained classical musicians—Ingo as a pianist and flutist, Henrik as an organist and conductor. In 2014, they created the cultural refuge Garden of Lemons, south of Granada, Spain, where they focus on food and community around the table. Today Henrik and Ingo are chefs who work year-round developing recipes inspired by the large kitchen gardens at Garden of Lemons. Their five cookbooks are *Lemons* (2019), *Olives and Olive Oil* (2020), *Avocado* (2021), *Tomatoes* (2023), and *Onions* (2025).

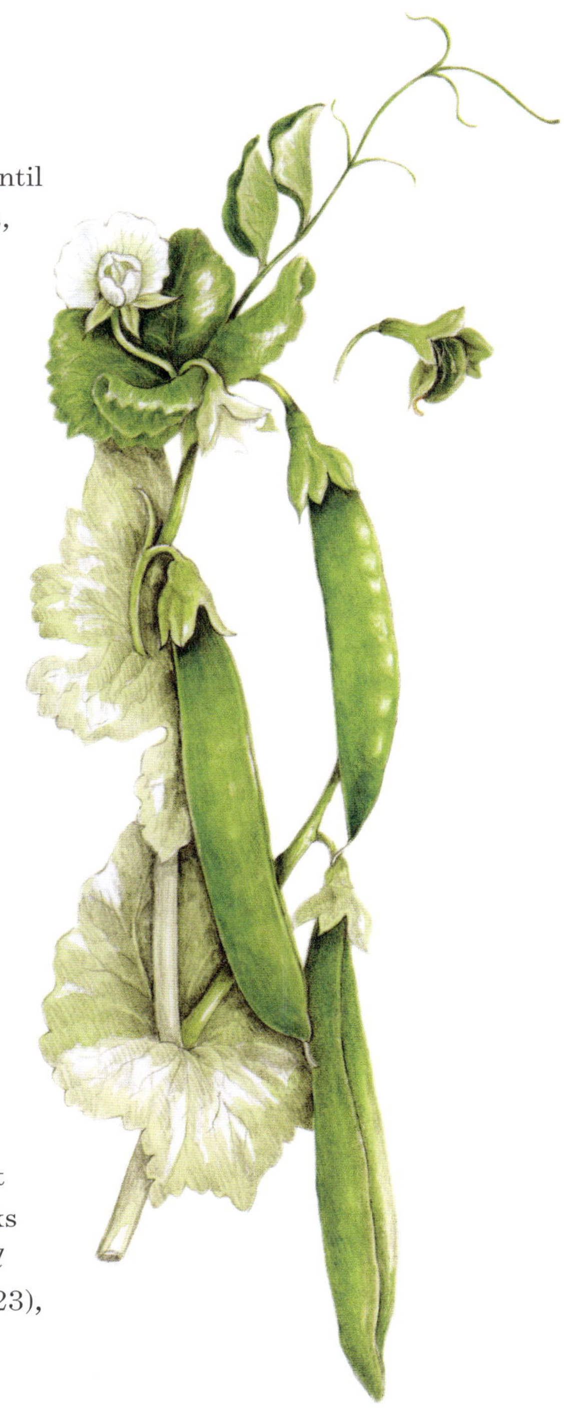

Shelling Pea (*Lathyrus oleraceus*)

# A *Fab*ulous Family

BY DUSTIN WOLKIS

Ah, Fabaceae, what a *fab*ulous family! Fabaceae, or Leguminosae, is one of the largest plant families on earth, represented by around 730 genera and over 19,000 species, distributed globally. The family is so large that taxonomists have grouped these species into six subfamilies, which vary widely in their morphology (even within a subfamily, species can have radically different flowers). Compare, for example, the banner and keel flower morphology of Wendy's partridge pea (*Chamaecrista fasciculata*), a member of the subfamily Caesalpinioideae, to the small pom-pom or powder-puff flower clusters of a *Mimosa* species, in the Mimosoideae subfamily. In each "pom-pom," composed of many individual flowers, the petals and sepals are greatly reduced, revealing the long, threadlike stamens that give it the soft, feathery appearance. As striking as the flowers of this family can be, the most interesting aspect of it is the seeds.

Most of us are familiar with this family through its culinary role, as we commonly eat Fabaceae seeds. While delicious and packed with protein, these seeds also reveal a fascinating anatomy with implications for seed conservation physiology.

As a plant conservation biologist, I study the longevity of seeds stored in seed banks and the factors that influence them: moisture, temperature, lipids, and time. "A seed is a baby plant in a box with its lunch." That is a quote from one of my seed heroes, Carol Baskin. As Professor Baskin explains, the "baby plant" is the embryo, the "lunch" is the fleshy endosperm (providing nourishment to the embryo), and the "box" is the seed coat. It's the seed coat that makes Fabaceae so intriguing.

Many Fabaceae seeds have a water-impermeable seed coat. In scientific terms, this puts them in a class of seed dormancy called "physical dormancy." In physical dormancy, the seed coat protects the seed and keeps it from sprouting. The seed coat must then become water-permeable before the seeds can imbibe (drink) water and thus germinate. Many a gardener knows to cut or sandpaper the water-impermeable seed coat in Fabaceae and other families. One unique aspect of seeds with physical dormancy is that they can lose (desorb) but not gain (adsorb) atmospheric moisture. This turns out to be important for *ex situ* (off-site) seed conservation.

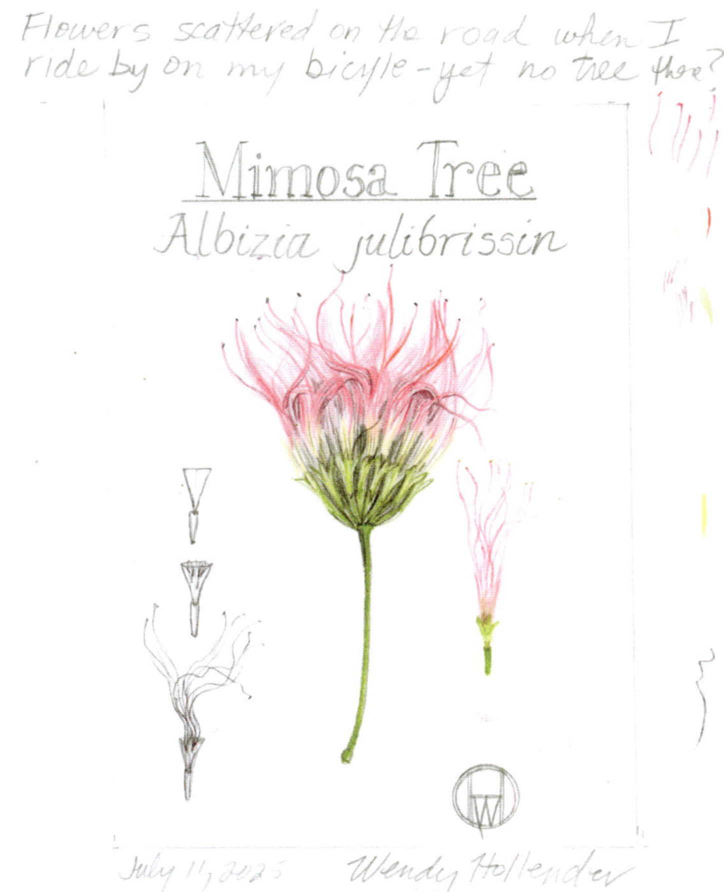


*Mimosa* species sprout pom-pom or powder-puff flower clusters.

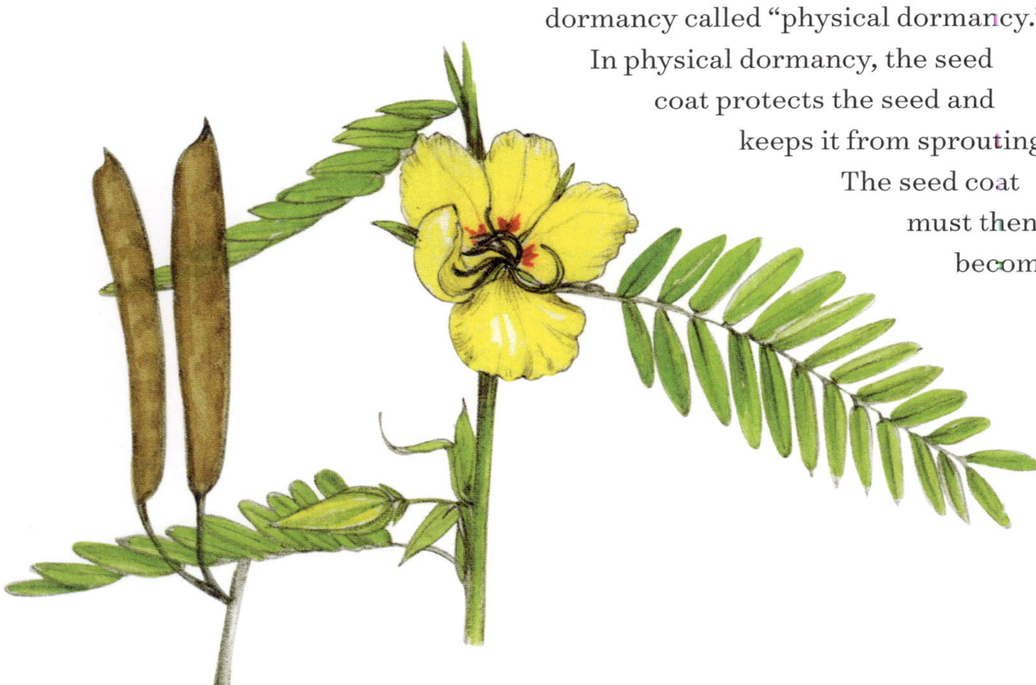

Partridge Pea (*Chamaecrista fasciculata*)

Plant extinctions are happening tens to hundreds of times faster than the background extinction rate (the normal, nonurgent extinction rate on earth). Seed banking is the most efficient and low-risk method of ex situ plant conservation. Seed banking involves drying seeds to low moisture levels, hermetically sealing them, and then storing them at cool temperatures. It is estimated that seed lifespan doubles for every 1 percent decrease in moisture content. Drying seeds also protects them from lethal ice crystallization when they are stored at temperatures below freezing. However, not all species' seeds tolerate desiccation. Fabaceae seeds, or at least the vast majority with physical dormancy, are excellent examples of those that can be dried, frozen, and safely stored long-term in seed banks.

Some of the longest-lived seeds identified are in Fabaceae. Using seeds harvested from herbaria (repositories for dry, pressed plant specimens), researchers have germinated *Prosopis juliflora* and five species of *Acacia* after fifty and thirteen years, respectively. From the herbarium of the National Tropical Botanical Garden in Hawaiʻi, we successfully germinated seeds of the federally endangered Fabaceae species ʻo-hai (*Sesbania tomentosa*) after thirty-plus years at ambient conditions. We attribute this to the unique anatomy of its seeds, in which the moisture content inside the seed would have stayed at whatever minimum was reached, regardless of ambient conditions, thanks to the property of physical dormancy.

Fabaceae really is a fabulous family.

DR. DUSTIN WOLKIS is the Scientific Curator of Seed Conservation at the National Tropical Botanical Garden. He is fascinated with the physiology of ex situ seed and pollen conservation, and focuses on longevity and the factors that influence it—moisture, temperature, lipids, and time. Based in Hawaiʻi, where 90 percent of vascular plants are endemic, but the extinction rate is the highest globally, he is especially interested in how ex situ conservation can help safeguard biodiversity for the future. In his role as the NTBG's Scientific Curator of Seed Conservation, Dr. Wolkis curates the conservation collection of endangered Hawaiian seeds, pollen, and fern/bryophyte spores and is the principal investigator for germplasm research. He earned his master's of science in plant biology and conservation at Arizona State University and his PhD in biodiversity from the University of Copenhagen. He is affiliate graduate faculty in the botany graduate program in the School of Life Sciences at the University of Hawaiʻi at Mānoa, and he serves as the deputy chair for the International Union for the Conservation of Nature, Species Survival Commission, Seed Conservation Specialist Group.

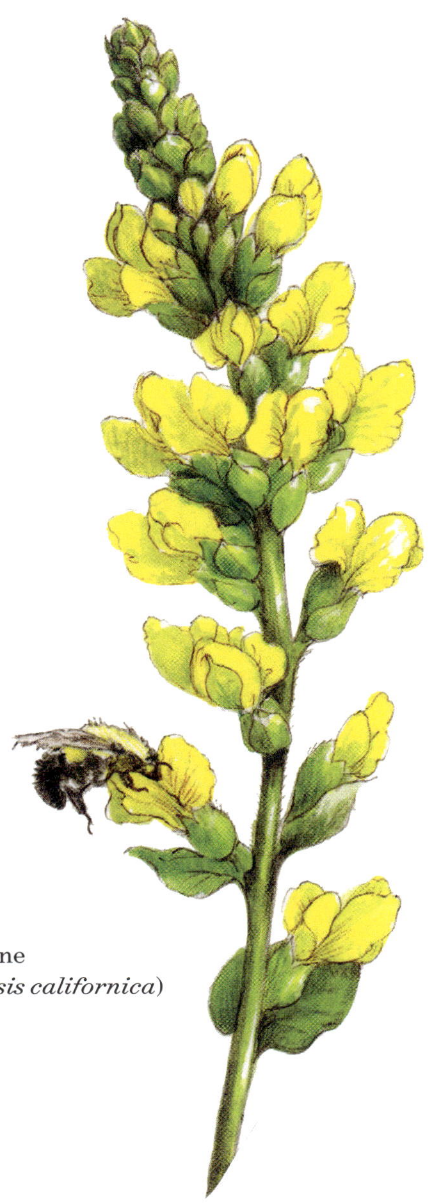

False Lupine (*Thermopsis californica*)

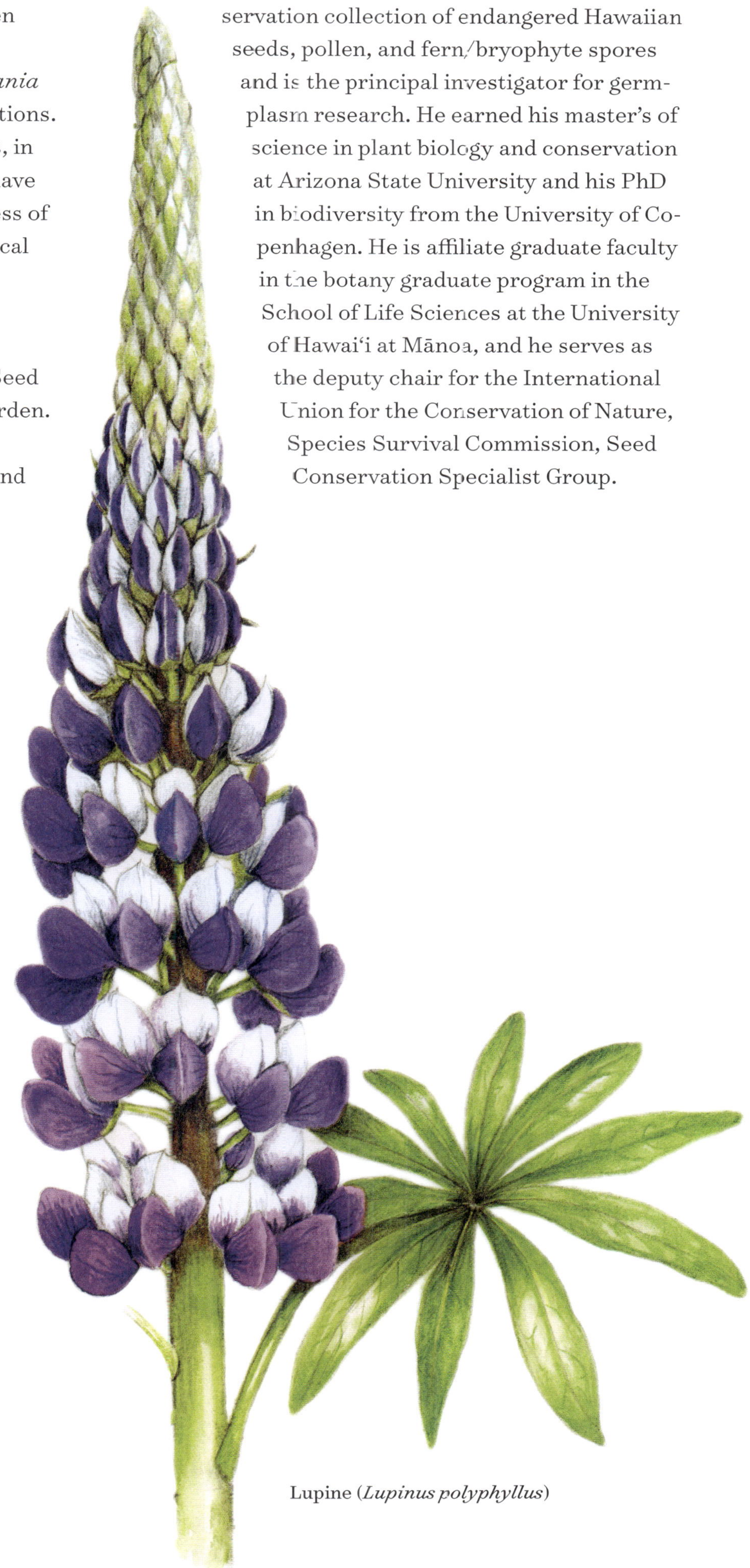

Lupine (*Lupinus polyphyllus*)

# 20
# Rose Family

## Rosaceae
## 3,000+ species

My fascination with botanical illustration began when I was a home-furnishings textile designer in New York City. I was mesmerized by vintage cabbage-rose fabrics and used them for inspiration in my designs. My dream was to be able to draw and paint florals as detailed as those in a Pierre-Joseph Redouté (often called the father of botanical illustration) botanical print or an antique textile design. That desire led me to study botanical illustration.

My rose family painting is influenced by those antique fabrics, and the center of this painting is composed to look like one. In the center is my homage to hybrid tea roses, and I selected a beautiful rose I received in my birthday bouquet while I was working on this painting in the summer months. In addition are some roses I grow in my garden and the prolific saltspray rose that grows all season long from the mountains to the coastline in the eastern United States.

But the rose family is so much more than roses. Who knew that so many of our edible and delicious fruits are in the rose family? By drawing these fruits I was also able to discover why almonds and peaches originally were considered the same species. Crack open a peach pit, and there is a tiny, almondlike kernel inside!

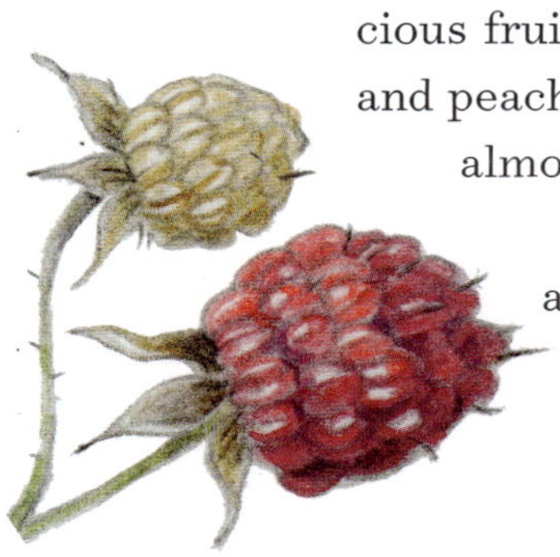

I have many of these fruits growing in my own garden, and even so, I did not know they were all in the rose family until I started this project. I watched while each plant started to flower, and I realized that they didn't all flower at once. In my garden, the almond trees bloomed first, followed by plum, peach, pear, cherry, apple, and wild strawberries. Notice how the flowers

## Plants Illustrated

*Cydonia oblonga*, Quince
*Fragaria × ananassa*, Strawberry
*Malus domestica*, Apple
*Prunus avium*, Cherry
*Prunus domestica*, Plum
*Prunus dulcis*, Almond
*Prunus persica*, Peach
*Pyrus communis*, Pear
*Rosa carolina*, Carolina Rose
*Rosa × hybrida*, Grandiflora Hybrid Tea Rose, Queen Elizabeth
*Rosa × hybrida*, Floribunda Hybrid Tea Rose, Julia Child
*Rosa × hybrida*, Hybrid Tea Rose, Miracle on the Hudson
*Rosa rugosa*, Saltspray Rose
*Rubus allegheniensis*, Blackberry
*Rubus idaeus*, Red Raspberry

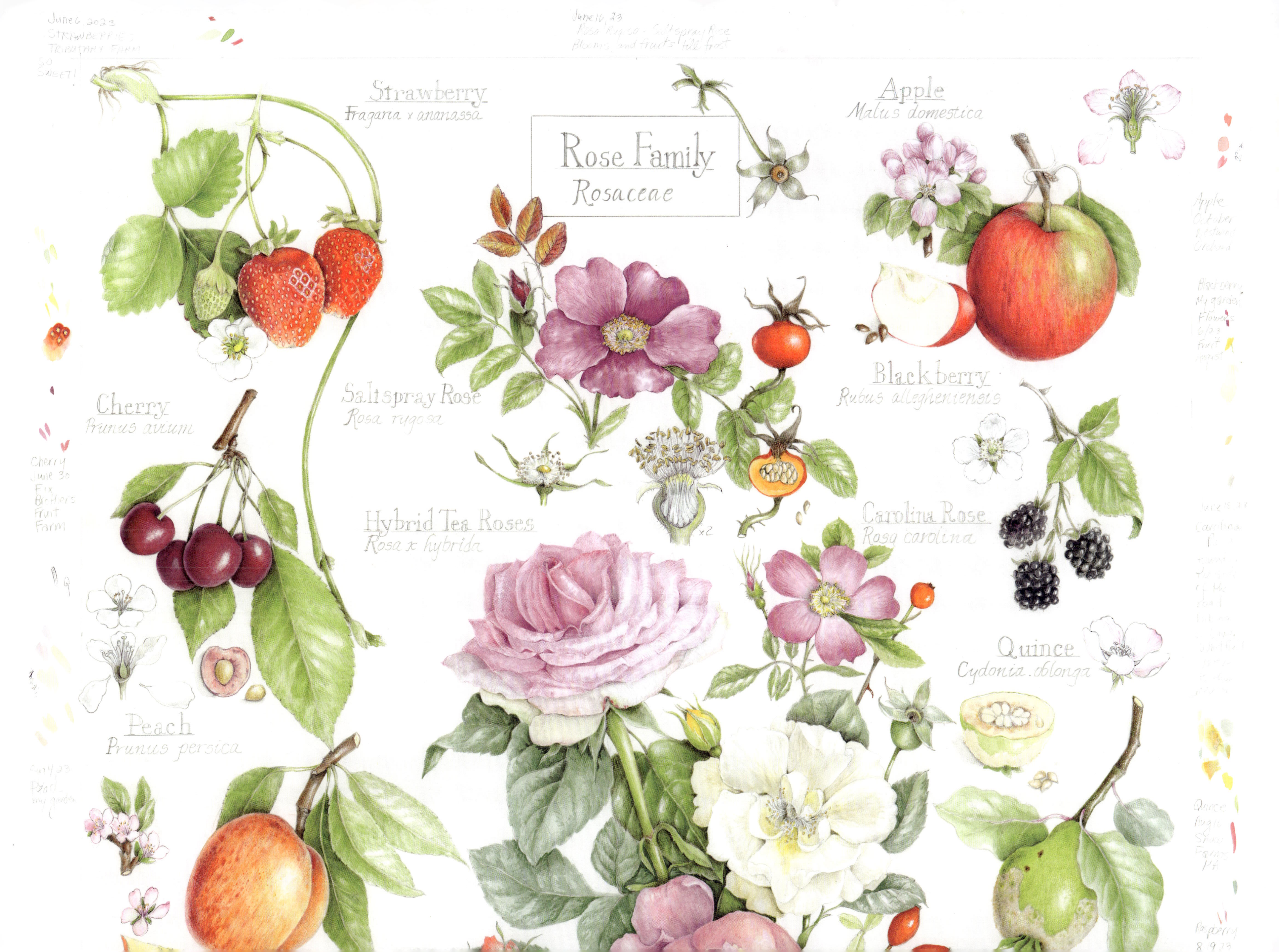

June 6, 2023
STRAWBERRIES
TRIBUTARY FARM
SO SWEET!
June 16, 23
Rosa Rugosa - Saltspray Rose
Blooms and fruits till frost
Rose Family
Rosaceae
Strawberry
Fragaria x ananassa
Apple
Malus domestica
Saltspray Rose
Rosa rugosa
Blackberry
Rubus allegheniensis
Cherry
Prunus avium
Cherry
June 30
Fix
Brothers
Fruit
Farm
Hybrid Tea Roses
Rosa x hybrida
Carolina Rose
Rosa carolina
x2
Quince
Cydonia oblonga
Peach
Prunus persica

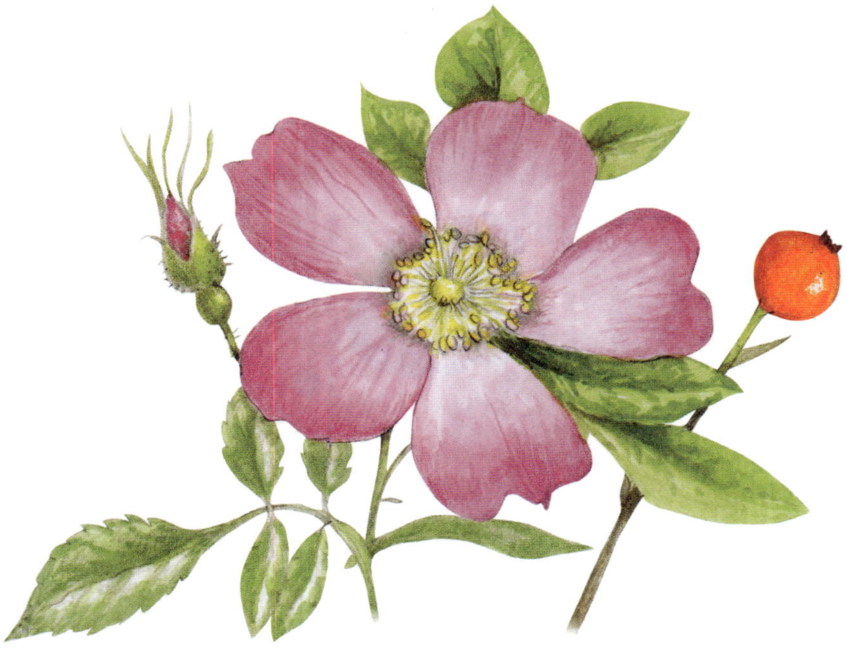

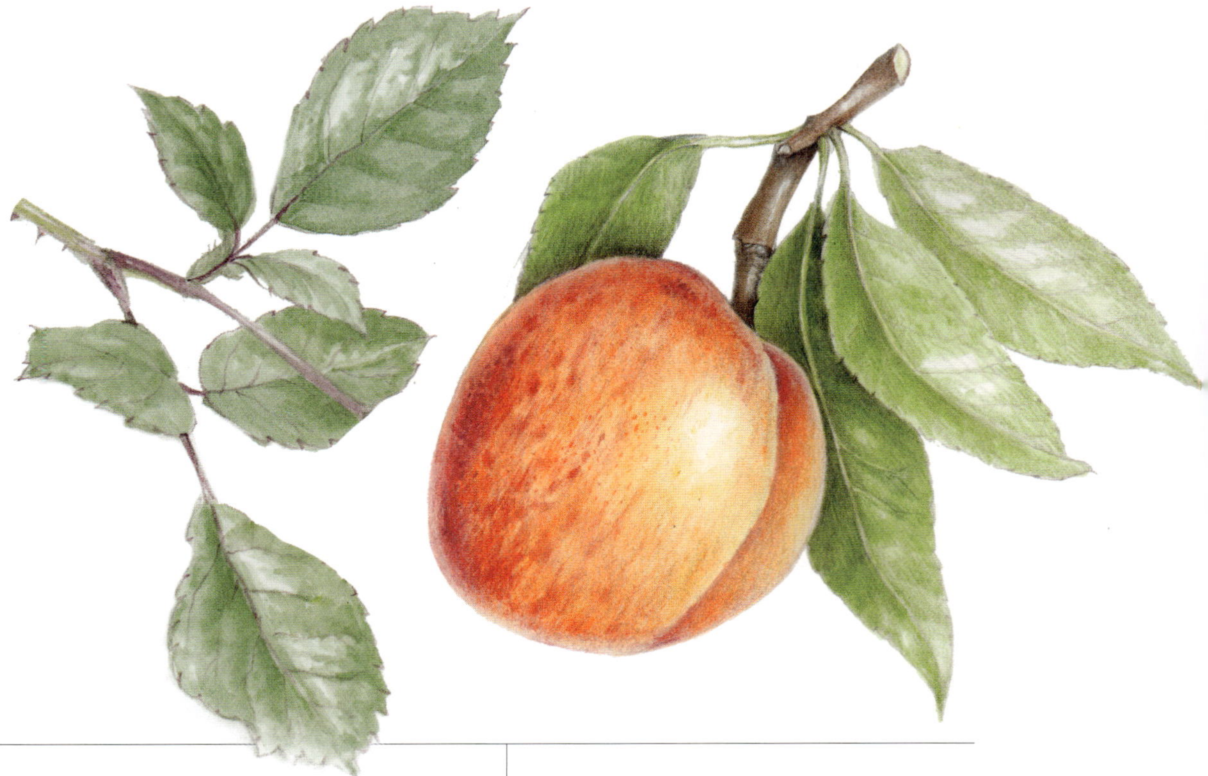

| FLOWERS | LEAVES | STEMS AND FRUITS |
|---|---|---|
| Separate petals in multiples of 5<br>5 sepals<br>Numerous, distinct stamens<br>Pistils often fused<br>Delicious scent! | Alternate or opposite along the stem<br>Oval or elliptic in shape<br>Serrated edges | Stems may have prickles (which help roses climb)<br>Fruit often edible |

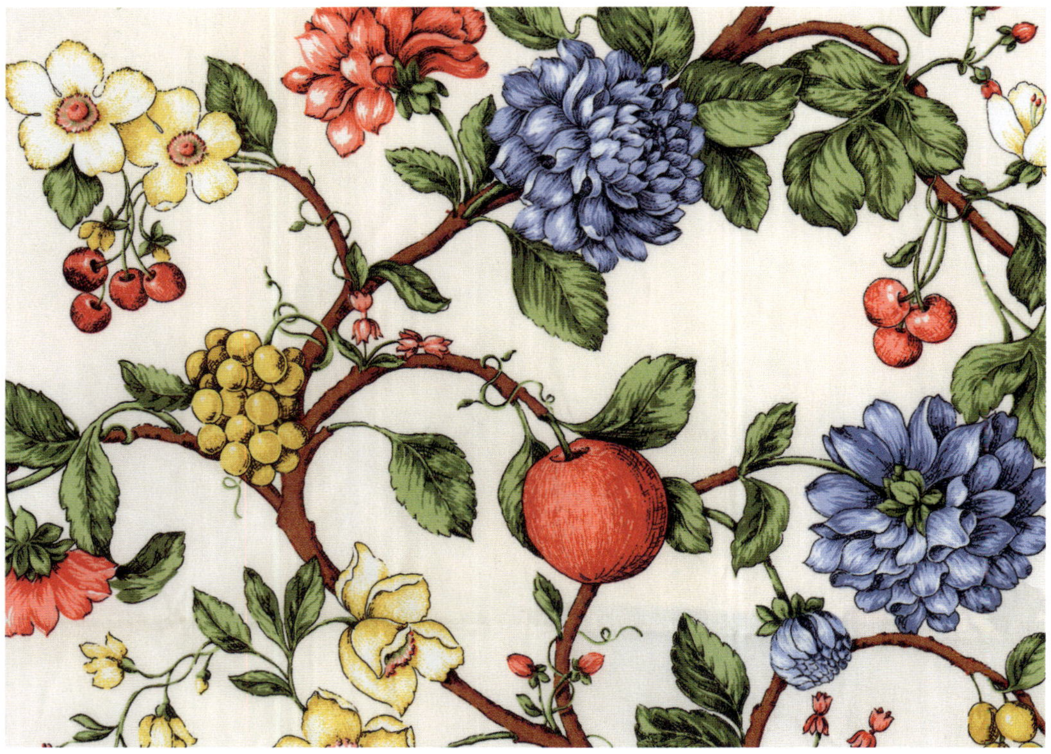

One of my textile patterns. As a textile designer, I wanted to create fabrics covered in roses and other flowers in the spirit of Redouté. The "tree of life" layout is often used in textile design, combining realistic and imaginary elements, as shown in this textile design that I created for Duralee Fabrics around 1994. The fruits and flowers appear to be growing on the same branch, but in reality, you wouldn't find grapes, roses, dahlias, and cherries all together. By the way, there are several plant families in this design. Can you identify some of them?

of all these plants are very similar, with five individual petals, numerous stamens, and often a nice sweet smell.

I was able to gather purply-blue plums from a local orchard called Jenkins-Lueken Orchards in New Paltz, New York. Plums ripen all season, but these purple ones ripen in September. I kept calling the farmer to ask if they were ready. I didn't want to miss them. Finally, the farmer took me out into the orchard in a golf cart and allowed me to pick from any tree. I chose the plumpest and most colorful plums I could find. I asked him why there is a whitish waxy bloom on plums, and he said it is a form of protection. I also can't help but wonder if the word "plump" is related in some way to the word "plum."

# Rose, Family of Desire

BY CHRIS BAKER

The tea rose, as a flower cut from all earthly context, signifies one thing most clearly: desire. Its glossy triad of leaves provides the perfect contrast to the blush of its buds. Its petals curl while they unfurl, revealing and concealing in the same gesture. Its sharp thorns and unearthly fragrance push and pull in tandem, putting tension on the line of our complex, storied attraction to this plant. While we no longer speak a whole language of flowers as the Victorians did, sending one bloom or another to signal our intentions, still this one word remains to us: *rose*, flower of desire.

Now some may reasonably feel the rose has been overplayed—that it tipped over the delicate line from emblem to cliché around the time one billion stems per year came bursting from bodegas and cluttered up the entrance of every pharmacy. But with reflection, I would argue, there is still no better-suited plant to carry its significance. Observe the graceful little vine bearing a sea of speckled strawberries in spring, the bush so laden with luscious raspberries in summer it must be provided an architecture of supports, the tree that drips with crimson apples in autumn: these too are roses. How many plant families have been so profoundly shaped—so sweetened, so plumped, so scented, so densely bloomed, so richly colored—by the short-lived whims and deep currents of human desire?

Before their evolutionary history intersected with humanity's, most flowers in family Rosaceae followed a highly similar blueprint: five sepals and five petals set in radial symmetry and fused at their base into a single hypanthium (the part that swells, in *Rosa* species, into those notorious scarlet "hips" after flowering). Inspect the illustration's perimeter, showing many of our favored fruit- and nut-bearing Rosaceae, all species whose mouthwatering produce is more typically the focus of our desire than the blossom itself. In these, the flowers remain similar, more or less, as time, nature, and their many dalliances with insects shaped them.

However, where the flower has been the object of our global obsession, as in the central *Rosa*, we have rendered them unrecognizable in a millennia-long, high femme makeover. Historical evidence from China suggests that people have been adoring, breeding, and thereby shaping the lineage of roses for some 5,000 years. The gene for the lavishly "doubled" blossoms of our modern Valentine's Day icons emerged later, from the long labors of Syrian gardeners. The damask rose (*Rosa damascena*), with pale pink petals like enticingly rumpled bedsheets and a fragrance that launched a thousand perfumeries, is said to have made its way to Europe via the bloody, defeated retreat of the Second Crusade—a small, thorny, highly consequential spoil.

As the passion for roses spread across the world along warpaths and trade routes, taking on innumerable cultural, spiritual, culinary, medicinal, and quite often political significances along the way, so too did the desire for novelty, for longevity, for more, more, more. The China rose (*R. chinensis*), prized for its repeat flowering, was crossed with the prodigious *R. gigantea* to produce the first tea rose. Damask roses were later crossed with these floriferous roses to produce hybrid perpetuals. These were, inevitably, crossed with tea roses, creating hybrid tea roses, which have since conquered our hearts, gardens, and wallets in popularity.

This ardent profusion of rose cultivation has never ceased: patient plant breeders produce new hybrids; diligent antiquarian gardeners tend the cuttings of Old World roses dating back to medieval apothecary gardens; and still the fragrant wild roses continue on their own

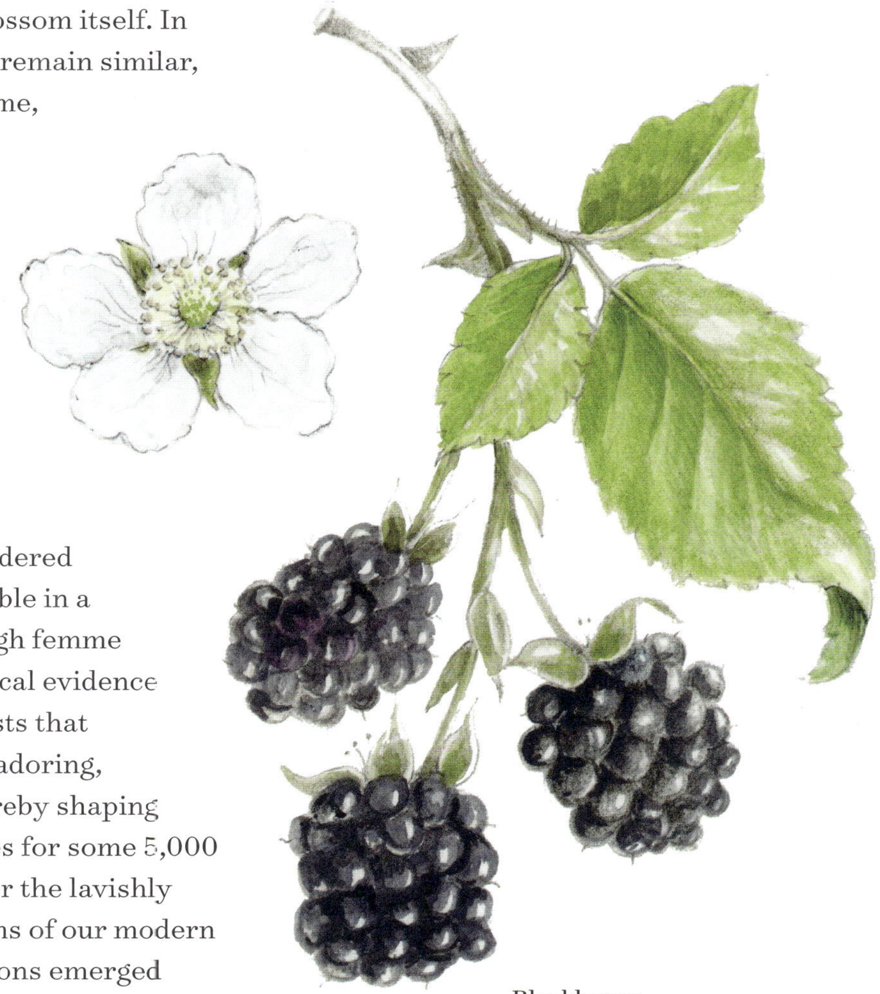

Blackberry (*Rubus allegheniensis*)

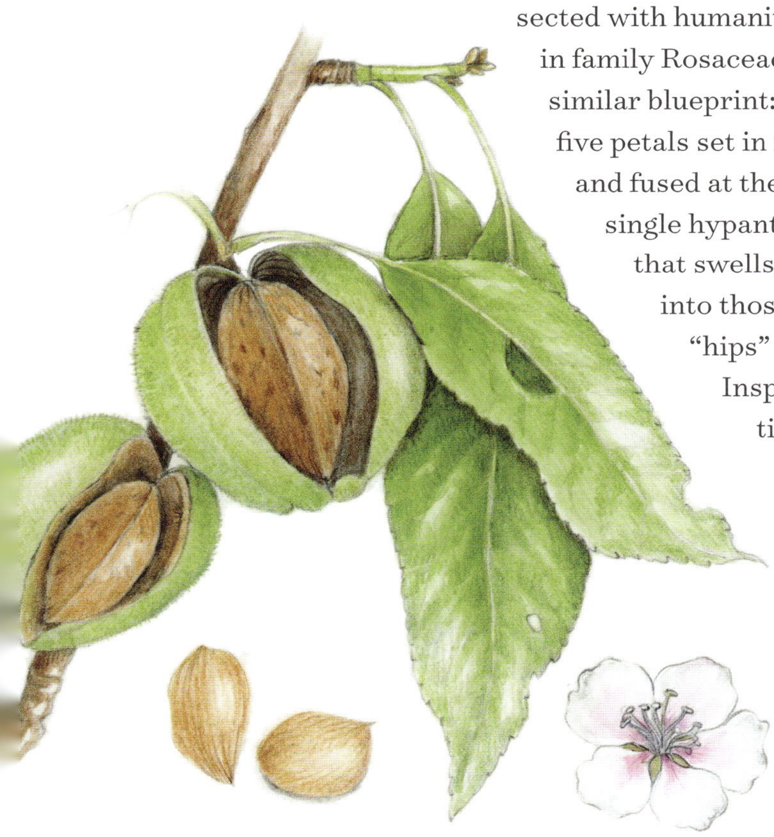

Almond (*Prunus dulcis*)

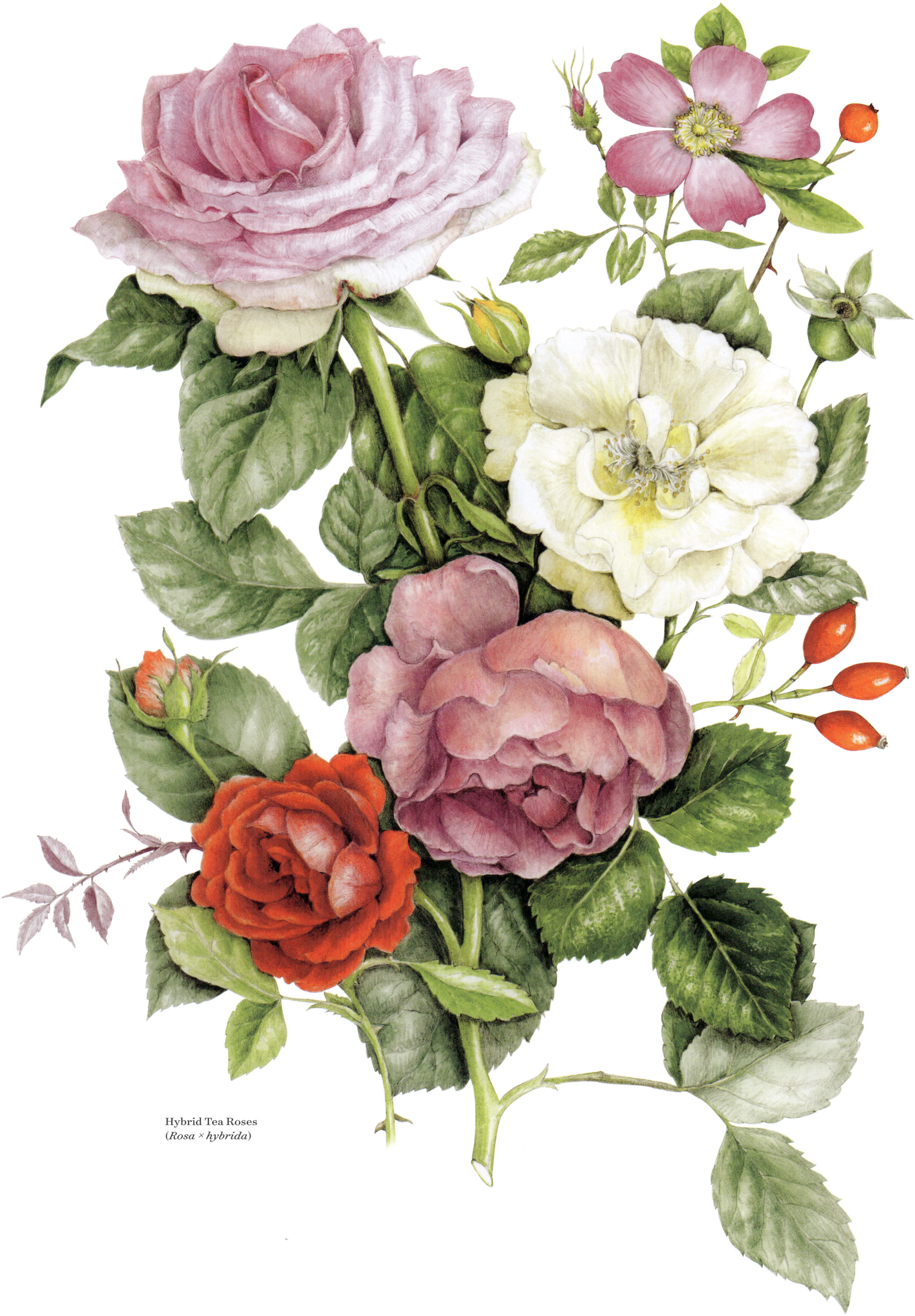

Hybrid Tea Roses
(*Rosa × hybrida*)

wild ways. And what of the fruits of family Rosaceae? Many people are shocked to learn that so many of the sweet fruits we grow in temperate regions today, as diverse in appearance and flavor as their flowers are similar, spring from the same branch of the tree of life. A majority of the drupes (including cherries, peaches, almonds, and plums), pomes (apples, quinces, and pears), and berries (blackberries, raspberries, and—not technically a berry—strawberries) that dominate the prototypical American's appetite for fruit are all in the family. And like their cousin *Rosa*, so too have these plants undergone umpteen transformations to suit our human desires.

In truth, anyone who has been lucky enough to browse on vine-ripened wild blackberries in a dappled understory knows that very little could improve upon the experience. However, take one bite of a wild crab apple followed by a bite of literally any apple you'll find at a farmstand, and you'll find our tastes have massively improved the irresistibility of Eve's temptation. Wild plums make delicious preserves (with sweetener to offset their powerfully sour punch) but can hardly compare to cultivated varieties—deep purple orbs dusted with waxy bloom that smudges like lipstick when you rub your thumb across it, so bursting with juice that the best culinary use for a pint's worth may be simply to devour them over the sink, one by one.

We have long understood other animals to be integral to the reproduction of many plants. Yet our demure deferral to "the birds and the bees" often obscures our own enduring passions at work. Does the desire we feel to feast upon a sun-warmed, peak-season peach at the pick-your-own (the interdiction only heightening the need) stem from our desire for the body parts the peach stands in for in lyric or glyph? Or did we work together, cross-culturally, generation after generation, millennium after millennium, to align our staple fruits and favorite flowers with our objects of desire? We have made ourselves indispensable in the sex lives of roses—most cultivars in our farms, orchards, and gardens could never grow true to type without us, and many have forfeited sexual reproduction altogether in favor of petals upon petals. But so too have they come to dominate us in mind and body, our hearts and appetites alike. As it has been to us from time immemorial, so it remains: rose, family of desire.

CHRIS BAKER is an author, professional mushroom lover, and founder of the Chicory Naturalist shop in Kingston, New York. She is a certified forager and enthusiastic guide to the realm of fungi, leading forays in the Hudson Valley bioregion. Running through all of Chris's professional endeavors is a passion for reconnecting broken chains of generational knowledge about the natural world. She lives a few blocks from the Hudson River with her spouse, child, and two calico cats.

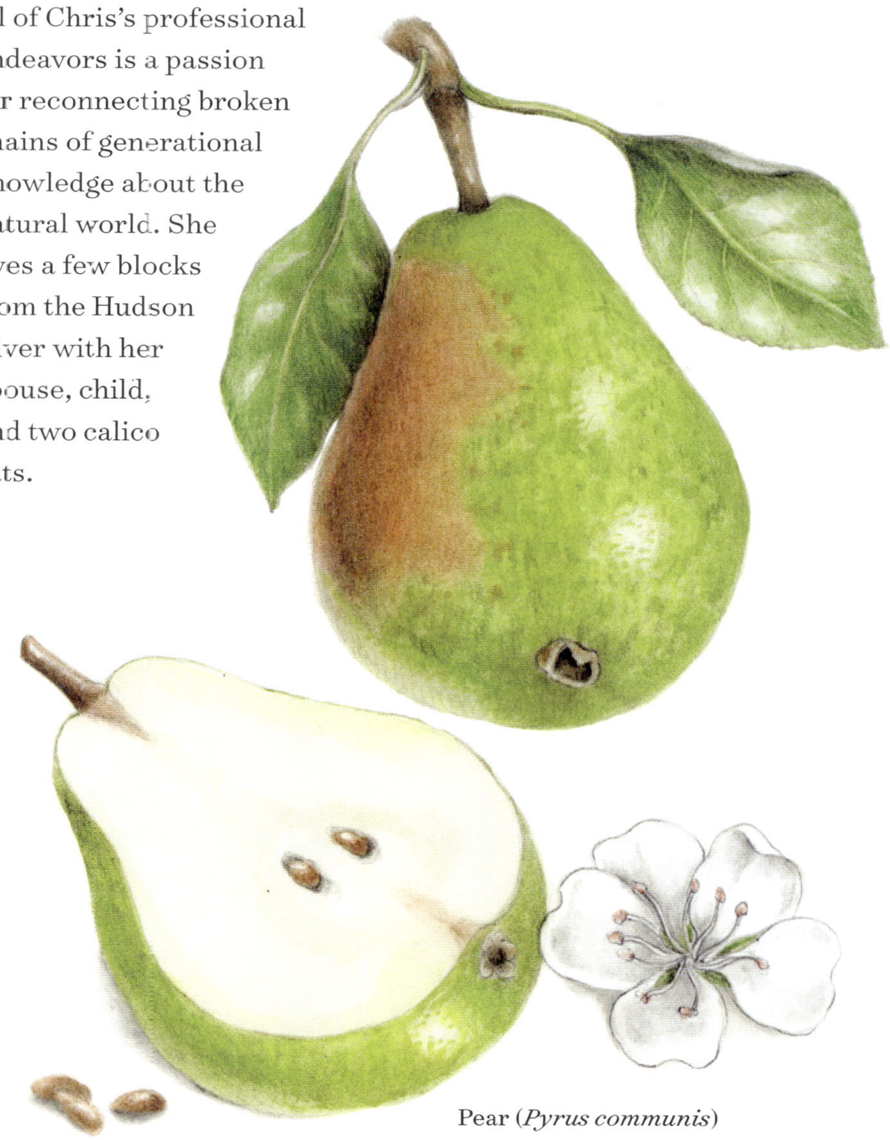

Pear (*Pyrus communis*)

# 21
# Soapberry Family

## Sapindaceae
## 1,700+ species

My journey with the soapberry family, Sapindaceae, started in early spring in the Hudson Valley. I think the very first sign that spring is about to erupt is the appearance of a red blush of color in the treetops visible when driving along highways or country roads of the region. These red maple flowers, tiny and hardly visible to the naked eye, are one of the earliest blooms of spring, emerging before the leaves. I enjoyed studying their decorative-looking red buds and then flowers as they opened under the microscope. Some trees have only male flowers, some only female, and some have bisexual flowers. The winged seedpods, which appear later, are called samaras. As kids we would open up a samara and stick it on our nose with the aid of the sticky sap on its surface. Many kids refer to these wind-borne seeds as helicopters for the way they spin while floating toward the ground. In the fall, red maple leaves turn a deep red and are glorious to behold.

Sugar maple is valuable for its sap, which is tapped in late winter to make maple syrup. It also yields a valuable hardwood, used for furniture and string instruments such as violins and guitars. I waited patiently for the sugar maple to flower, but last year, for whatever reason, the trees near me did not flower. I searched for the flowers and then later the winged samaras, but when no flowers appear, there will be no seedpods in the fall. I saw an Instagram post from a fellow botanical artist of her drawing of the sugar maple samaras in Virginia. I got excited and asked her to send me some to add to my painting. When the subject arrived, I studied the leaves and realized they were slightly different from my sugar maple leaves. I researched and discovered that this was not a sugar maple but a close relative, the black maple—so close that some botanists classify it as a sugar maple subspecies, *Acer saccharum* subsp.

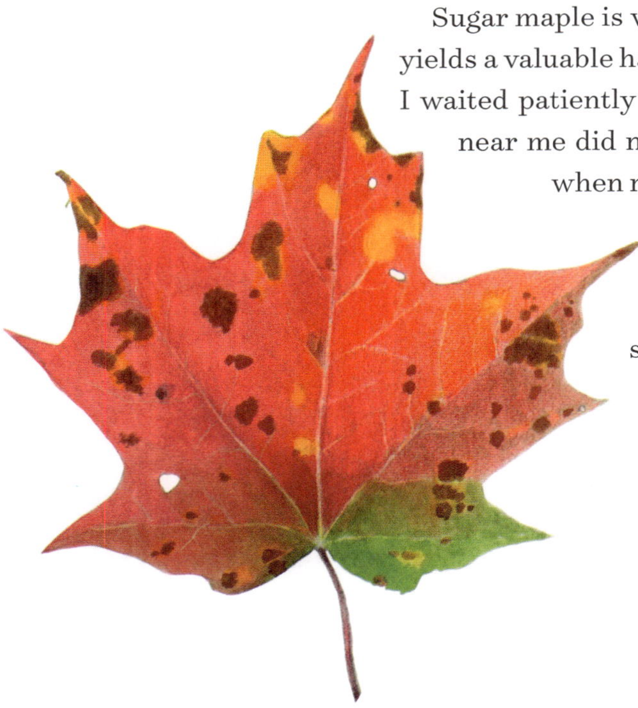

## Plants Illustrated

*Acer nigrum*, Black Maple
*Acer pectinatum*, Snake Bark Maple
*Acer rubrum*, Red Maple
*Acer saccharum*, Sugar Maple
*Aesculus hippocastanum*, Horse Chestnut
*Blighia sapida*, Ackee
*Dimocarpus longan*, Longan or Dragon's Eye
*Koelreuteria paniculata*, Golden Raintree
*Litchi chinensis*, Lychee
*Majidea zanguebarica*, Mgambo Tree
*Melicoccus bijugatus*, Spanish Lime
*Nephelium lappaceum*, Rambutan
*Sapindus saponaria*, Soapberry

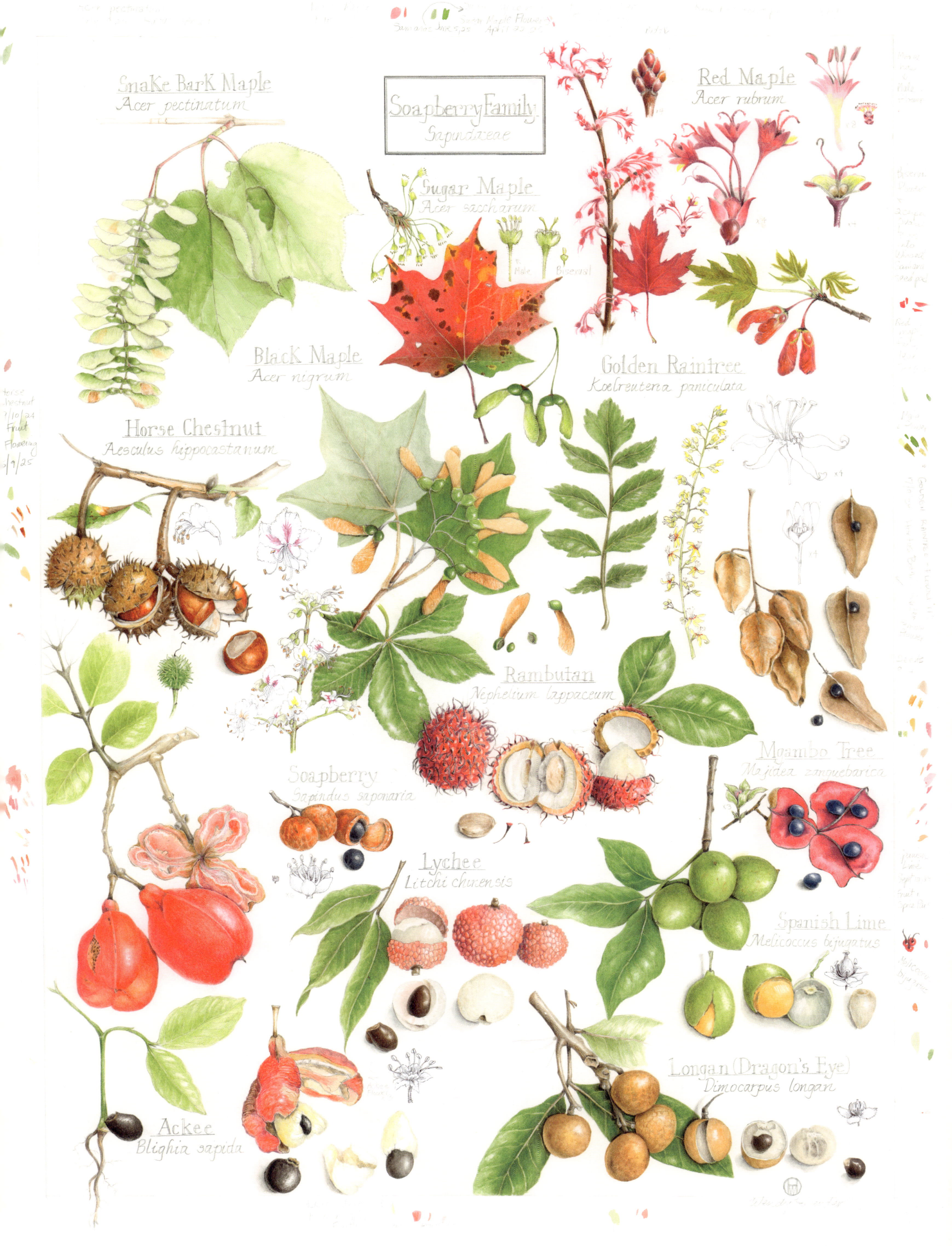
Soapberry Family
Sapindaceae
Snake Bark Maple
Acer pectinatum
Sugar Maple
Acer saccharum
Male
Bisexual
Red Maple
Acer rubrum
Black Maple
Acer nigrum
Golden Raintree
Koelreuteria paniculata
Horse Chestnut
Aesculus hippocastanum
Rambutan
Nephelium lappaceum
Soapberry
Sapindus saponaria
Mgambo Tree
Majidea zanguebarica
Lychee
Litchi chinensis
Spanish Lime
Melicoccus bijugatus
Longan (Dragon's Eye)
Dimocarpus longan
Ackee
Blighia sapida

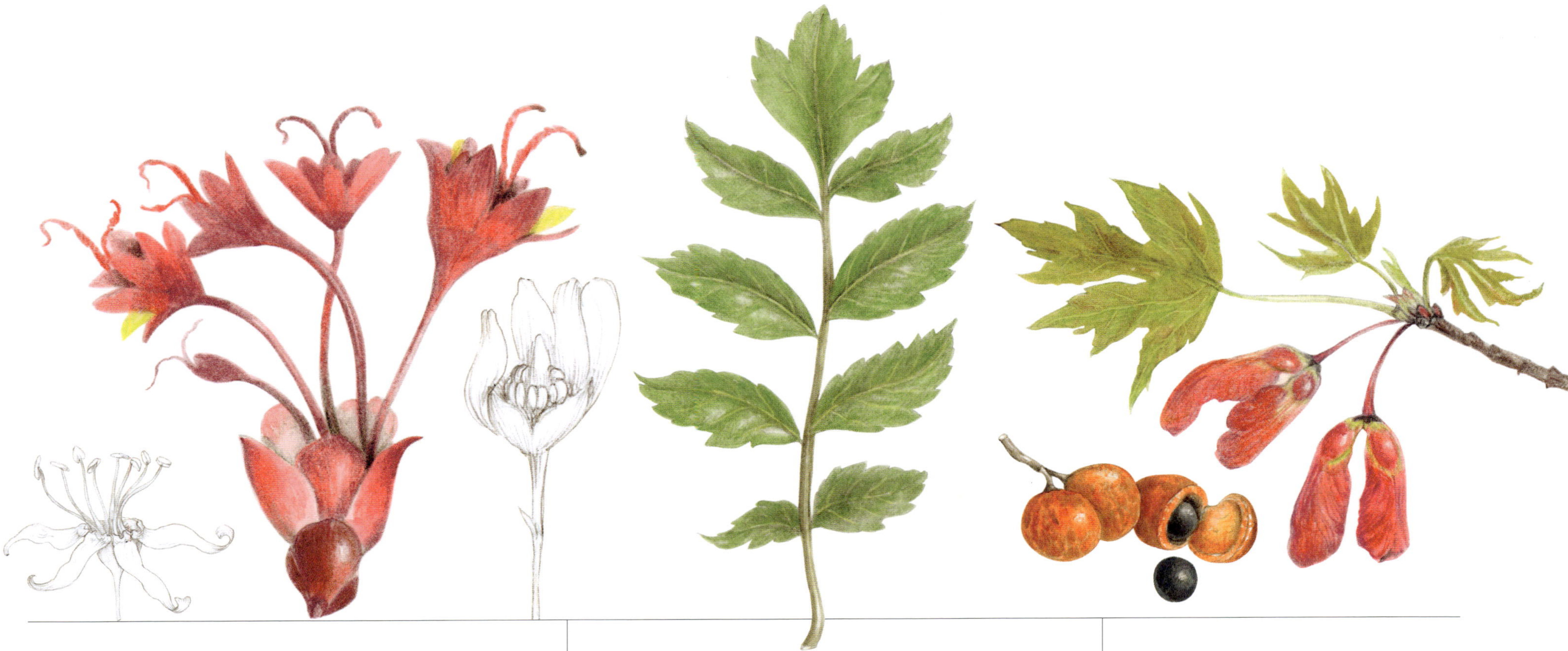

FLOWERS

4 or 5 sepals
4 or 5 petals
4–10 stamens
Bisexual or unisexual

LEAVES

Usually compound

FRUITS

Variable; sometimes winged samaras or nutlike seeds

*nigrum*. I was proud that I was able to notice the difference. As I write, it is the beginning of March, so I am going to study the sugar maples again this spring and see if they flower and fruit. Sugar maples are also known for their beautiful fall color, so at least I was able to include a leaf here.

Horse chestnut trees grow in my area and produce a spiky pod with a shiny seed inside. This is another fruit that I imagine kids love to play with. In Great Britian and Ireland they play a game called conkers using the seeds.

The top half of my painting was reserved for the temperate plants in this family, and the bottom half was home for some of the tropical species. In late September, I went to Canal Street in Chinatown in New York City in search of tropical fruits in the Sapindaceae family. They are available there regularly, often with accompanying stems and leaves from the trees. I was looking for rambutan, lycee, and longan.

The first market I saw had many unusual fruits and vegetables but only one of these fruits. I approached the grocer and asked if he had any of the other fruits, showing him photos on my phone as he did not speak much English. He said no, this is not the season, and you won't find them. I left determined to check further. Soon I reached the heart of Canal Street, where many outside vendors' stands were laden with fruits piled high. Across the wide avenue I spied the bright red and fuzzy exterior of what I hoped was rambutan. Years of searching for plants at great distances came in handy here. I remember asking a botanist once known for climbing cliffs and rappelling down on ropes how he looks for plants. He said that it starts with searching for color, and I now know that to be true with my own experience. Sure enough, when we crossed the street, there they were: rambutan, snuggled up next to lychee, and the longan nearby. Back in the studio, with the precious fruits preserved in a cooler, I got to work. These fruits have unusual patterns, each different from the other but in some ways similar. Sometimes, after several layers of color and texture, I notice the pattern emerging on the paper, as if the paper is drawing the pattern and not me. It can start to appear, and then I just have to emphasize it in places. It feels like magic.

Florida is a good place to find tropical fruits that have origins all over the world. The Fruit and Spice Park in Homestead grows over 500 different tropical fruits, the plants arranged in regions from which they originated. The trees are accessible to all, and while you are

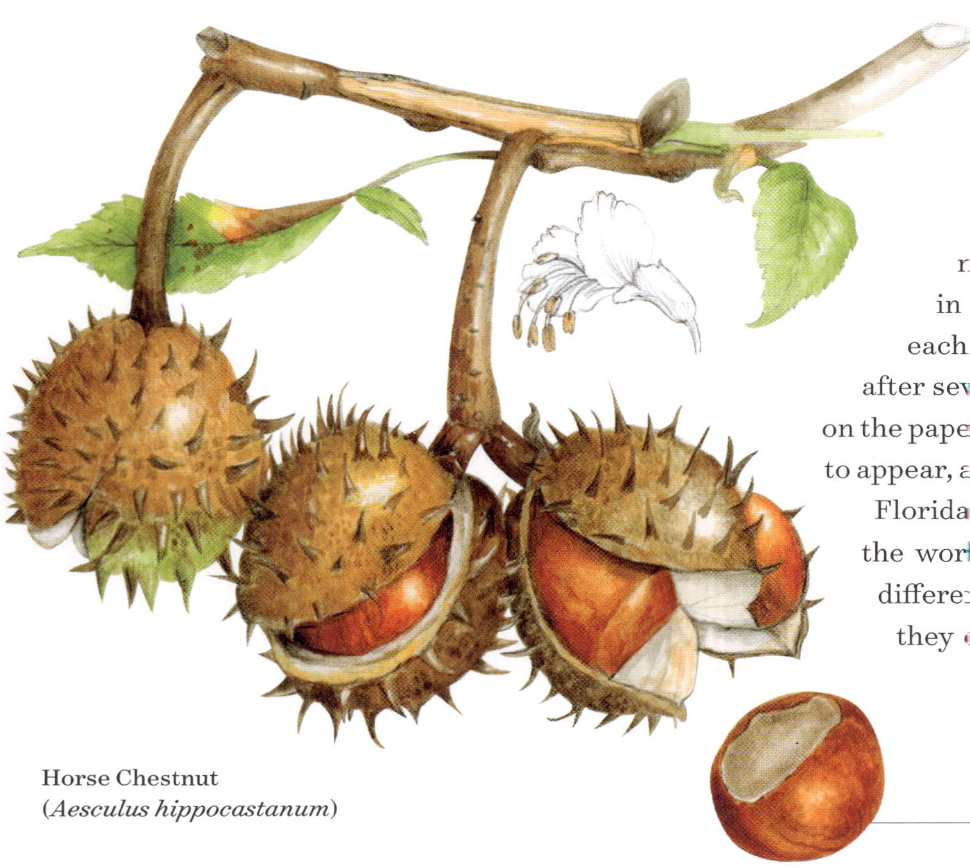

Horse Chestnut
(*Aesculus hippocastanum*)

allowed to pick up fruit on the ground to try, you are not supposed to cut any off the trees. But a shake of a tree is encouraged. The garden's tour leaders take you around on a tram and give out interesting tidbits of ethnobotany about the plants in two languages, Spanish and English. Spanish is a big part of the culture in south Florida, as this region is one of the agricultural centers of the state, with a large number of immigrants who do the hard work of farming. The land is very flat, and you can see for miles and miles in every direction. What I notice is the diversity of crops—not all one crop growing but many, interspersed between annuals and perennials, herbs and trees. Plant nurseries providing trees, shrubs, and orchids abound. There are also tropical fruit stands where you can sample many of the unusual varieties of fruits grown here.

I started my journey to south Florida with a family vacation off the west coast on a small island called Captiva. In the past two years, three tropical storms/hurricanes have gone through and ravaged this part of Florida, leaving patches without vegetation. When we arrived, the island was mostly up and running, but it was hard not to notice how barren the landscape looked and felt without the usual abundance of lush tropical plants. From Captiva, I drove three hours across the state to the agricultural region of Homestead. As I drove across the Everglades, the tropical plants started to reappear, and it was a joy to be back in a lush vegetative environment. This is something I am grateful for every day and do not take for granted. Without plants, we could not survive; they are the structure upon which we humans rely for food, sustenance, resources for building and clothing, and the oxygen we breathe.

Once in Homestead, I was able to see some of the tropical fruit trees in this family. The ackee was fruiting, and I added it to my painting. This unusual fruit is poisonous until the moment it naturally splits open on the tree. Then you can harvest the fruit and cook the creamy, brainlike flesh that covers the black seeds. It is prepared with salt fish in the national dish of Jamaica.

After Florida, I headed to Hawai'i, where I was able to add a plant from the genus that gives the soapberry family its name. The fruit of soapberries exudes a liquid that can be used as a soap. If you are curious about the purpose of this soapy liquid, the plant uses it to protect the seeds from insects and microbes. You've got to love nature's solutions to problems. I also was able to get herbarium specimens of some of the family's tropical fruits from the National Tropical Botanical Garden's herbarium to add to my paintings. Herbarium specimens are plants that are collected, pressed, dried, and mounted onto paper to preserve the materials for future use. One use is to help the botanical artist to illustrate a plant when fresh material is not available.

I patiently awaited one more growing season to finish this painting—first to see if the sugar maple produced flowers this year, and secondly to add in the gorgeous horse chestnut flowers when they bloomed in the spring. These flowers start out yellow and turn red when they are pollinated—just another amazing example of ingenious Mother Nature at work, directing the insects to the unpollinated flowers.

As spring blossomed in 2025, I watched my sugar maple trees closely. It was exciting because indeed the flowers emerged, and I was able to study them in detail under the microscope. The flowers hang down on long skinny stems in clusters reminding me of dangling earrings. I seemed only to have male flowers, but many nearby trees had bisexual flowers as well. As my tree's flowers withered, and the leaves started to appear, I could see that samaras were indeed developing as well, meaning that my tree must have had both male and female flowers. I certainly learned a lot about these tiny flowers that we often don't notice. I am sure that each year I will continue to look closely at these trees as spring emerges.

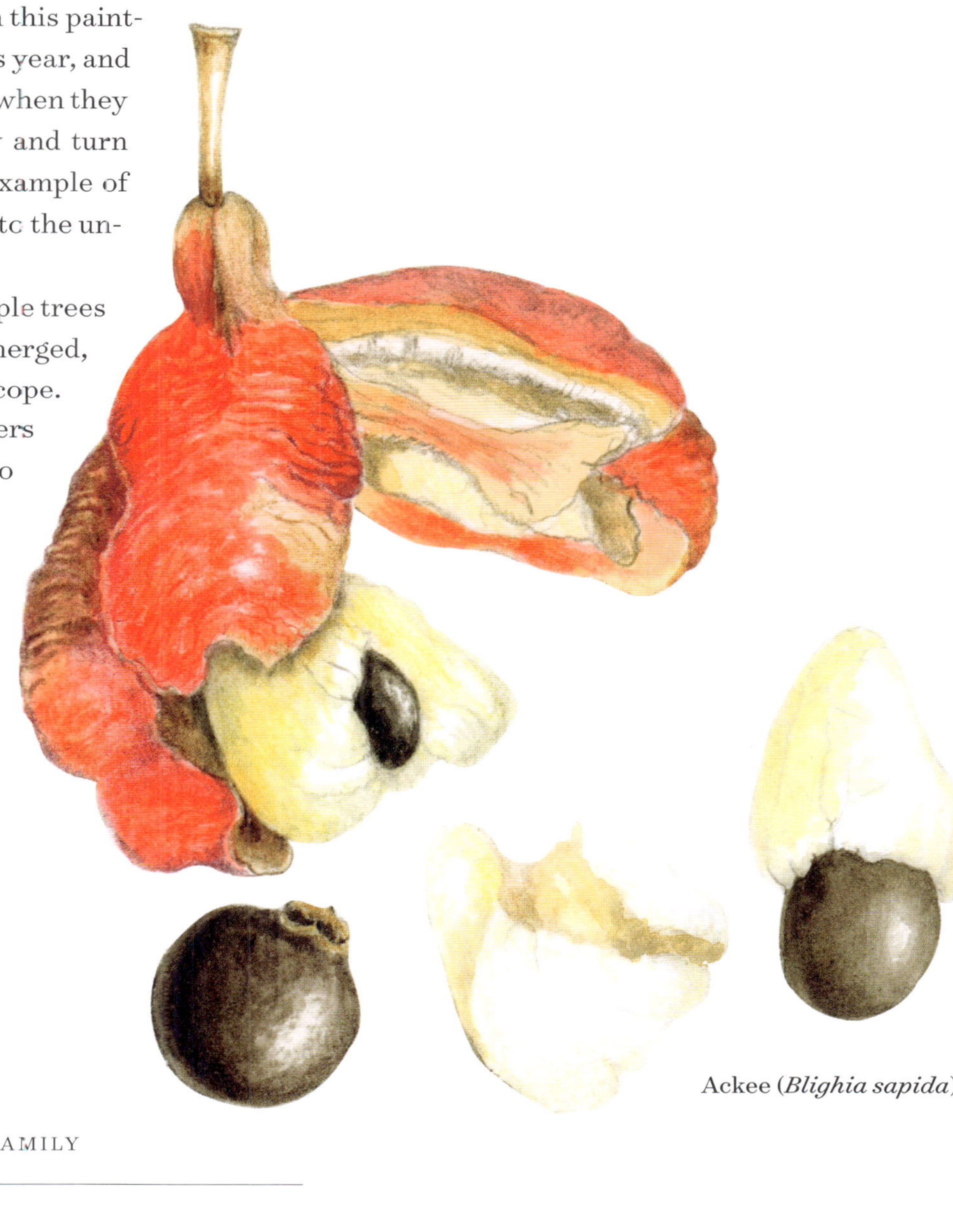

Ackee (*Blighia sapida*)

# 22
# Sunflower Family

## Asteraceae
## 30,000+ species

The sunflower family (Asteraceae) is the largest family of flowering plants, with over 30,000 individual species, surpassing the orchid family (Orchidaceae), with over 28,000 species. Some of the smallest flowers and largest flowers are in the Asteraceae family. Sunflowers can grow as tall as 30 feet, and one composite flower head can be 12 inches in diameter and weigh close to 5 pounds when laden with seeds.

This is a good family to start with for identification purposes because of its distinctive flower head, which basically looks like a daisy or a dandelion. When I think "daisy," I picture a simple flower, but in reality this family has some of the most complex flowers out there. What most of us consider a single daisy flower, with its round center surrounded by petals, is a complex composite flower head comprising often hundreds of tiny individual flowers. The center is composed of tightly packed disc flowers (or florets), while the outer "petals" are individual ray florets. Each disc floret has a male and a female part and turns into an individual seed, allowing the whole flower head to house hundreds of seeds, each capable of becoming a new plant.

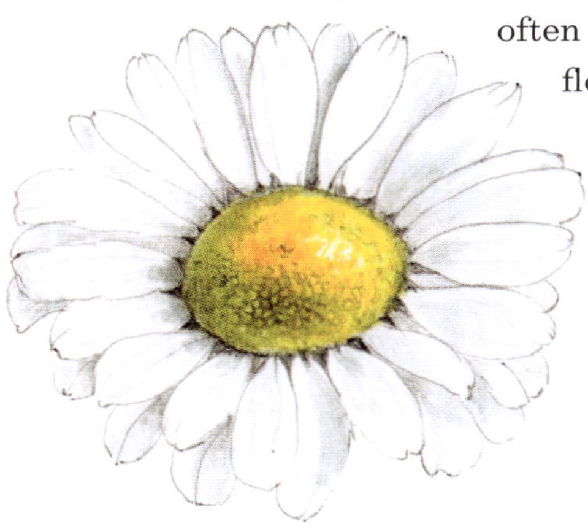

The life cycle of the dandelion can be quite rapid, from my observation, progressing from flower to seed within five days. Sometimes I would cut a dandelion in flower and bring it into my studio, and within a few days it would have turned into a seed head. The dandelion seed head is the feathery, fuzzy ball that many of

## Plants Illustrated

*Artemisia vulgaris*, Mugwort
*Cichorium intybus*, Chicory
*Coreopsis lanceolata*, Coreopsis
*Cynara cardunculus*, Cardoon
*Cynara scolymus*, Artichoke, Purple Globe
*Dahlia*, Dahlia, Natalie G
*Echinacea purpurea*, Purple Coneflower
*Helianthus annuus*, Sunflower
*Helianthus tuberosus*, Jerusalem Artichoke
*Lactuca canadensis*, Wild Lettuce
*Lactuca sativa* var. *longifolia*, Romaine Lettuce
*Leucanthemum vulgare*, Oxeye Daisy
*Liatris spicata*, Blazing Star
*Taraxacum officinale*, Dandelion
*Tragopogon dubius*, Yellow Goat's Beard
*Tussilago farfara*, Coltsfoot
*Zinnia elegans*, Zinnia

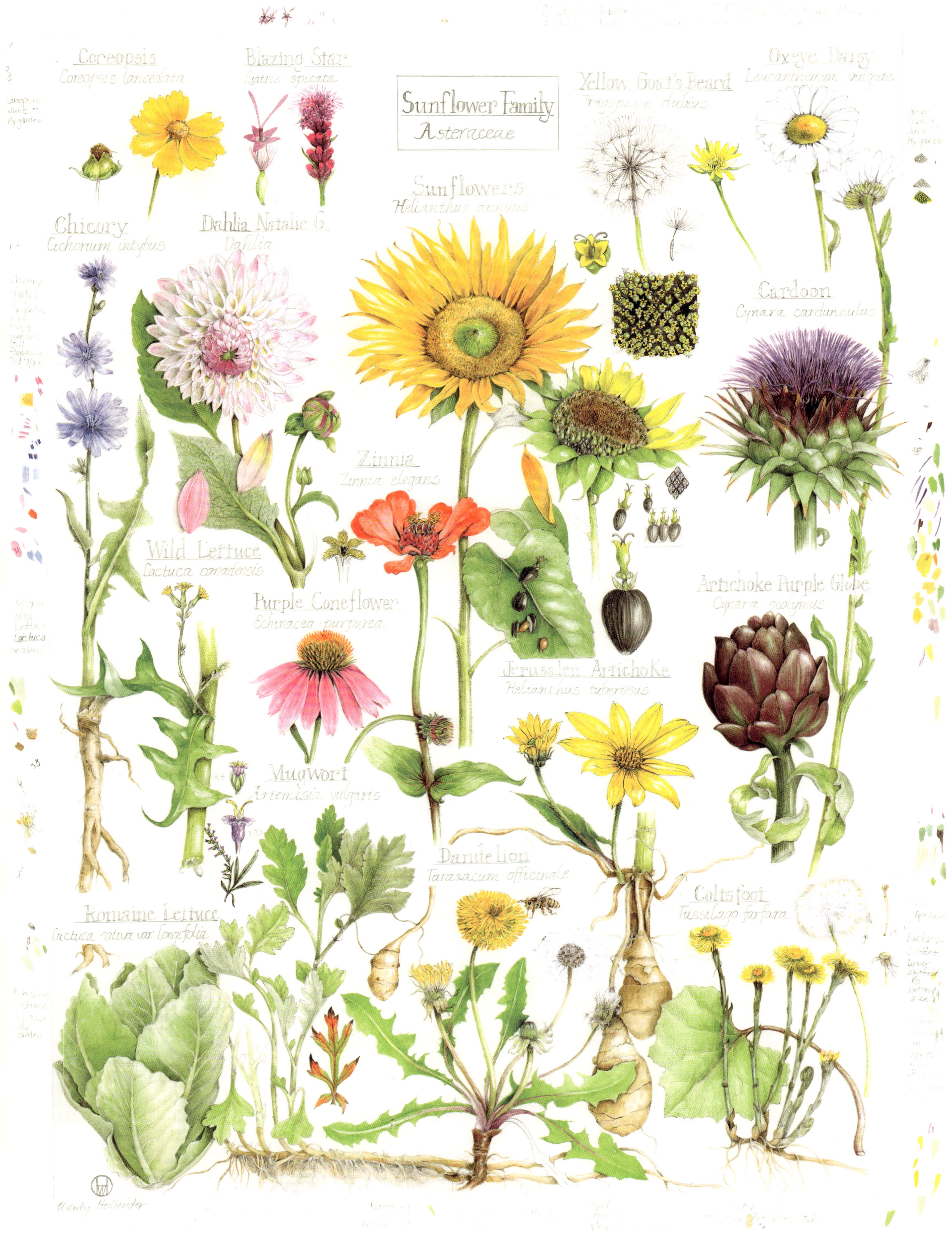

Sunflower Family
Asteraceae
Coreopsis
Coreopsis lanceolata
Blazing Star
Liatris spicata
Yellow Goat's Beard
Tragopogon dubius
Oxeye Daisy
Leucanthemum vulgare
Chicory
Cichorium intybus
Dahlia Natalie G
Dahlia
Sunflowers
Helianthus annuus
Cardoon
Cynara cardunculus
Zinnia
Zinnia elegans
Wild Lettuce
Lactuca canadensis
Purple Coneflower
Echinacea purpurea
Artichoke Purple Globe
Cynara scolymus
Jerusalem Artichoke
Helianthus tuberosus
Mugwort
Artemisia vulgaris
Dandelion
Taraxacum officinale
Coltsfoot
Tussilago farfara
Romaine Lettuce
Lactuca sativa var longifolia
Wendy Hollender

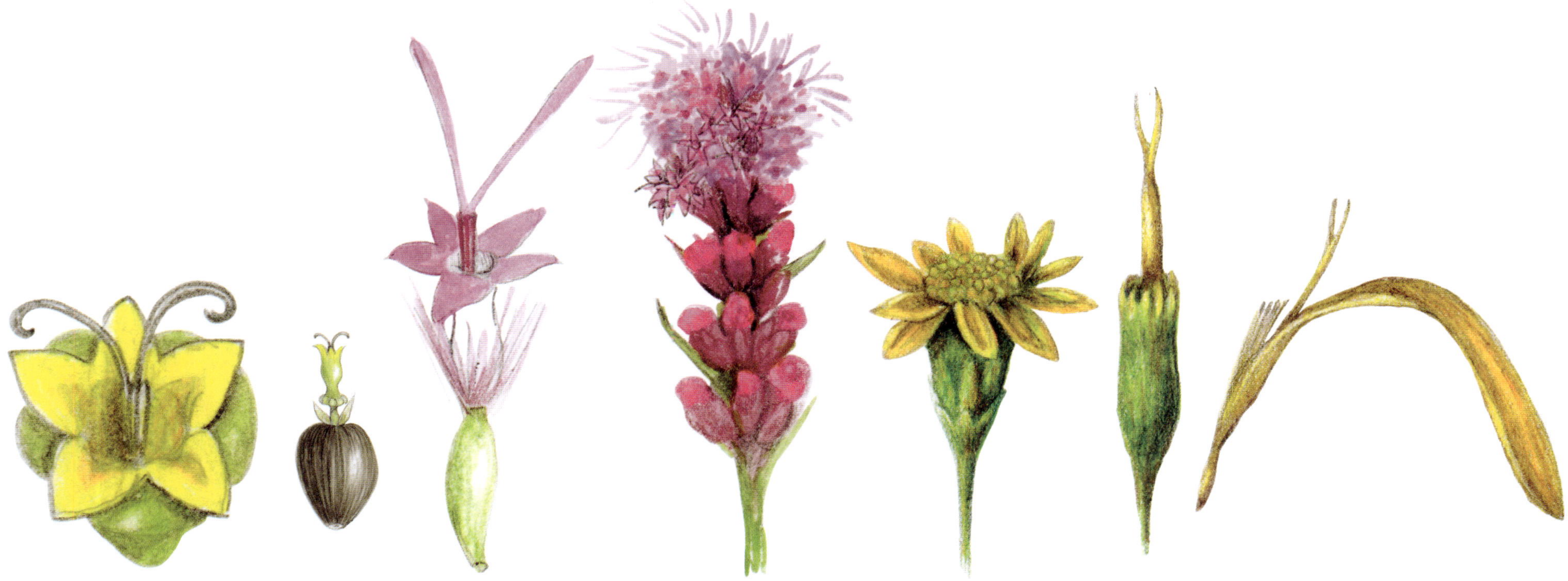

FLOWERS

In composite flower heads
Central flowers (disc florets) develop into seeds

Large outer petals are ray florets
Flower heads may be all disc or all ray florets

us pick to make a wish with. Is this the dandelion's way of getting us to help it spread its seeds? The seeds are wind-dispersed, each one attached to its own "parachute."

Coltsfoot comes up first in my area. It is easy to confuse coltsfoot with dandelion, but they are different looking if you get up close. The coltsfoot flower is not attached to the leaves; its leaves come up later, whereas the dandelion's basal rosette of leaves comes first, and the flower develops in the center of these leaves (which, by the way, are tender and delicious to eat in early spring).

I started my painting at the bottom of the sheet with the low-growing coltsfoot and dandelion, highlighting the runner-style root system of the coltsfoot and the center taproot of the dandelion. From there I moved to the top of the paper to add in the meadow wildflowers that I knew would bloom soon. Coreopsis and the classic-looking oxeye daisy arrived. I planned ahead on this painting, knowing I would be including artichoke, cardoon, and the late summer and fall–blooming zinnias, dahlias, coneflowers, and sunflowers. I knew sunflowers were going right in the center. The first sunflower I chose was a hybrid that did not develop seeds, so I left room for a sunflower head that would go to seed, adding the developing seeds in the center of its complex flower head.

I drew my artichoke and cardoon in Spain, using plants growing at the magical Garden of Lemons in Andalusia, where I take students on a botanical drawing adventure in the spring (see a sample recipe, p. 97). It is a privilege to eat directly out of

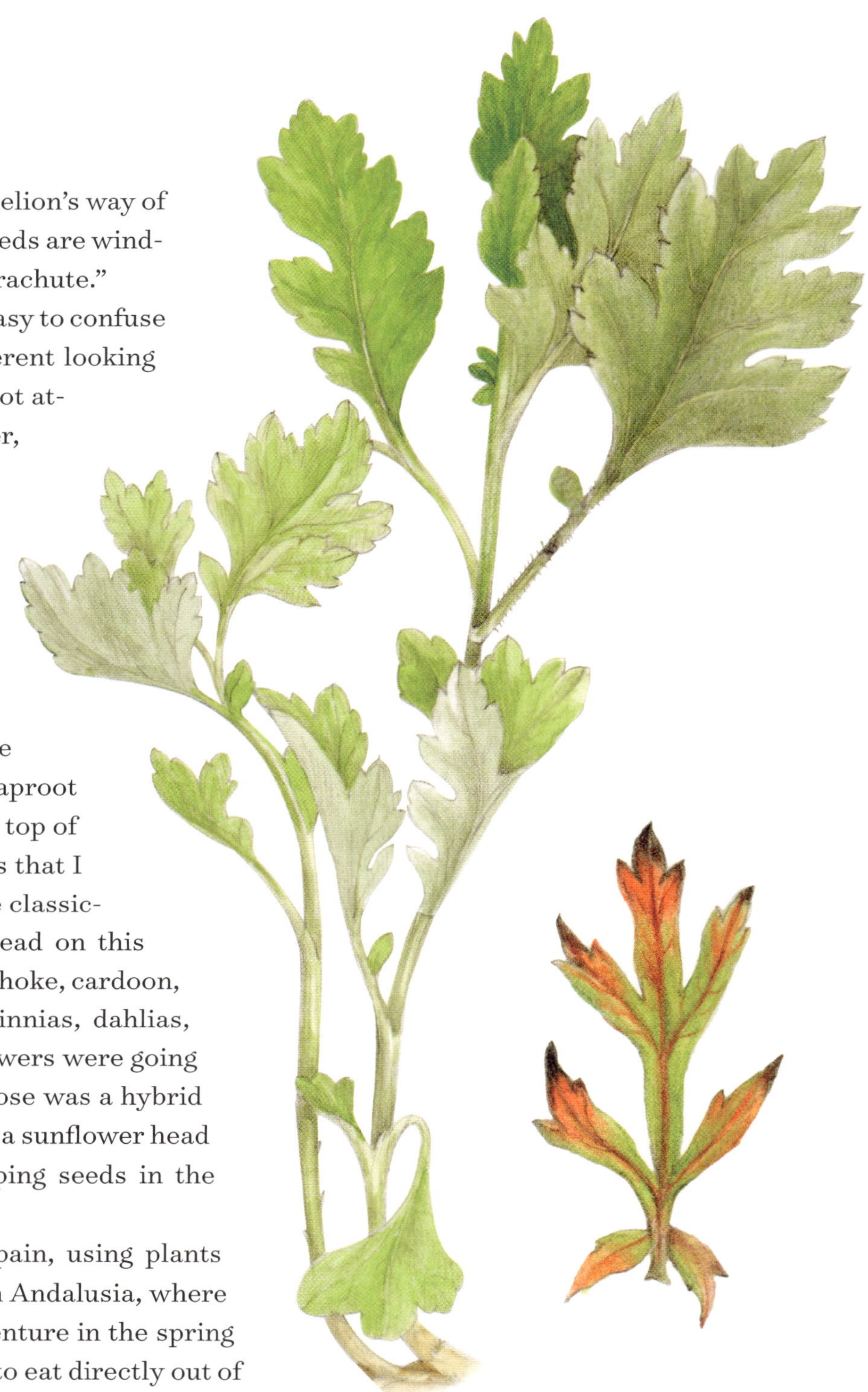

Mugwort (*Artemisia vulgaris*)

a garden, and I am thankful to get this opportunity not only when I go to Spain but at home from my own garden and local farms.

The roots and tubers of chicory, Jerusalem artichoke, and dandelion are used as food and medicine. Lettuce is also in this family, which may come as a surprise, but if you let some of your cultivated lettuces go to flower, you will see the tiny daisylike flowers appear. These gifts are readily available everywhere in my garden, and if nothing else, the bees are able to get so much nourishment from these plants.

I carefully chose my flowers by color for this painting. Yellow is prevalent in the sunflower family, and I consciously decided to include purplish flowers to contrast with the yellow, as exhibited by cardoon, blazing star, and dahlia. Yellow and purple are complementary, meaning they are opposite each other on the color wheel. Complementary colors are used together to show contrast. Fields of yellow goldenrod mixed with purple asters offer a beautiful version of this in fall. The wildflower fields are abundant where I live, and it is a treat to see the continuously changing march of wildflowers blooming throughout the seasons.

One of the most invasive plants in my garden, and everywhere else it grows, is mugwort. I can't avoid it, but recently it surprised me when I was looking for flowers in October. In my garden it grows alongside goldenrod, and sometimes it is hard to distinguish between the two plants. When the goldenrod goes to seed, and its flowers lose their distinctive bright yellow color, they start to look like mugwort. I notice, though, that the mugwort's distinctive leaves are a clue to the difference. Today my mugwort is putting on a late season show as its leaves start to turn a beautiful pinkish-peachy color, making it very pretty to look at. It is as if the mugwort is saying to me, *we aren't all bad*. The mugwort is referring to how I often feel about it, as the mugwort and I have been having an ongoing discussion for years now.

Me: *Mugwort, you just don't know your place in my garden. You keep pushing everyone else out. Why can't you just get along better and allow other plants to grow near you? If you didn't take over so much, I would be happy to let you stay.*

This discussion is starting to remind me of when I am with my grandchildren, especially the two siblings. When we go out together for our weekly Taco Tuesday dinner, Lucy (six years old) and Toby (four) are always competing for my attention or just letting the world know that they want to be first, and they don't want to share that toy. I am wondering if there is a lesson here for either the grandchildren or the mugwort. . . .

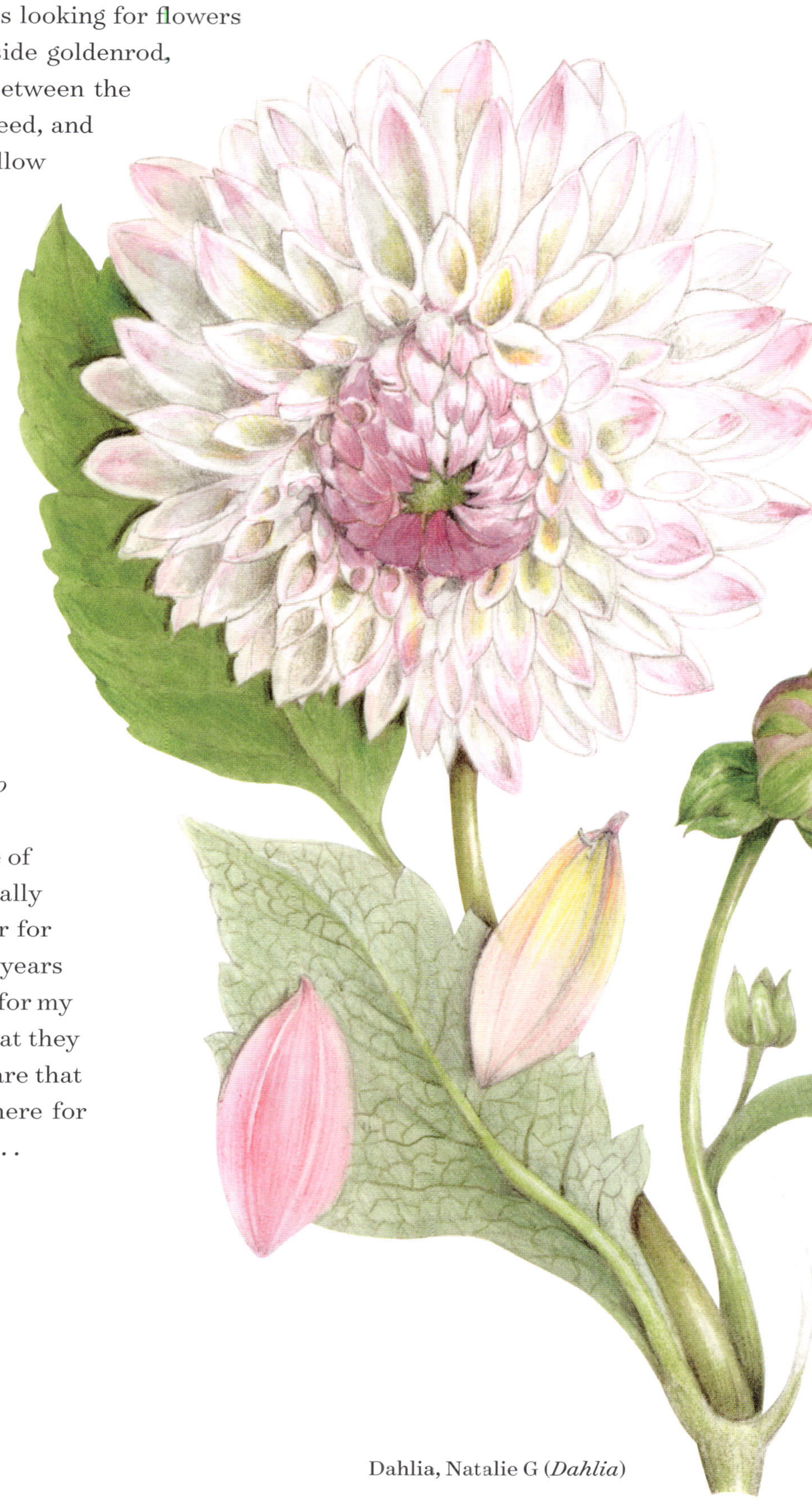

Dahlia, Natalie G (*Dahlia*)

# An Asteraceae Reunion

BY K GREENE

All plants have families. Some have really big families, like the Asteraceae. Commonly known as the daisy or sunflower family, Asteraceae is one of the largest families of flowering plants, encompassing around 30,000 species across more than 1,600 genera. It's a sprawling, diverse clan that's been around for over 100 million years, each member offering something unique to the world.

Imagine yourself at the Asteraceae family reunion. There's your roaming cousin Ambrosia, wild and free like ragweed. Your fabulous uncle is as elegant as ever with a Queen Lime zinnia adorning his lapel. Your great-grandparents give you chamomile-scented hugs that feel like good medicine, while your sibling's child, an aspiring young culinarian, brings a bowl of homegrown baby radicchio salad to the picnic. There's even your third cousin twice removed, who finds joy in spreading poisonous snakeroot rumors, but fortunately, they're easy to avoid. Despite their differences, all these relatives share something remarkable in common. Each seemingly individual bloom of a zinnia, lettuce, dandelion, or ragweed holds an often overlooked tight-knit internal community. I love this commonality because it allows me to say one of my favorite flower expressions: *inflorescence*.

Botanically speaking, an inflorescence is the particular arrangement of a cluster of flowers on a floral axis. In the Asteraceae, this arrangement is called a capitulum—a cluster of sessile (stalkless) flowers springing up from a level base. While it may look like a single flower to our eyes, each bloom is actually made up of many smaller flowers—disc florets and ray florets—nestled together in a way that gives the impression of one cohesive whole.

Take *Helianthus annuus*, for example, the common sunflower, which is anything but common. The sunflower would be one of the tallest and most visible plants at the reunion. Everyone would want that radiant face turned in their direction. Its broad flower heads beam like sun rays containing a miniature cosmos of interconnected life. Its unique centripetal arrangement of florets spirals inward, each one full of pollen and the potential to nurture a new seed. The 1,000 or more disc florets at the center resemble uncountable stars, tightly clustered in a golden embrace. Surrounding this celestial core, the ray florets stretch outward like sunbeams, inviting the world in. Each individual floret is a testament to the enduring strength of creation, a seamless, nonbinary perfection blending male and female, enabling self-pollination and extending invitations to pollinators. Beneath this vibrant display, the bracts cradle, protect, and support. The peduncle, or stalk, sturdy and steadfast, connects the vibrant crown to the earth.

It's no wonder the sunflower has become a hopeful symbol for so many. From one seed, its journey of seeking light brings joy, nourishment, and the promise of many more seeds. All the elements convene as a whole so that one seed can blossom into exponential generosity in the form of hundreds, if not thousands of seeds to sow next season. And each of those seeds radiates out and creates the same abundance, again and again.

There are lessons in seeds we can seek to emulate. A sunflower, or the inflorescence of any one of the guests at our expansive Asteraceae family reunion, transcends the idea of the individual. These plants and the ways they've thrived over millions of years remind us that we are all part of something greater—a collective spirit composed of families, chosen families, and communities that support and nurture us. Just as the Asteraceae family, despite its vast diversity, finds harmony through its collective spirit, we too can find strength in community. Our webs of support thrive when we come together, each of us contributing something invaluable to the whole. Host an Asteraceae reunion in your garden. Look closely at the parts, greet them lovingly, and then revel in the beauty of the whole.

K GREENE is a seed being. And so are you. K's seed journey began twenty years ago when they created the first seed library in a public library. Over time, Greene and their partner, Doug Muller, turned the library into the Hudson Valley Seed Company—a national seed company focused on the art, culture, and improvement of open-pollinated and organic seed. Greene also created the Seed Growing program at the Hudson Valley Farm Hub, where they partner with Indigenous and cultural organizations to support seed rematriation and sovereignty work. Along with these seed relationships, Greene is developing a nonbinary approach to botany to question and reimagine how we relate to plants and participate in generational seed cycles. Greene continues to teach seed-saving skills and enjoys bonding with new plants every season.

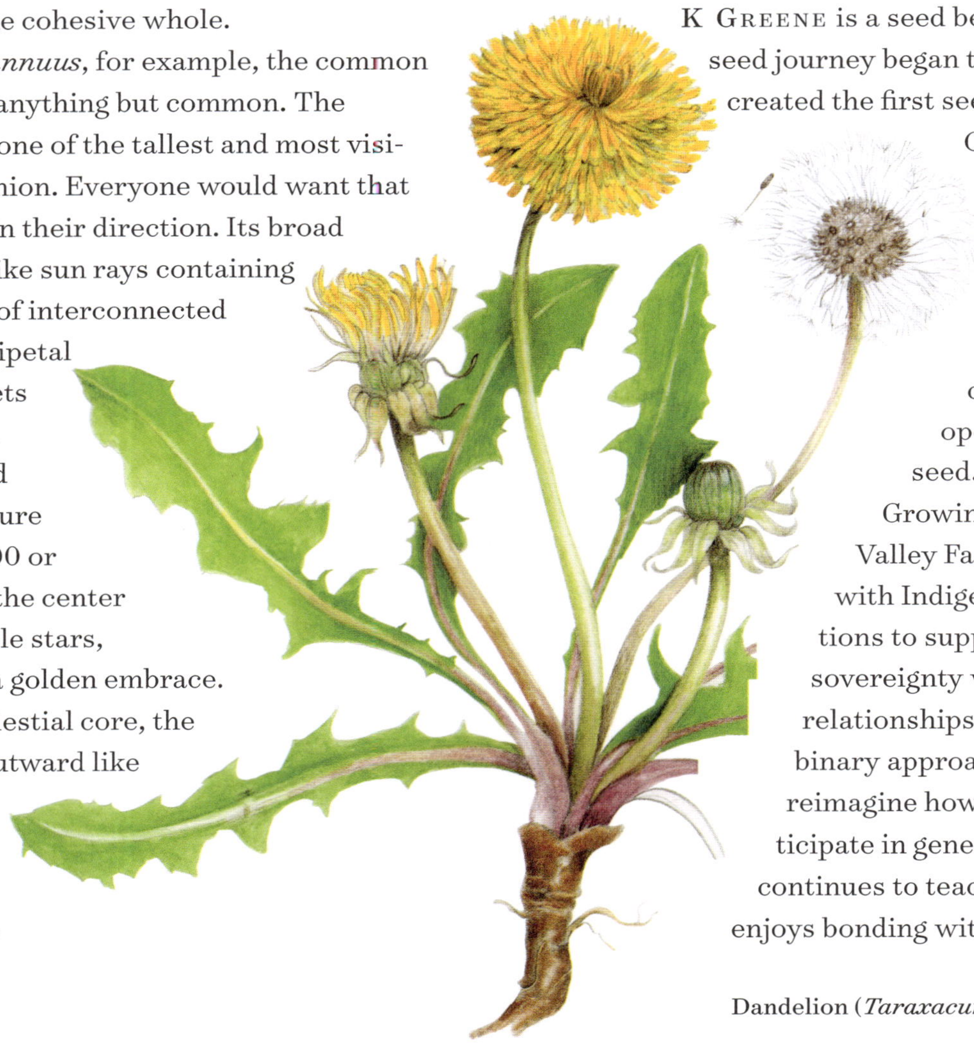

Dandelion (*Taraxacum officinale*)

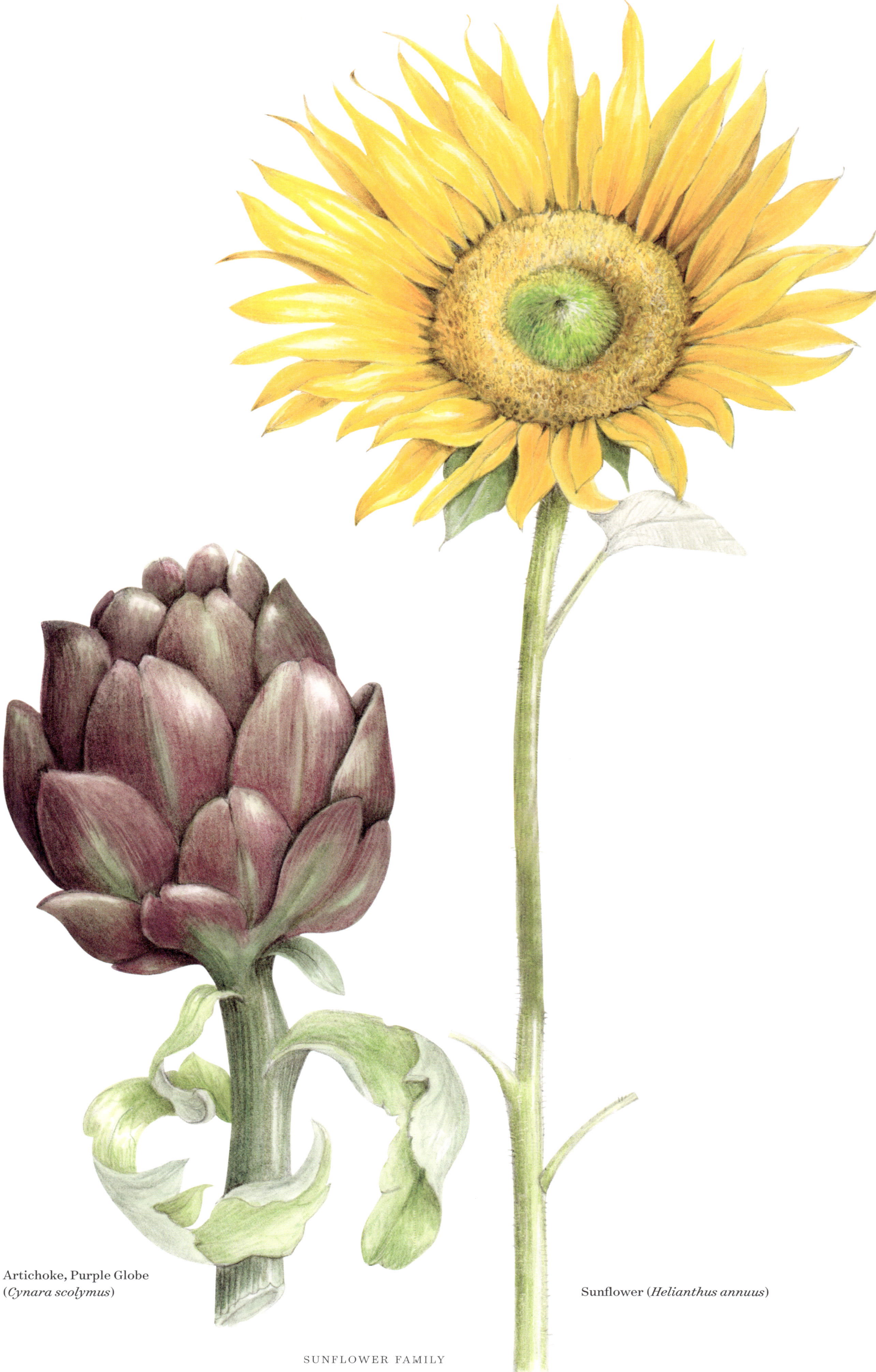

Artichoke, Purple Globe
(*Cynara scolymus*)

Sunflower (*Helianthus annuus*)

# 23

# Mushrooms of New York

## April–August 2023

### Fungi (Fungus Kingdom)

Mushrooms, part of the Fungi, or fungus kingdom, are not considered plants, but they are closely tied to the plant kingdom, literally, as mushroom mycelia intertwine with the roots of trees and plants underground. The mushrooms presented in my two paintings are not organized by family because I did not know what mushrooms I might find in a particular year. I followed mushrooms through one complete growing season in my temperate environment and added all the interesting ones I found, or that were found for me by the local mushroom enthusiasts who helped me track down these most elusive subjects. Fortunately, it was a rainy year, and there was an abundance of mushrooms to choose from.

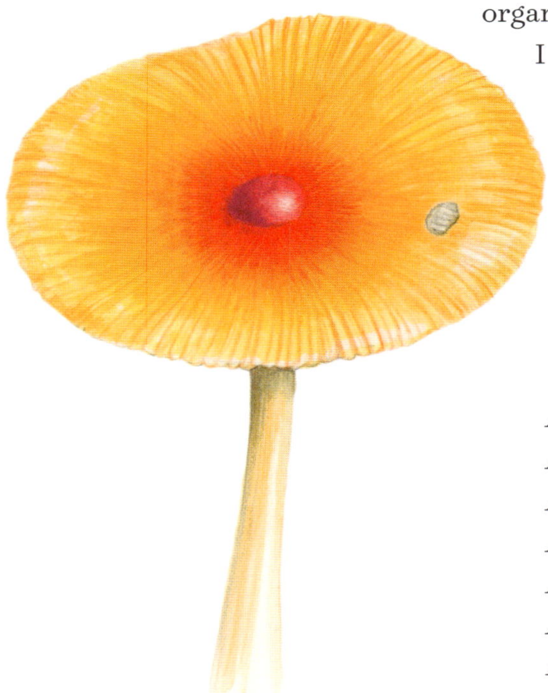

## Mushrooms Illustrated

*Amanita bisporigera*, Destroying Angel, found 7/12/23. Toxic
*Amanita ceciliae*, Snakeskin Grisette, found 8/8/23. Edible with caution
*Amanita flavoconia*, Yellow Patch Amanita, found 7/10/23. Toxic
*Amanita flavorubens*, American Blusher, found 7/10/23. Edible with caution
*Amanita gemmata*, Gemmed Amanita, found 7/9/23. Toxic
*Amanita jacksonii*, American Caesar, found 7/12/23. Edible with caution
*Boletus subluridellus*, Red-mouth Bolete, found 7/4/23. Edible with caution
*Cantharellus flavus*, Golden Chanterelle, found 6/30/23. Edible
*Cantharellus lateritius*, Smooth Chanterelle, found 7/19/23. Edible
*Cerioporus squamosus*, Dryad's Saddle, found on 4/23/23. Edible
*Crucibulum laeve*, Bird's Nest, found 7/8/23. Inedible
*Galerina paludosa*, Bog Bell, found 6/14/23. Toxic
*Galiella rufa*, Rubber Cup, found 7/8/23. Inedible
*Hypomyces hyalinus*, Ghost Amanita, found 7/9/23. Toxic
*Lactifluus hygrophoroides*, Milk Caps, found 7/12/23. Edible
*Microstoma floccosum*, Pink Fringed Fairy Cup, found 7/8/23. Inedible
*Mitrula elegans*, Swamp Beacon, found 6/14/23. Inedible
*Morchella americana*, Yellow Morel, found 4/25/23 and 5/10/23. Edible
*Neofavolus alveolaris*, Hexagonal Polypore, found 7/8/23. Inedible
*Resupinatus applicatus*, Smoked Oysterling, found 7/10/23. Inedible
*Russula mariae*, Purple-bloom Russula, found 7/6/23. Toxic
*Sarcoscypha occidentalis*, Stalked Scarlet Cup, found 7/1/23. Inedible
*Strobilomyces strobilaceus*, Old Man of the Woods, found 7/12/23. Edible
*Stropharia rugoso-annulata*, Wine Cap, found 5/22/23. Edible
*Tylopilus felleus*, Bitter Bolete, found 7/10/23. Inedible
*Xylaria polymorpha*, Dead Man's Fingers, found 7/4/23. Toxic

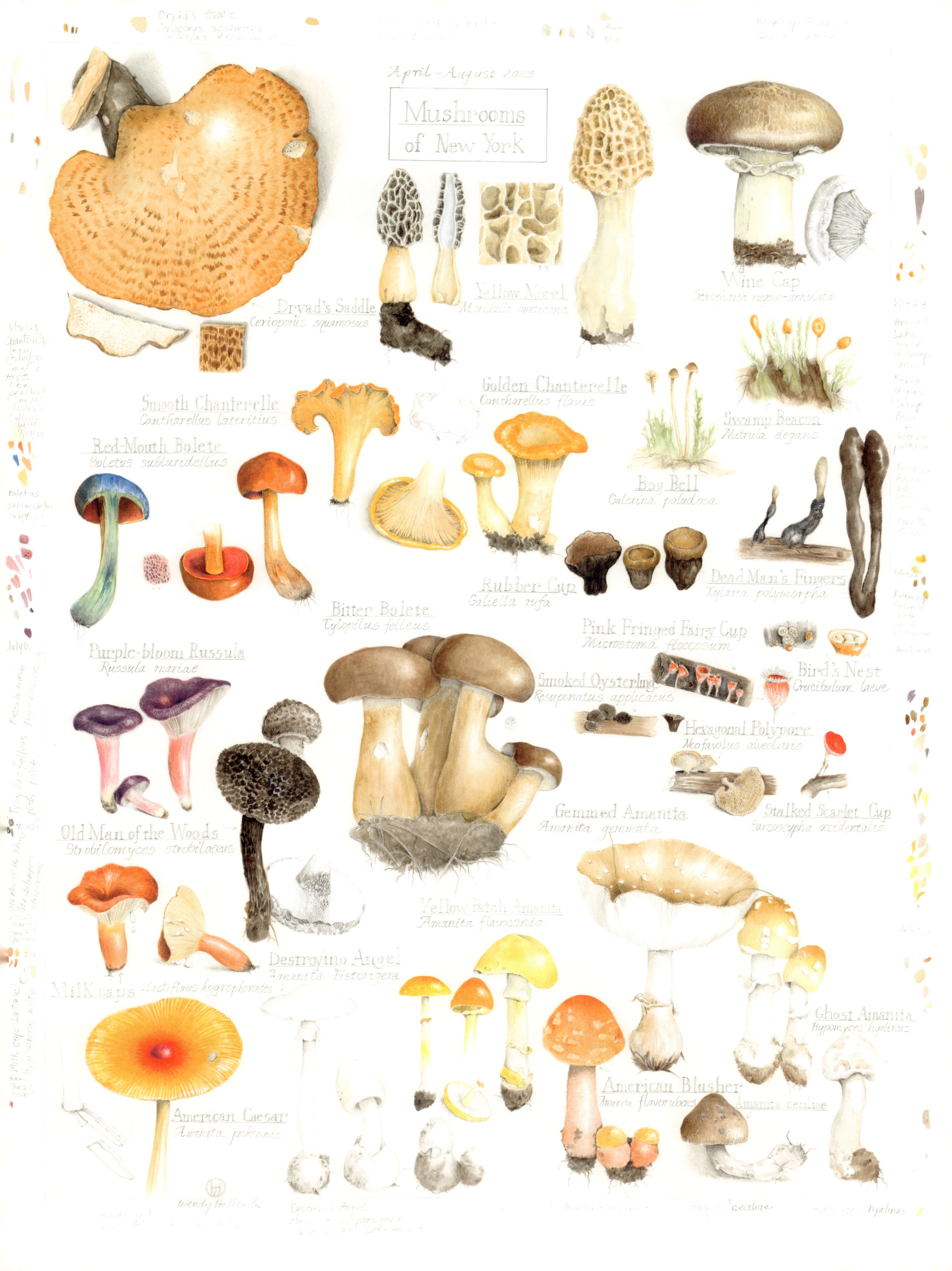

April - August 2023
Mushrooms of New York
Dryad's Saddle
Cerioporus squamosus
Yellow Morel
Morchella americana
Wine Cap
Stropharia rugoso-annulata
Smooth Chanterelle
Cantharellus lateritius
Golden Chanterelle
Cantharellus flavus
Swamp Beacon
Mitrula elegans
Red-Mouth Bolete
Boletus subluridellus
Bog Bell
Galerina paludosa
Rubber Cup
Galiella rufa
Dead Man's Fingers
Xylaria polymorpha
Bitter Bolete
Tylopilus felleus
Pink Fringed Fairy Cup
Microstoma floccosum
Purple-bloom Russula
Russula mariae
Bird's Nest
Crucibulum laeve
Smoked Oysterling
Resupinatus applicatus
Hexagonal Polypore
Neofavolus alveolaris
Gemmed Amanita
Amanita gemmata
Stalked Scarlet Cup
Sarcoscypha occidentalis
Old Man of the Woods
Strobilomyces strobilaceus
Yellow Patch Amanita
Amanita flavoconia
Destroying Angel
Amanita bisporigera
Milk caps - Lactifluus hygrophoroides
Ghost Amanita
Hypomyces hyalinus
American Blusher
Amanita flavorubens
Amanita ceciliae
American Caesar
Amanita jacksonii
Wendy Hollender

I find mushrooms so mysterious because of all the work they do underground, away from our prying eyes, and then miraculously, one day or not, pop up out of the earth. If you have ever tried foraging for mushrooms, especially for certain elusive edible mushrooms such as morels, it almost feels as if they know you are there and they decide whether you will be able to see them or not. I have often walked over one spot looking for morels and found nothing, yet on my return trip an hour later, after I have put in lots of hard work looking and talking to the morels, they appear. I describe it as "getting your mushroom eyes on." Once you do, you start to see more and more mushrooms, or so it seems.

The year 2023 was very rainy in the Hudson Valley/Catskills region of New York. Fortunately, this was the year I decided to find and draw mushrooms for my project. I needed help, because finding mushrooms takes time, knowledge, and the willingness to travel around looking for them. I worked with Luke Sarrantonio, a local mycologist, as well as Dina Falconi, a local herbalist. I sometimes went on a walk with Luke to look for mushrooms, but more often than not, Luke would send a text to let me know he had found a subject for my painting or to invite me to come see what was growing somewhere in the area. I also took countless walks and bike rides scanning the landscape in pursuit of mushrooms. I was able to find so many just by moving slowly and looking at the ground. Subsequently, 2024 was extremely dry, and mushrooms were hard to find, but fortunately I had completed the two paintings in 2023, finishing one page halfway through the season and the second page by the season's end. I organized the mushrooms on page 1 by date, just adding whatever I found to the page when I found it. I organized page 2 a bit by type of mushroom, keeping the rosellas and amanitas together, for example.

Some mushrooms are known to be toxic, others to be edible; these are indicated in the mushroom lists below. Novice mushroom hunters should always have their identifications confirmed by a reliable, experienced mushroom guide.

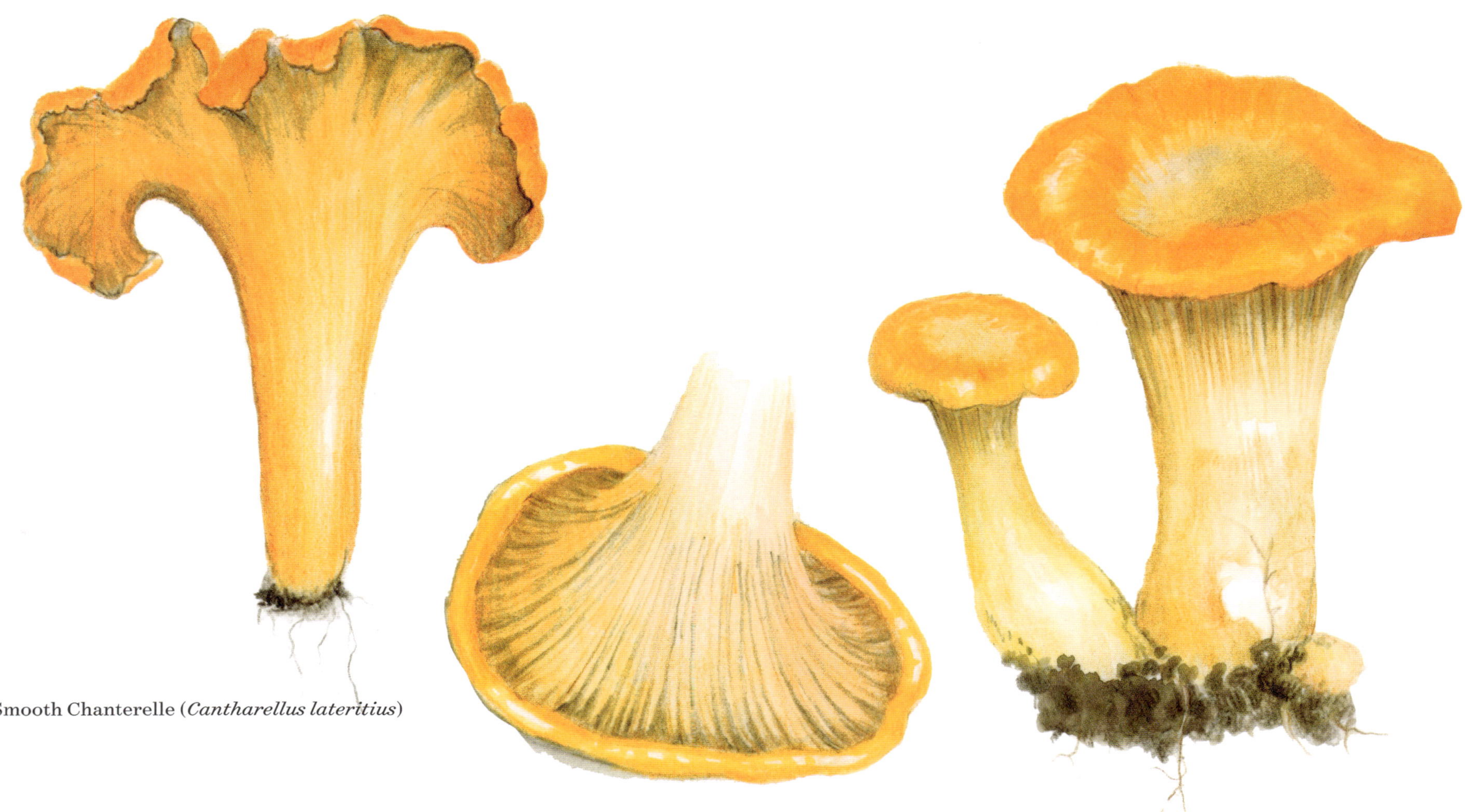

Smooth Chanterelle (*Cantharellus lateritius*)

Golden Chanterelle (*Cantharellus flavus*)

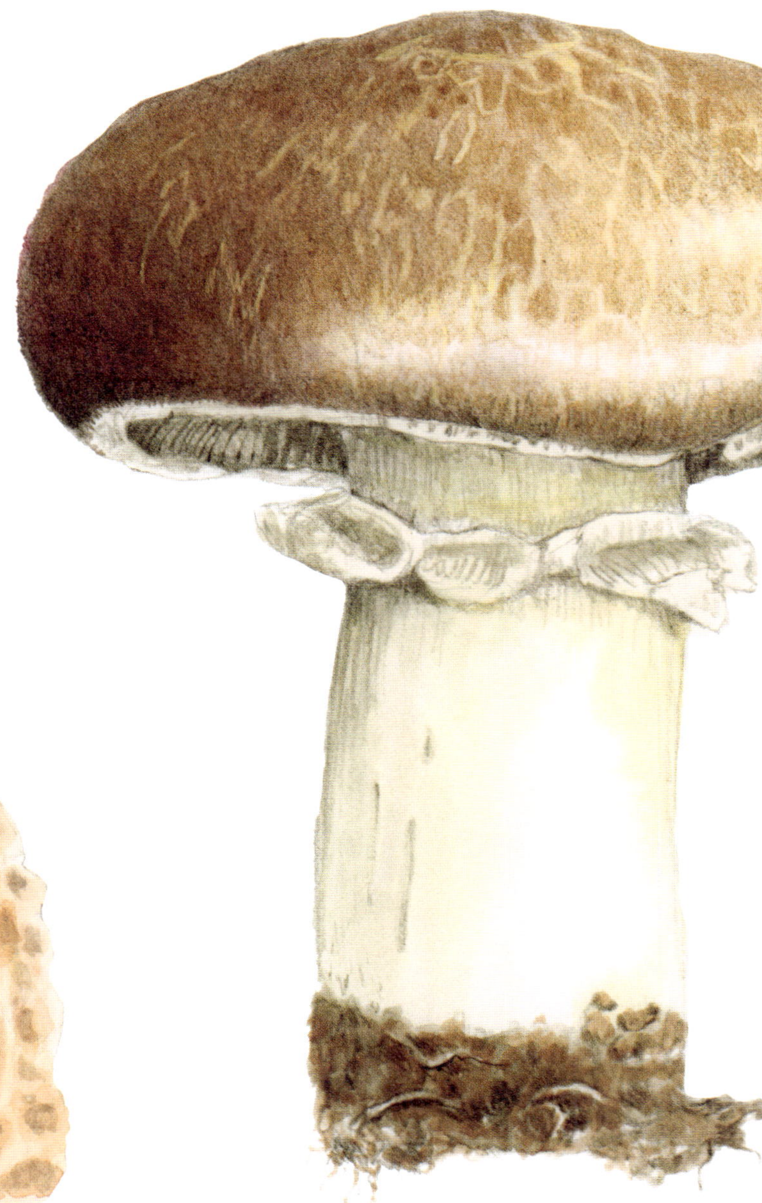

Wine Cap (*Stropharia rugoso-annulata*)

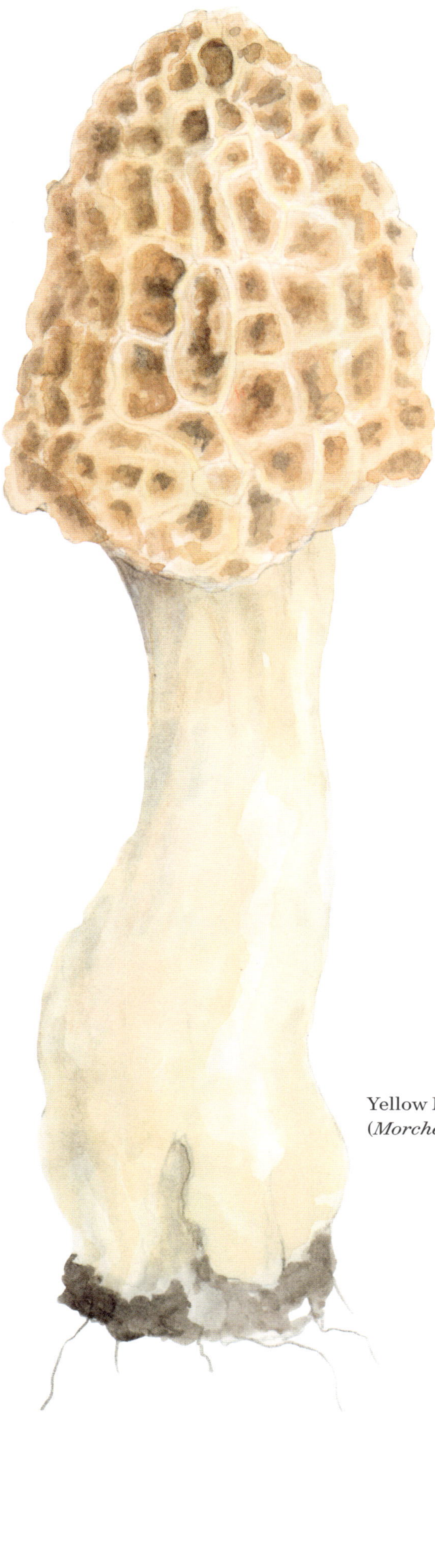

Yellow Morel (*Morchella americana*)

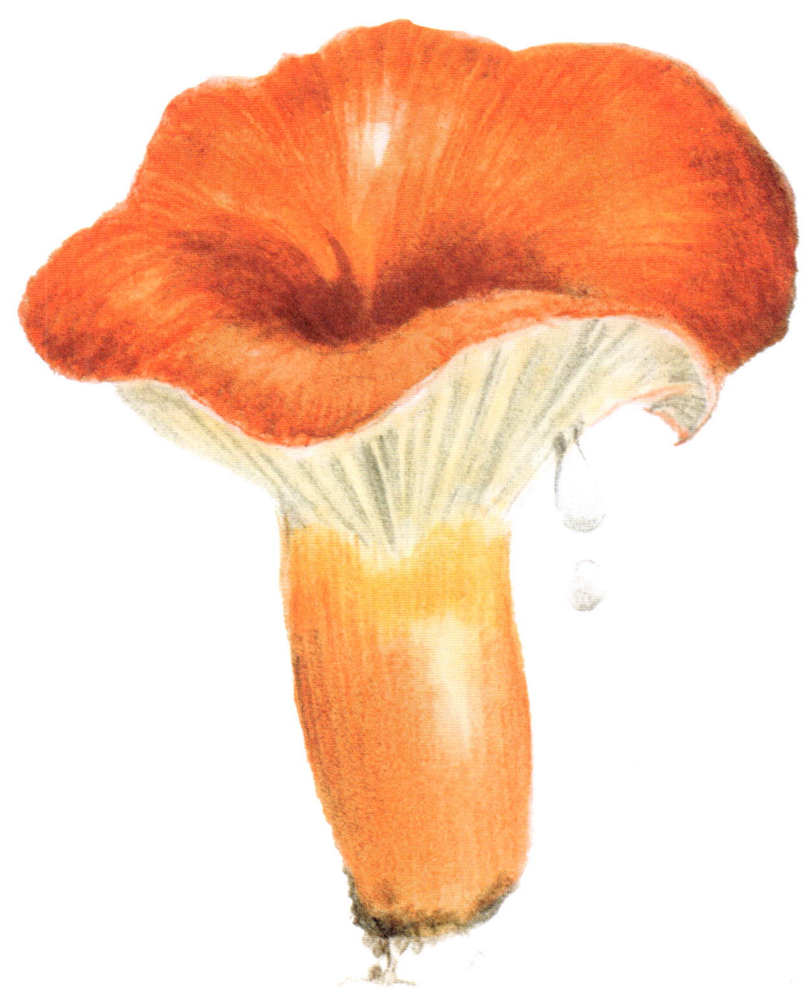

Milk Caps (*Lactifluus hygrophoroides*)

24

# Mushrooms of New York

## July–October 2023

### Fungi (Fungus Kingdom)

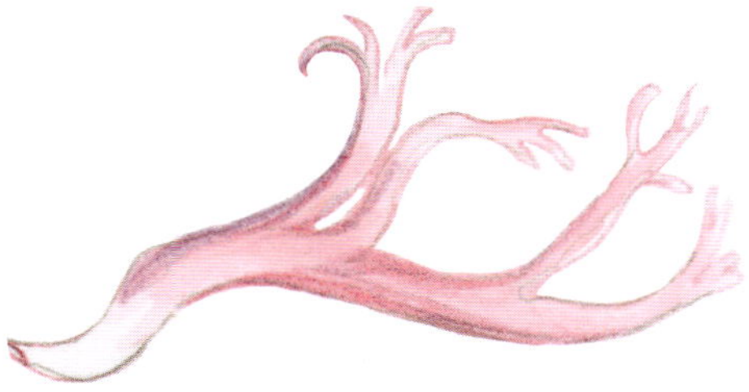

### Mushrooms Illustrated

*Amanita muscaria*, Fly Agaric, found 8/16/23. Toxic
*Boletus separans*, Lilac Porcini, found 7/18/23. Edible
*Cantharellus cinnabarinus*, Cinnabar Chanterelle, found 7/17/23. Edible
*Cantharellus lateritius*, Smooth Chanterelle, found 7/17/23. Edible
*Cantharellus minor*, Minor Chanterelle, found 8/9/23. Edible
*Chlorophyllum molybdites*, Green-spored Parasol, found 10/16/23. Toxic
*Clavaria fumosa*, Smoky Coral, 8/8/23. Edible
*Clavaria zollingeri*, Violet Coral, found 8/9/23. Inedible
*Clitocybe nuda*, Blewit, found 10/4/23. Edible
*Cortinarius iodes*, Violet Viscid Cortinarius, found 8/9/23. Edible with caution
*Craterellus fallax*, Black Trumpet, found 8/1/23. Edible
*Ganoderma applanatum*, Artist's Conk, found 10/4/23. Edible
*Grifola frondosa*, Hen of the Woods or Maitake, found 10/5/23. Edible
*Hericium americanum*, Bear's Head, found 10/11/23. Edible
*Hydnum umbilicatum*, Hedgehog, found 10/6/23. Edible
*Lactarius chelidonium*, Celandine Milky, found 10/1/23. Toxic
*Lactarius indigo*, Indigo Milky, found 10/1/23. Edible
*Lactifluus corrugis*, Corrugated-cap Milky, found 7/18/23. Edible
*Laetiporus cincinnatus*, Chicken of the Woods, found 7/29/23. Edible
*Lenzites betulina*, Gilled Polypore, found 7/17/23. Inedible
*Lycoperdon perlatum*, Gem-studded Puffball, found 10/2/23. Edible with caution
*Lycoperdon pyriforme*, Pear-shaped Puffball, found 10/1/23. Edible with caution
*Omphalotus illudens*, Jack O'Lantern, found 10/2/23. Toxic
*Russula aeruginea*, Quilted Green Russula, found 7/20/23. Edible
*Russula claroflava*, Yellow Russula, found 7/20/23. Edible
*Russula silvicola*, Red Russula, found 7/19/23. Toxic
*Sarcomyxa serotina*, Olive Oysterling, found 10/23/23. Edible
*Scleroderma citrinum*, Pigskin Puffball, found 10/4/23. Toxic

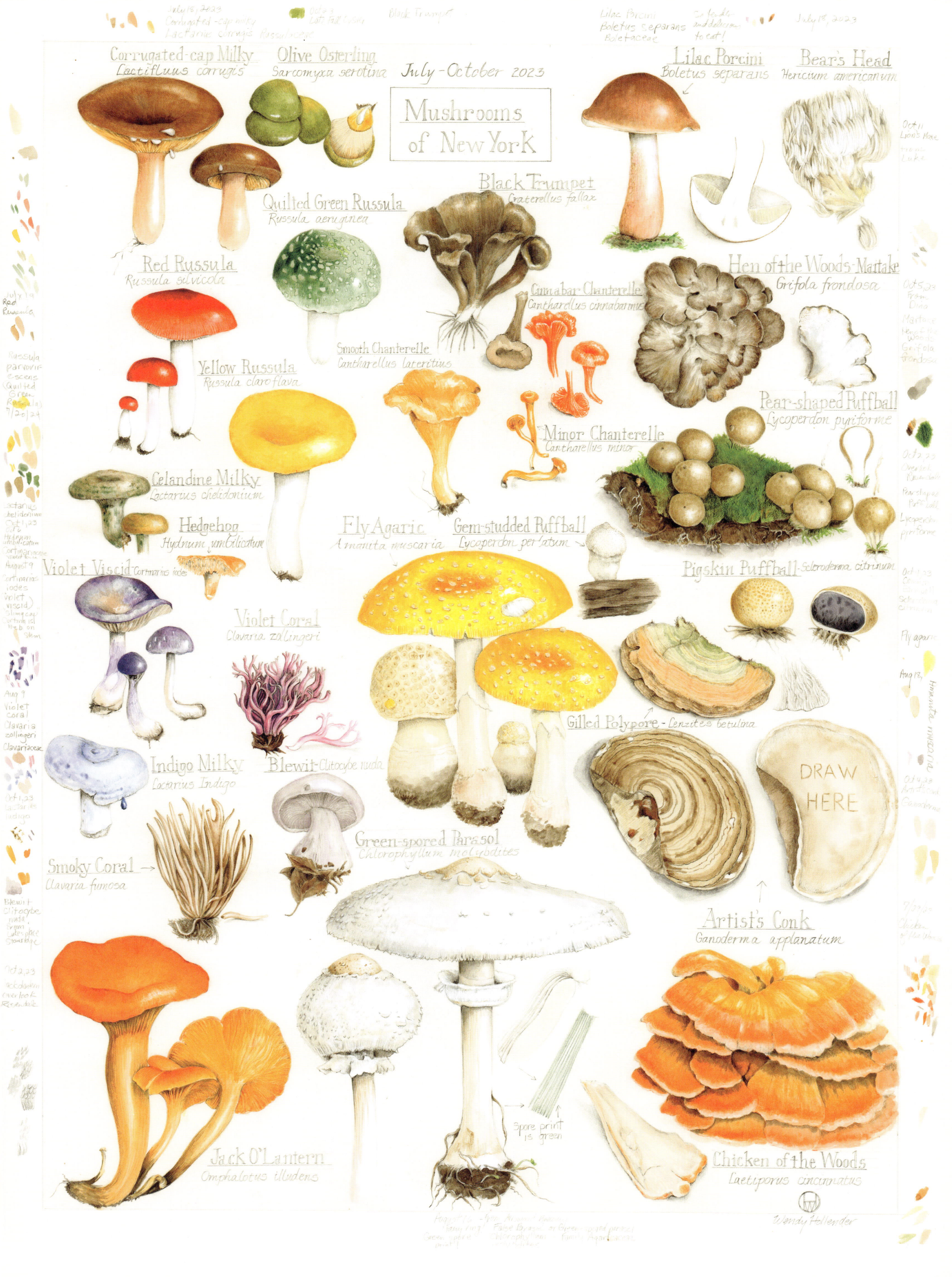
July-October 2023
Mushrooms of New York
Corrugated-cap Milky
Lactifluus corrugis
Olive Oysterling
Sarcomyxa serotina
Lilac Porcini
Boletus separans
Bear's Head
Hericium americanum
Black Trumpet
Craterellus fallax
Quilted Green Russula
Russula aeruginea
Red Russula
Russula silvicola
Hen of the Woods-Maitake
Grifola frondosa
Cinnabar Chanterelle
Cantharellus cinnabarinus
Smooth Chanterelle
Cantharellus lateritius
Yellow Russula
Russula claroflava
Pear-shaped Puffball
Lycoperdon pyriforme
Minor Chanterelle
Cantharellus minor
Celandine Milky
Lactarius chelidonium
Hedgehog
Hydnum umbilicatum
Fly Agaric
Amanita muscaria
Gem-studded Puffball
Lycoperdon perlatum
Violet Viscid-Cortinarius iodes
Pigskin Puffball-Scleroderma citrinum
Violet Coral
Clavaria zollingeri
Gilled Polypore-Lenzites betulina
Indigo Milky
Lactarius Indigo
Blewit-Clitocybe nuda
DRAW HERE
Green-spored Parasol
Chlorophyllum molybdites
Smoky Coral
Clavaria fumosa
Artist's Conk
Ganoderma applanatum
Spore print is green
Jack O'Lantern
Omphalotus illudens
Chicken of the Woods
Laetiporus cincinnatus
Wendy Hollender

# A Glorious Year for Mushrooms: Mysterious and Elusive Fungi

BY LUKE SARRANTONIO

What a glorious year for mushrooms! Those words came to mind as I arrived at Wendy's house in Accord, New York, in the spring of 2023. We had just received record-breaking precipitation in the Hudson Valley, and some areas south of us were hit by devastating floods. The rain was threatening again that afternoon, but fungi love the rain, so it was more welcoming than a threat to us. Rain provides the moisture they need to carry out essential ecological functions, and allows fungi to create the beautiful, fascinating, and in some cases delicious fruiting bodies we call mushrooms! Not all fungi produce mushrooms, though—only about 20,000 of the ~155,000 described species do. Current estimates of the total number of unique fungi on earth range from 2.2 to 13.2 million, depending on calculation methods. That's quite a big range, but even if it were only 2.2 million, it would mean we've described only about 6.8 percent. We'd better get on it! The loss of species to habitat destruction is reaching an alarming rate. Conservation is important not just for ecological reasons; fungi also produce essential nutrients, novel compounds, and useful structures that may help us navigate obstacles of the future.

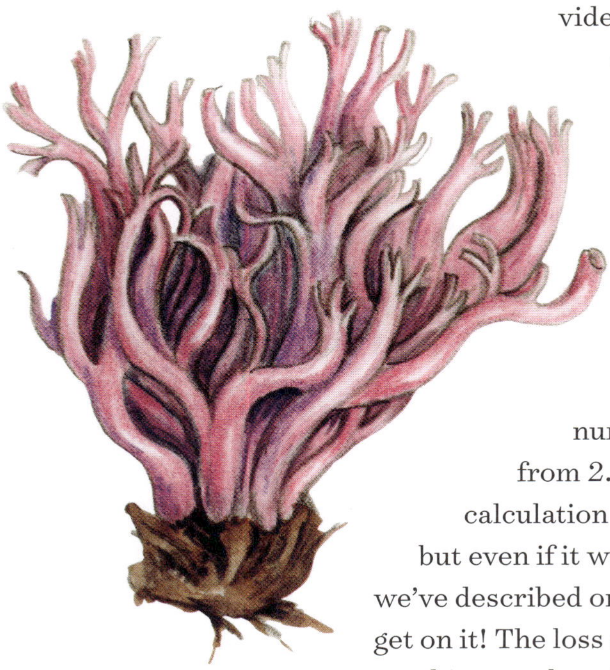

Violet Coral
(*Clavaria zollingeri*)

I was excited to hear that Wendy wanted to include mushrooms in this beautiful collection of plant families; they were considered plants up until 1969 by Western science, so it seems fitting. They also form essential relationships with plants: some fungi (saprotrophs) release nutrients through decomposition; others (mycorrhizal fungi) shuttle them (as well as water) to plant roots in exchange for sugars the plants make through photosynthesis. Mycology, the study of mushrooms, is seen as a young science by many, but if you look around the world, it seems clear that we have had a deep ancient connection with fungi throughout history that we are now working to rekindle.

As Wendy and I stepped off the trail that rainy day into the damp undergrowth, we were greeted by a vast landscape of mushrooms in many shapes and sizes. The ghostly white parasitized amanita (*Hypomyces hyalinus*), which appeared to be exceptionally abundant that year, loomed into view, and the black, gnarled old man of the woods (*Strobilomyces strobilaceus*) exhibited some true contrast against the forest floor. Mushroom collection is not always easy; some species are very elusive, especially coveted edibles like morels (*Morchella* species and relatives). A well-known mycologist once told me that he stares at honeycomb before going out morel hunting to train his eyes because the pattern is similar. "Sometimes certain mushrooms just don't seem to want to be found," I've heard people say. When it comes down to it, I think it just takes time to train the eyes or, as Wendy puts it, "get your mushroom eyes on." You might go a whole day looking for morels or another particular mushroom, only to find your first specimen as you're getting back to the car, in a spot next to the trail that you were sure you checked! As we strive to learn more, safety is always something to keep in mind when foraging. When identifying a mushroom for edibility, we often utter the phrase "if in doubt, toss it out." It's just not worth taking a chance with a mushroom you're not 100 percent sure about.

We found an amazing diversity of growth forms that season, from fascinating structures like the brilliant, towering castlelike appendages of the vibrant violet coral (*Clavaria zollingeri*), a favorite of mine, to delicious trumpet-shaped golden chanterelles (*Cantharellus flavus*) and black trumpets (*Craterellus fallax*). The later season brought in cold-temperature species like the spiny hedgehog (*Hydnum umbilicatum*), bear's head (*Hericium americanum*), and maitake (*Grifola frondosa*). Many structures in the fungal world seem nearly identical, and we have traditionally grouped species together on this basis, yet with recent DNA work, we're finding many cases of convergent evolution within the origins of fungi. We have polypores, gilled

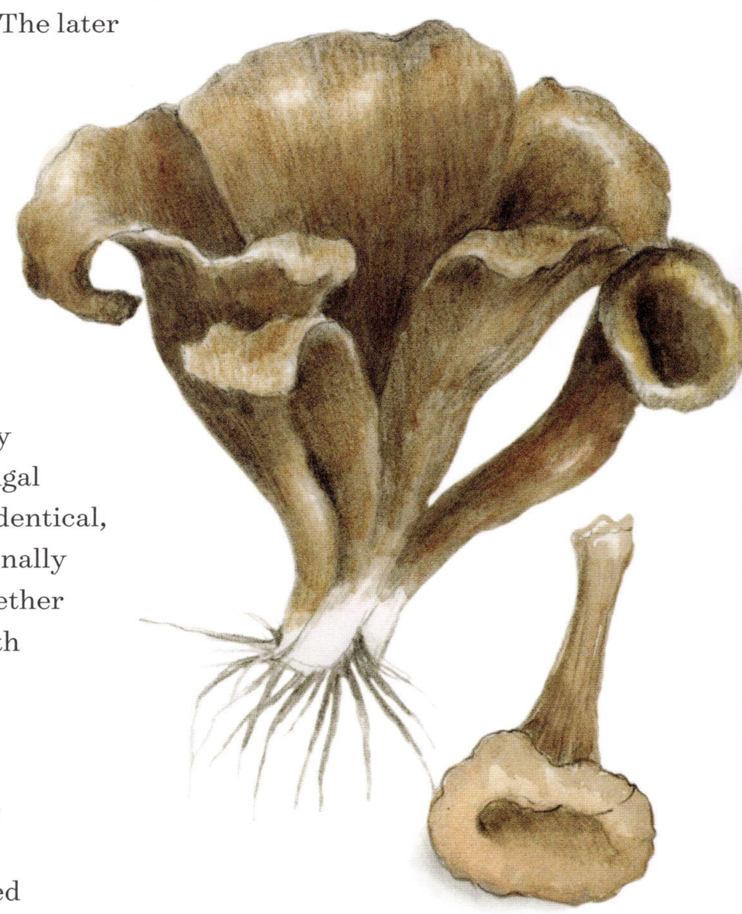

Black Trumpet
(*Craterellus fallax*)

and spined mushrooms, boletes, cups, clubs, puffballs, earthballs, crusts, rusts, smuts, coral shapes . . . Convergent evolution is when similar forms or characteristics arise in organisms that are not closely related (like wings in bats and birds), often due to similar ecological situations in different geographic locations. For example, many mushrooms have been grouped together by the presence of gills for spore dispersal, but we now understand that gill structures have evolved at least eight different times in separate groups throughout fungal evolution.

Wendy and I each did a bit of collection on our own for this project and helped each other with specimen choice and identification, but it was especially nice to get out in the forest with a kindred spirit, exploring our fascinations with form, function, and color. It is truly an honor to be part of this project!

Luke Sarrantonio grew up in Rosendale, New York, exploring the local landscape and its unique ecology. He truly discovered fungi while studying at SUNY ESF (State University of New York College of Environmental Science and Forestry) and immediately fell in love! He has since developed a dynamic career based around his passion for fungi. Luke's work includes educational programming focused on ecological understanding and cultivation, organizing an annual mushroom arts festival in the Catskill region (For the Love of Fungi), facilitating community projects (Eat More Mushrooms Project), and creating a line of functional mushroom products (under the name Mycophilic). He also consults for outdoor mushroom farms and manages his own in Accord, New York. Luke's main goal is to be an accessible resource for people who want to learn more about this fascinating group of life-forms.

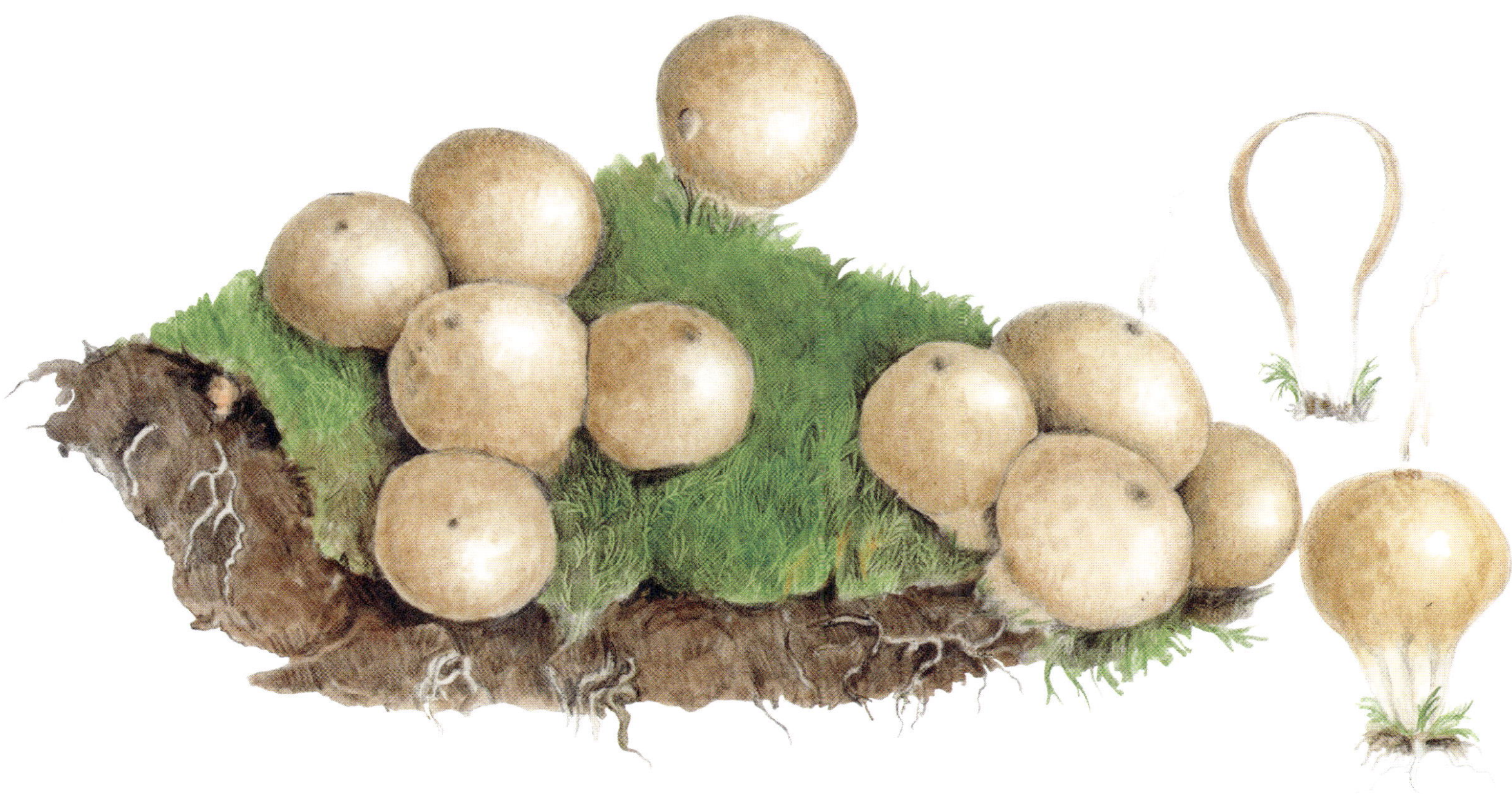

Pear-shaped Puffball (*Lycoperdon pyriforme*)

# Afterword and Acknowledgments

It is a late October day in 2025 when I am reading the final version of my edited book manuscript in preparation for publication in the spring of 2026. It seems fitting to be writing this paragraph while flying home from an annual conference of the American Society of Botanical Illustrators, of which I am a member. As I reflect on my now four-plus-year journey in creating *The Album of Plant Families*, I am grateful to have been given this opportunity to take a deep dive into the plant world. Sometimes people, especially fellow botanical artists (who know how time-consuming creating botanical art can be), ask me how I was able to take on such a large project. I don't think I was in charge of this project, nor did I decide to take it on. Rather, it found me. I was inspired to try one painting of alliums, and once it was completed, I just had to continue. It often felt as if the plants were holding my hand and leading me on a journey of discovery and documentation.

I wanted to sum up some of my discoveries while on this journey at this time, but honestly, I think the publication of this book and coordinated exhibitions and programming will allow for this project to continue to grow and develop. I don't know yet how the world will respond to this book, and thus far, I can only imagine what might take place, so for now I am looking forward to what that might be. So, please, let me know if this project has helped you, excited you, or anything else!

I am so grateful to the many plants and people that have helped me on this journey. I am deeply grateful to Abbeville Press, specifically my editor, Lauren Orthey, and publisher David Fabricant, for believing in this project as a large-format art book. I want to thank the copy editor, Amy K. Hughes, for her thoughtful editing, and Misha Beletsky, for his exquisite design. It was a privilege to finally have representation by an exceptional agent, B. G. Dilworth, and to have his wealth of publishing experience for this book. And, finally, I am grateful to my family and friends who support and encourage me.

Nature and especially the plant world are truly my inspiration and guiding force. I hope that in my own small way, this project allows for many people to deepen their own understanding and appreciation of our plants, and a desire to keep this world healthy and abundant. Let's all take a moment to thank and celebrate this amazing planet!

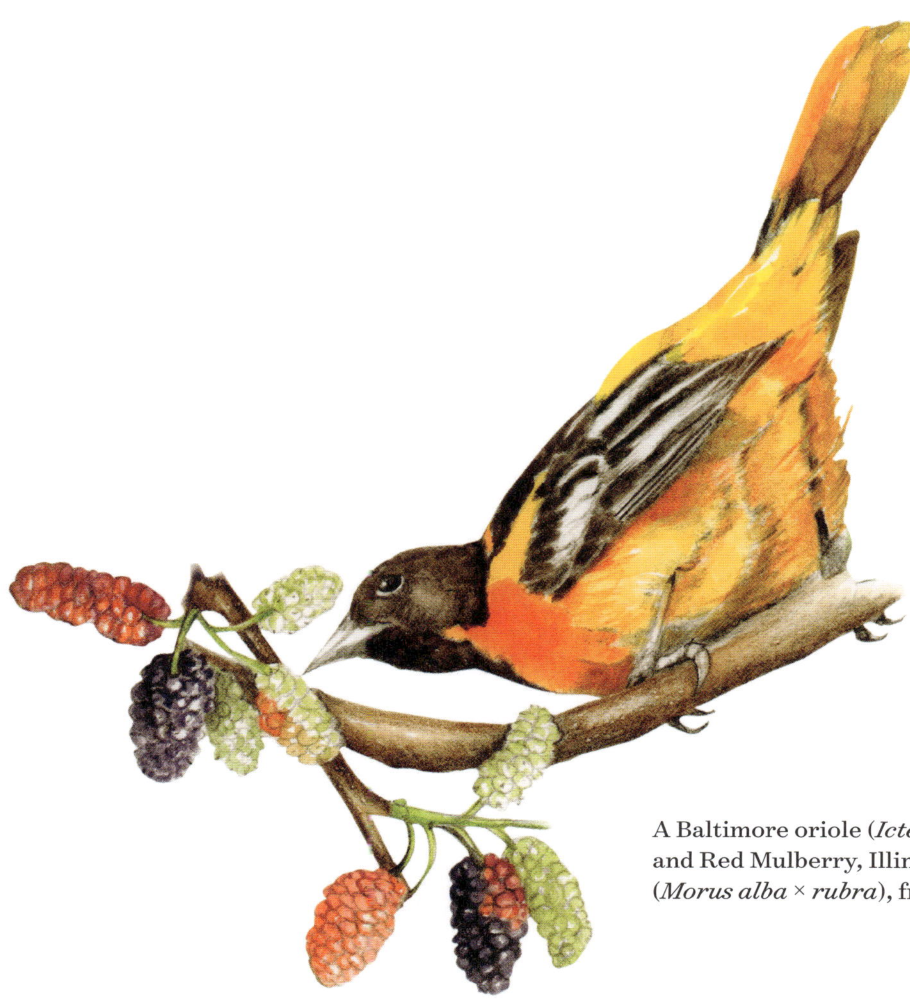

A Baltimore oriole (*Icterus galbula*) and Red Mulberry, Illinois Everbearing (*Morus alba* × *rubra*), from the fig family

# Farms, Growers, and Gardens Where I Gathered Plants

I would like to thank all the growers who gave me plants to draw for my project. I am in awe of all the hard work that goes into cultivating plants. I think all growers—from the backyard gardener growing for their own personal use to the farmer providing vegetables to small or large groups of people—have a love and appreciation of the importance and fragility of the environment and the need to preserve what nature has provided. The following is a list of the many places I collected plants for my paintings. Most of these places I visited myself, but in a few instances I had to rely on the postal system to bring me the plants from far away. Some of these growers provided one plant or type of plant to include in my painting, and those plants are listed by their name below. A few growers provided several plants and countless hours of advice to help me on my journey. If a particular person at a farm was especially helpful, I have included their name.

Hortus Arboretum and Botanical Gardens provided many plants in the magnolia family as well as maples and the only citrus fruit that grows in a temperate environment. The proprietors sent me out into the landscape, to gardens and rail trails that they know, in search of poison hemlock and other elusive subjects. The New York Botanical Garden invited me into a studio, and Marc Hachadourian brought me plants from the greenhouses to work from. Tributary Farm, the organic vegetable grower in my area that provides food for families all year long, is my son Jesse's farm. He allowed me to traipse out into the fields in search of the perfect subject, and I was able to fill the nightshade, amaryllis, gourd, mustard, and carrot families with lovely and delectable subjects to draw and of course eat. I was excited to discover a farm growing rice nearby; Ever-Growing Family Farm, based in Ulster Park, New York, was started by the Badjie family from Gambia in West Africa. Fortunately for them, a local landowner in Kerhonkson wanted her land to be farmed, but it was too wet for conventional growers—turns out it is perfect for rice. I visited the Fruit and Spice Park in Homestead, Florida, which has a large and varied collection of tropical fruits for people to explore and sample. The owners encourage visitors to eat the fruits they provide or those on the ground around the trees, and they provide excellent information about many of the tropical and unusual fruits growing there. The National Tropical Botanical Garden provided me access to most of my tropical plants, along with expertise on where to find them and what I might consider including. Dave Lorence provided me with hours of education while we took fitness walks throughout the hilly garden. I can't think of a better way to learn, and I encourage everyone to combine your curiosity about nature with some old-fashioned walking outside. Tim Flynn brought me many herbarium samples to help with flower or seedpod details. Limahuli Garden, part of the NTBG, allowed me to go back in time to experience plants growing in a traditional *pu'uhonua*, where Indigenous traditions and flourishing relationships between plants and people continue today. Lastly, I must give a shout out to my own garden, which I developed alongside my two grown children and countless other people to create an abundance of a variety of plants that nourish many living creatures, including myself, rabbits, and woodchucks. It provided many subjects to use in this project.

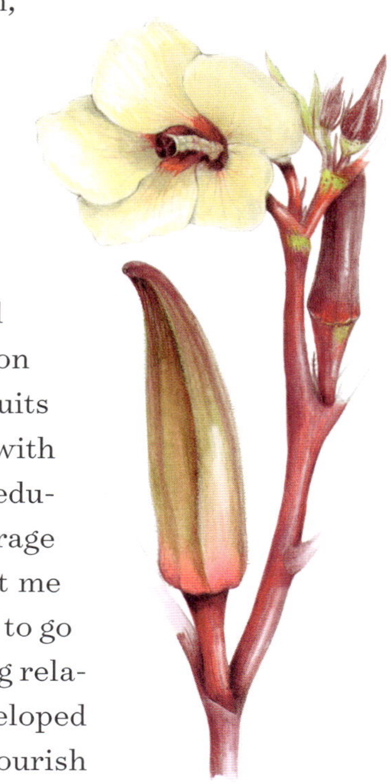

Okra (*Abelmoschus esculentus*), from the mallow family

# List of Sources for Plants

Alma Roots Farm, High Falls, NY (luffa gourd)
Ashokan Center, Olivebridge, NY (mulberries, mushrooms)
Chase Farms (Donald Chase), Oglethorpe, GA (peanuts)
Ever-Growing Family Farm, Ulster Park, NY (rice)
Dina Falconi garden, Stone Ridge, NY
Fruit and Spice Park, Homestead, FL (tropical plants)
Garden of Lemons, Andalusia, Spain (figs, peas)
Katie Grove garden, High Falls, NY (sweetgrass)
Christopher Hitz garden, Rosendale, NY (Osage orange)
Hortus Arboretum and Botanical Gardens, Stone Ridge, NY
Hudson Valley Farm Hub, Hurley, NY (wheat, corn, dried beans)
Hudson Valley Seed Company, Accord, NY (corn, Ruby Moon hyacinth bean)
Jenkins-Lueken Orchards (Eric James), New Paltz, NY
Kaua'i Coffee Company, Kalāheo, HI (coffee)
Kaua'i Farmacy, Kaua'i, HI (cardamom)
Kelder's Farm, Accord, NY (sunflowers)
Mountain Top Arboretum, Tannersville, NY (golden alexander)
National Tropical Botanical Garden, Kaua'i, HI
Neu Mana Hui Farm, Kaua'i, HI (cashew)
New York Botanical Garden, Bronx, NY (orchids, gingers)
Rail trail, Stone Ridge, NY (water hemlock)
Sam Cobb Farms, Desert Hot Springs, CA (dates)
Saunderskill Farm, Accord, NY (sweet corn)
Soroa Orchids, Homestead, FL (orchids)
Starling Yards (Kimberley Hart), Red Hook, NY (ginger, turmeric, roselle)
Stoke Albany, Northamptonshire, England (lupine, hogweed)
Top-Notch Pistachios (Sarah Ogilvie), Benson, AZ (pistachios)
Tributary Farm, High Falls, NY
Upperville, VA (black maple)
The Vanillery of Kaua'i, HI (vanilla orchid)
Westwind Orchard, Accord, NY (apples)

# Resources

Websites and apps for plant identification:
iNaturalist, https://www.inaturalist.org/
iPhone camera "Look Up": take a picture, click the "i" icon beneath the image, click the "Look Up" prompt.
Picture Mushroom, https://picturemushroom.com/
PictureThis: Botanist in Your Pocket, https://www.picturethisai.com/
Plants of the World Online, Royal Botanic Gardens at Kew, https://powo.science.kew.org/
USDA Plants Database, https://plants.usda.gov/
Herbarium collections at botanical gardens, universities, and nature centers

# Bibliography

Balick, Michael J. *Rodale's 21st Century Herbal: A Practical Guide for Healthy Living Using Nature's Most Powerful Plants.* New York: Rodale, 2017.

Bayton, Ross, and Simon Maughan. *Plant Families: A Guide for Gardeners and Botanists.* Chicago: University of Chicago Press, 2017.

Castner, James L. *Photographic Atlas of Botany and Guide to Plant Identification.* Gainesville, FL: Feline Press, 2004.

Elpel, Thomas J. *Botany in a Day: The Patterns Method of Plant Identification.* Pony, MT: HOPS Press, 1996.

Falconi, Dina. *Foraging and Feasting: A Field Guide and Wild Food Cookbook.* Accord, NY: Botanical Arts Press, 2013.

Hollender, Wendy. *The Practice of Botanical Drawing.* Accord, NY: Botanical Arts Press, 2017.

Madison, Deborah. *Vegetable Literacy.* Berkeley, CA: Ten Speed Press, 2013.

Pell, Susan K., and Bobbi Angell. *A Botanist's Vocabulary: 1300 Terms Explained and Illustrated.* Portland, OR: Timber Press, 2016.

Rauh, Dick. *The Science behind Flowers: Plant Morphology for Botanical Artists.* Westport, CT: Dick Rauh, 2020.

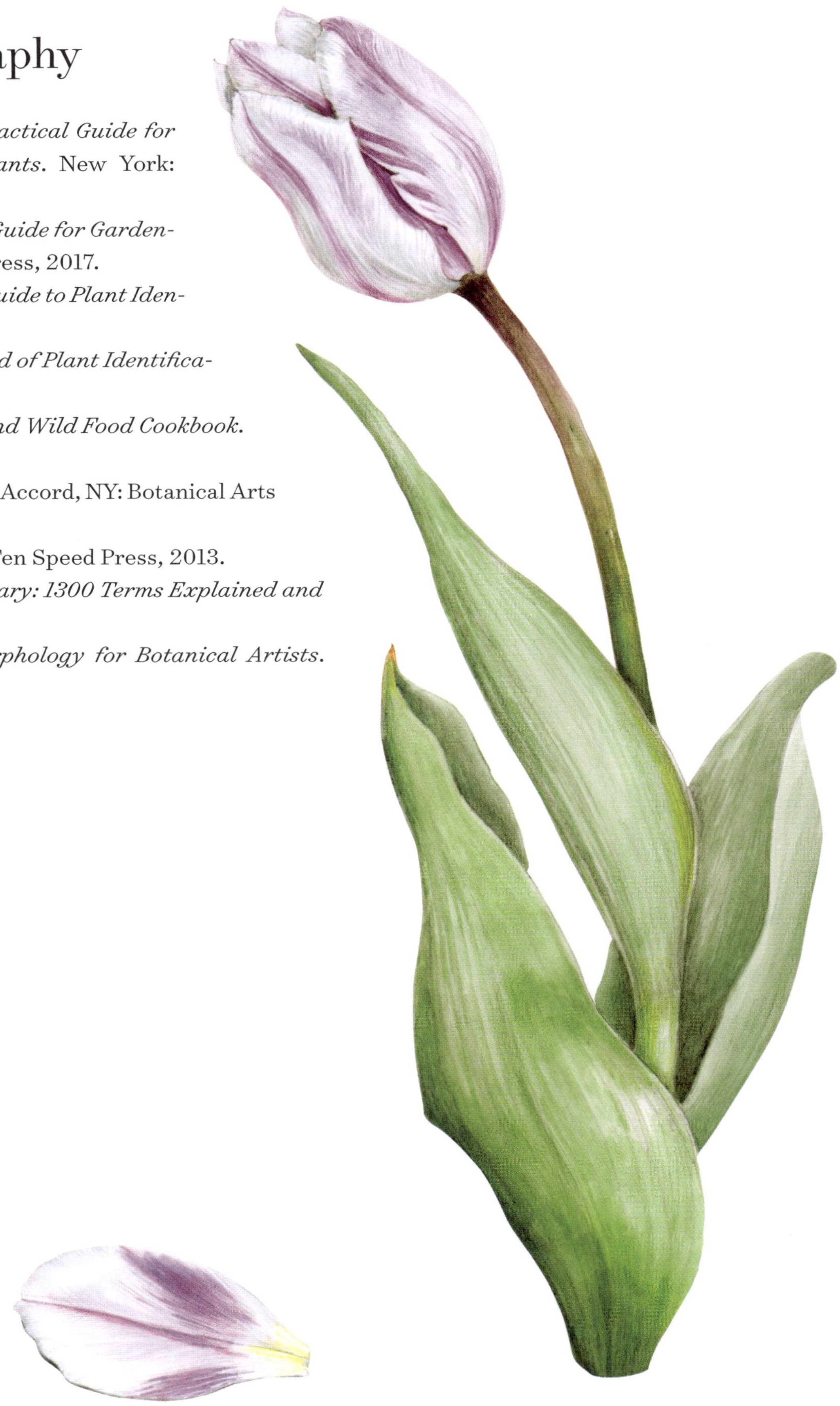

Tulip, Rem's Favourite (*Tulipa*), from the lily family

# Index

*Page numbers indicate the plate(s) where the species is illustrated.*

## PLANTS BY COMMON NAME

## PLANTS BY SCIENTIFIC NAME

## MUSHROOMS BY COMMON NAME



## MUSHROOMS BY SCIENTIFIC NAME



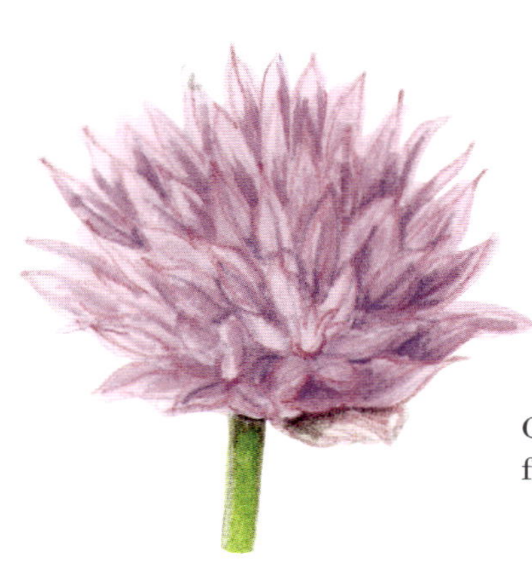

Chive (*Allium schoenoprasum*), from the amaryllis family

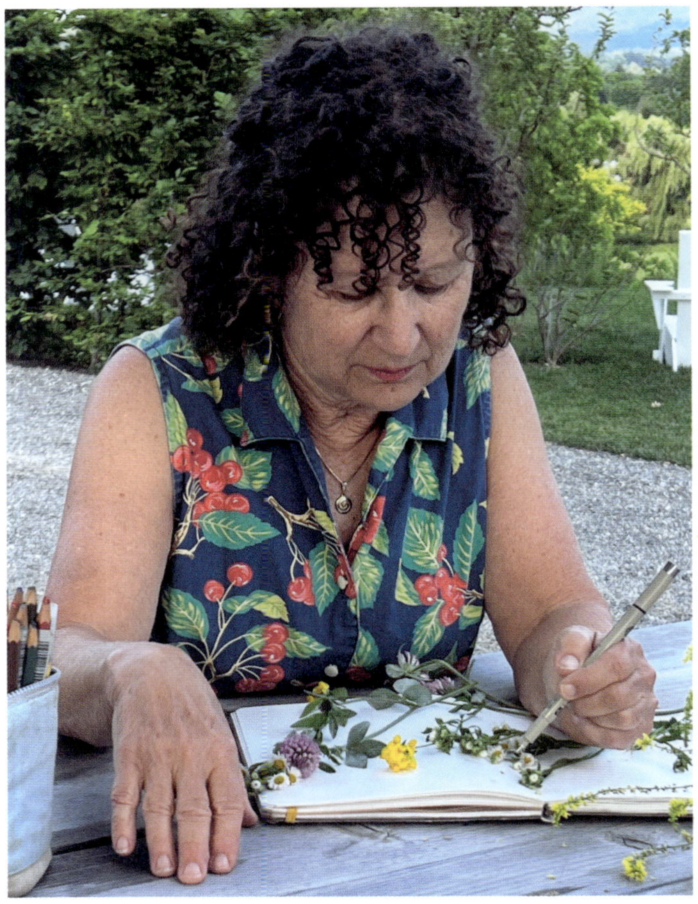

PHOTO BY STEVE BERNSTEIN

## ABOUT THE AUTHOR

Wendy Hollender is a world-renowned botanical artist and author whose work has been exhibited at the Royal Botanic Gardens, Kew, the New York Botanical Garden, and the Smithsonian National Museum of Natural History, and published in the *New York Times*, *O: The Oprah Magazine*, and numerous other outlets. A graduate of the Rhode Island School of Design, Hollender is the author of the best-selling books *The Joy of Botanical Drawing* and *Botanical Drawing in Color*, and teaches botanical drawing at the New York Botanical Garden and other beautiful venues around the world. See more about Wendy at www.wendyhollender.com and her educational website www.drawbotanical.com.

ILLUSTRATIONS
Front cover: Sunflower family (Asteraceae) (detail). See page 113.
Back cover: Amaryllis apple blossom (*Hippeastrum*) (left), Natal Lily (*Clivia miniata*) (center), and Daffodil (*Narcissus*) (right). See page 19.
Endpapers: *Dahlias, from the Sunflower Family* by Wendy Hollender
Pages 8–10: The illustrations on these pages have also been reproduced in *The Joy of Botanical Drawing: A Step-by-Step Guide to Drawing and Painting Flowers, Leaves, Fruit, and More* by Wendy Hollender (Watson-Guptill, 2020).
Page 102, bottom: "Tree of Life" (c. 1994) for Duralee Fabrics, by Wendy Hollender

Editor: Lauren Orthey
Copy editor: Amy K. Hughes
Production editor: Kayla Hassett
Design: Misha Beletsky
Production manager: Louise Kurtz

 The text of this book was set in Surveyor. Printed in China.

First edition
10 9 8 7 6 5 4 3 2

ISBN 978-0-7892-1541-3

Library of Congress Cataloging-in-Publication Data available upon request

For bulk and premium sales and for text adoption procedures, write to Customer Service Manager, Abbeville Press, 655 Third Avenue, New York, NY 10017, or call 1-800-ARTBOOK.

Visit Abbeville Press online at www.abbeville.com.